FROM NAKED APE TO SUPERSPECIES

DAVID SUZUKI

HOLLY DRESSEL

from

NAKED

➤ APE ➤

to

SUPERSPECIES

humanity and the global eco-crisis

David
Suzuki
Foundation

GREYSTONE BOOKS
Douglas & McIntyre Publishing Group
Vancouver/Toronto/Berkeley

05 06 07 08   5 4 3 2

Greystone Books
A division of Douglas & McIntyre Ltd.
2323 Quebec Street, Suite 201
Vancouver, British Columbia
Canada V5T 4S7
www.greystonebooks.com

David Suzuki Foundation
2211 West 4th Avenue, Suite 219
Vancouver, British Columbia
Canada V6K 4S2

National Library of Canada Cataloguing in Publication Data
Suzuki, David, 1936–
    From naked ape to superspecies: humanity and the global eco-crisis / David Suzuki and Holly Dressel. — Rev. ed
    Co-published by the David Suzuki Foundation.
    Includes bibliographical references and index.
    ISBN 1-55365-031-X
    1. Nature—Effect of human beings on. 2. Human ecology. 3. Social ecology.
4. Renewable natural resources. I. Dressel, Holly Jewell. II. David Suzuki Foundation.
III. Title.
GF75.S99 2004   304.2'8   C2004-901157-X

Library of Congress information is available upon request.

Editing by Michael Carroll
Cover design by Jessica Sullivan
Printed and bound in Canada by Friesens
Distributed in the U.S. by Publishers Group West

We gratefully acknowledge the financial support of the Canada Council for the Arts, the British Columbia Arts Council, and the Government of Canada through the Book Publishing Industry Development Program (BPIDP) for our publishing activities.

Greystone Books is committed to reducing the consumption of old-growth forests in the books it publishes. This book is one step towards that goal. It is printed on acid-free paper that is 100% ancient-forest-friendly, and it has been processed chlorine-free.

We remember, with admiration and gratitude, the love of nature that motivated Rachel Carson to write *Silent Spring*. She was a role model for generations of writers, activists and concerned citizens.

H.D. also dedicates this book, with love, to her husband, colleague and friend, the late Gary Toole.

CONTENTS

ACKNOWLEDGEMENTS

This is a revised, updated and expanded edition of a book we wrote in 1999. *From Naked Ape to Superspecies* originally grew out of the eight-hour radio series of the same name, which Holly Dressel researched and field-produced and which we wrote together in 1998 for CBC. The book refined the series; it was the culmination of three and a half years of research, travel and thought, of delightful and distressing conversations with hundreds of wonderful and a few frightening people. We would like to express our lasting gratitude to Jane Lewis, the Montreal producer of the series, as well as to our executive producer, Bernie Lucht; and we would especially like to express our deep appreciation to the people at Greystone Books, whose faith in the continuing usefulness of this book has made this new edition possible: Rob Sanders, Chris Labonte, and particularly, our editors, Nancy Flight and Michael Carroll. We also remember with gratitude the people who helped shape the first edition at our former, and now defunct publisher, Stoddart: editors Jennifer Glossop, Marnie Kramarich, Jane McWhitty and especially Angel Guerra.

Lisa Hayden, my secretary at the time, was an enormous help to both authors, and Joel Silverstein completed the demanding task of transcribing hundreds of hours of taped interviews with grace and enthusiasm. Many other people went beyond the call of duty in sending information quickly and without complaint, no matter how often we asked. They include Andy Kimbrell of the International Center for Technology Assessment, everyone at the Sierra Club of Canada, Christine von Weizsaecker of Ecoropa, Beth Burrows of the Edmonds Institute, Danny Kennedy of Project Underground, and particularly

Brewster Kneen, publisher of the *Ram's Horn*. We are also especially grateful to the Third World Network.

On a personal level, my collaborator, Holly Dressel, would like to thank the following friends, who kept up with all the issues so brilliantly and tirelessly allowed her to bounce ideas off them, often far into the night: John Hughes and Martha Hughes; Jack Kofman; Gwynne Dyer and Tina Viljoen; and, for his invaluable post-mortems, Jim Latteier. And no such list would be complete without mention of her daughter, Thea; her son, Jack; and all the other friends and family who had to bear the highs and lows of such all-consuming work.

Finally, I want to acknowledge Jim Murray for his years of friendship and patience. He kept me in broadcasting when I wanted to quit and return to the lab, and he deepened my understanding of the human place in the natural world. The generous and uncomplaining support and enthusiasm of Tara Cullis, despite many weeks of absence, can never be fully acknowledged, but I make a meagre gesture of gratitude with a promise to put a lid on any new projects. I thank also Severn and Sarika, for being so supportive even though I was often away when they wanted me to be with them and too often responded, "Not now. I've got to finish this book."

David Suzuki
Vancouver, British Columbia

INTRODUCTION

Each chapter in this new edition contains fresh examples and analyses that update the original arguments. The book also presents a new first chapter that provides an overview of how the world has changed in the five short years since the first edition was published. The grave problems our sources describe in the current structure of Western society, such as its neglect of basic social structures, rapacious attitude towards natural systems and its concentration of resources on global economics and invasive technologies, are reviewed and updated. Within this fresh context, this edition also introduces ideas about what people can do to correct the excesses of material greed and technological optimism that have placed the natural systems that support this planet in increasing peril.

It is not an unusual contention to state that modern Western culture and its underlying belief systems have as their basic goals ever-increasing human populations, rising GDPs and the unlimited fulfillment of all human desires. It is perhaps less usual, in mainstream analyses at least, to maintain that this is the systemic problem that lies at the root of the continuing and dangerous depletion of the natural products and services on which physical life depends. With so much at stake in terms of our increasing loss of the basic materials required for survival, it has become an urgent matter to convince people that we need to reassess the direction in which we are headed. We also need to explore seriously whether, as our leaders so often claim, it really is our only choice.

In the new first chapter in this edition, we assess the extent to which we are still faced with the same or increasing challenges, while highlighting the many areas where significant progress has been made.

For example, "globalization" is no longer a new term that needs to be defined, as was the case in 1999; it has become a rallying cry that has polarized societies and individuals all around the world, and is now seen as one of the major issues of our time. We now have even graver political and military destabilization in the Middle East and the Third World than we did five years ago, as countries such as the United States, Britain, Japan and Canada try to move into these areas in order to exploit increasingly scarce natural commodities like oil, forests and gold. Real wars are now increasingly present in our lives as new flashpoints such as Iraq, Afghanistan, North Korea and Liberia have joined existing trouble spots like Israel, sub-Saharan Africa and the Balkans.

In the interim since we wrote about the dangers of placing too much credence in the more ethereal aspects of the Information Revolution, that bubble collapsed, leaving economic havoc and a legacy of lost ecological and social opportunities. Despite continuing hype, almost every product of genetic engineering has proven itself to be dangerous or unstable, and the continuing fight against this technology has created a diplomatic gulf between the United States, Canada and Australia, and the rest of the world. The new chapter also updates our analysis of what was happening to journalism as it became centralized under the power of a few private entities; today, this danger is so well understood that many alternatives, mostly on the Internet, have popped up to counteract a loss of basic freedom of speech. In those five years, we also wrote another book, *Good News for a Change,* which took off from where the last chapter of this one, "Complex Pleasures," left off. In this new edition, "Complex Pleasures," as well as the first-chapter overview, provides not just warnings of problems, but some guidance towards solutions.

We called the radio series and the book *From Naked Ape to Superspecies* in an attempt to articulate what has happened to our species in the past few hundred years. Human beings entered the second millennium more or less as we began our sojourn on Earth — as intelligent, inventive creatures with numerous cultures to our credit, but tied to the Earth in every way imaginable. Back in the year 1000, most human beings were still nomadic hunter-gatherers or traditional agriculturalists, completely dependent on the vagaries of weather, climate and the generosity of the web of living things. A tiny minority of people lived in towns and cities, where they had laws, markets, stores and the trappings of civilization, but little control over natural forces like climate and disease. Industrialization occurred only in the past two or three centuries, and then total human populations began to climb steeply as Europeans rapidly colonized most other places and peoples of the world.

When the twentieth century began, many countries were well along in industrialization, and there were more than a billion people overall. While most of nature remained untouched, we were beginning to have an impact on the forests, waterways and soils on a large scale. The second half of the twentieth century witnessed the emergence of most modern societies from economic collapse and world war into a period of spectacular economic growth and social change. The accelerating growth in population, technological prowess, consumer demand, transnational corporations and the global economy has created a new force on the planet. We have been transformed from naked apes into what can be called a superspecies. We are now transforming the biosphere — depleting the oceans, poisoning the air, levelling mountains and altering the composition of the atmosphere — and we are doing it all in a mere instant of geological time. In the nearly four billion years that life has existed on Earth, no species has possessed this capability for changing the biophysical makeup of the planet and thus affecting every other species on Earth.

For most of human existence, the natural world seemed to be the unlimited source for our physical, social and spiritual needs. Natural systems readily absorbed the impact of our activities, and even when we exceeded the finite limits in one area we were able to move to new spaces unoccupied by our kind. But those seemingly limitless vistas have disappeared far faster than our culture has proved able to comprehend and adapt. There are no new lands to be exploited, and the diversity and abundance of every natural system on Earth are being rapidly degraded by human activity. This crisis has to be faced collectively, yet the driving forces of our current culture are still aimed in the wrong direction, still pursuing endless growth in a world that is perceived as an infinite source of wealth and opportunity.

I've been working on environmental issues for almost four decades. I was first moved by Rachel Carson's 1962 book *Silent Spring*, which warned of unpredictable negative consequences of new technology. As a scientist, I still had a profound faith in the benefits of science and technology, but her book made me re-examine many of my assumptions. I didn't begin my career in broadcasting and journalism thinking that all natural systems were in peril. But I was concerned about the profound implications of advances in genetics, which is my area of expertise, and I felt a responsibility as a scientist. It was only after many years of observation, reflection and discussion with many people that I began to realize the magnitude and gravity of what was happening on the planet. For the past twenty-five years, I've been making films about people who are fighting to save the last local woods or endangered species or habitat; I have noticed the quality of air, water and soil rapidly diminishing and the productive capacities of forest and marine ecosystems being stretched beyond their limits. And I've watched our governments and businesses respond reluctantly and too little to these escalating crises.

Like most other people, I'd rather spend time with my family, go to the lab and do research in my field of genetics, or pursue my hobbies. But I have children and grandchildren. I have a profound stake in the future. I'm only one person, and I have no illusions or conceits about saving the world. But I hope my grandchildren will never look at me and tell me, "Grandpa, you could have done more for us." If we adults fail to put the environment on the front burner, our children and their children will not have any hope of experiencing the abundance and diversity of life's creatures that existed when we were still young. I have read the statistics on contaminated water, asthma and cancer rates, allergies and immune problems, climate change and the loss of topsoil. I realize that the very things that give our children health, and even life, are in jeopardy.

In 1996, I approached the writer and researcher Holly Dressel with the idea for another radio series. I had worked with her in the past and knew that, apart from being one of the top people in her field, she had a deep understanding of and concern for the global eco-crisis. Together we proposed to the CBC an eight-part series that would move beyond a simple description of the degradation of the environment. What we wanted to investigate was why, when there is so much consensus among scientists that the biosphere is being altered catastrophically by human activity, with results that will be disastrous for our own species, are governments and businesses failing to respond appropriately to reduce our ecological impact? After we interviewed scores of scientists, activists, politicians and businesspeople, it became clear that it is the deep-seated beliefs and values of modern culture that are both creating these problems and blinding us to their consequences.

We repeat like a religious mantra the unquestioned benefits and power of science, information and economics, without inspecting the structures and methodology on which they are built. Many of these beliefs are insupportable and dangerous. For example, the notion that human beings are so clever that we can use science and technology to escape the restrictions of the natural world is a fantasy that cannot be fulfilled. Yet it underlies much of government's and industry's rhetoric and programs.

Another dangerous assumption is that science can penetrate all of life's mysteries and extract the information we need to manage everything around us, thereby giving us total control over our own destiny. Science has generated many of the wonderful technologies that we enjoy today, but its methodology, which separates nature into narrow specialties, has prevented us from seeing the big picture. In real life, environmental issues cannot be separated from the economic, informational or cultural aspects of our lives. They are intertwined.

For all these reasons, the radio series looked at many aspects of culture, not just those considered to be "environmental." If we can see where we are now and where we are heading, then we might be able to assess whether we really want to continue on the same path. Some areas we investigated for the series, like the rapid and unregulated progress of biotechnology, have frightening implications, while others, like the way people in the Third World are recognizing and fighting the recolonization of their countries by the global economy, are inspiring.

Overall, we interviewed nearly 200 people and spent more than two years gathering tape and editing the results. Our ambitious eight-hour series attempted to provide a picture of the steady loss of our natural systems in the context of the global culture that is currently shaping all our lives. We tried not only to analyze what is happening to our physical support systems but also to suggest some areas of our culture that are ripe for change. Public response to the radio series was very gratifying. It attracted a large listening audience, and the audio package has sold more copies than any other such package in the history of the CBC. To us, this meant that people were indeed very concerned about environmental issues, and that they will go out of their way to get information about them.

The reality of radio, however, is that once each show was cut and assembled, we were able to include only a small fraction of all the interviews we'd done. The number of topics and ideas we could cover in each hour of radio was also restricted. But we had so much wonderful raw material that we wanted to get more information out. This book was, therefore, originally built around the radio interviews, both those that were included and those that for a variety of mostly technical reasons didn't make the final cut. Our intention was to look at all the things that affect the environment in our culture. We wanted to show that one reason we're in so much trouble is that our modern culture is paradoxically behind the times, still assessing the world the way it did in the nineteenth or even eighteenth centuries, as a place of inexhaustible resources, where humanity is at the pinnacle of creation, separate from and more important than anything around us.

This book is in many ways a departure for me, a discussion of modern media, economics, culture and history that might seem some distance from my real expertise, science and nature. But culture controls how we treat the Earth, so we interviewed many experts who could give us a glimpse of the big picture. The reason we're looking at such a big picture, however, is to assess the state of the natural world, the state of our life-support systems as described by the best scientific studies. For that reason, in the early chapters of this book, I felt we had to do an inventory of what we've done to nature, how much we use and how much we have left. That inventory is always at the core of our discussion of the belief systems — modern culture, economics, politics and media — that have led us to consume our natural resources so excessively, and so quickly. Hence the new first chapter in this edition.

The second chapter, "Bugs Я Us," is a kind of history of the cultural assumptions on which we have based our present use of the natural world. One major assumption we make is based on size. We continue to be impressed with megaflora and -fauna, the big things, when in fact it is the insect and microbial world that keeps this planet habitable. We are only just starting to realize that worms, ants and microorganisms are more important than humans or their technology when it comes to the continuance of life on Earth. We are also just starting to realize that ancient human cultures held belief systems that dovetail with the most recent discoveries of science.

The third chapter, "Bigfoot," takes an inventory of natural resources on Earth. How much was there before humanity's great expansion? How much is left? And how much would each of us have if resources were shared out equally? These basic questions must be addressed if we are to learn to live in balance with the natural world.

The fourth chapter, "Sez Who?" establishes just what is being conveyed through the media and how very destructive values, such as the glamour of consumption, are perpetuated. This chapter also suggests ways that concerned citizens can wade through the morass of information and use the media to present issues that shape the planet's future. The fifth chapter, "Unnatural Selection," explores how the desire to conquer and control nature has led to biotechnology and the deliberate redesigning of life through the manipulation of DNA. This chapter explains why this manipulation is not only dangerous but is also doomed to fail in the long term because of its reductionist base. The sixth chapter, "Your Money or Your Life Forms," shows how this search for gold inside our own bodies is tied to an increasing lack of capital in the world. Most of the world's true wealth has already been exploited. We investigate the implications and justice issues surrounding the commodification of life itself.

Chapter 7, "Follow the Money," examines the economy — an entity that late-twentieth-century culture has not only reified but also deified. What is the economy really and, as Lewis Carroll asked, "Which is to be master?" Chapter 8, "Globalization Blues," explores what it means to extend the paradigm of the ever-growing, ever-insatiable economy to the entire planet. How does globalization both help and harm the environment and local communities? Who reaps the benefits? Are there alternatives?

Chapter 9, "The Other World," includes many voices that were left out of the radio series. Ironically, people in the Third World, who are perceived as poor and in need of aid from industrialized nations, are themselves the focus of treasure hunters trying to cash in on biological and human diversity. In fact, it is the rich nations that are sometimes poor in varied approaches to resource conservation, and the lesser-developed societies that have much to teach us about sustainability and about enduring values.

The final chapter, "Complex Pleasures," tackles the logistics of sharing, taking action and working for change. In our research, we discovered that these strategies not only benefit the environment, but also satisfy basic human needs far more than the infantile ones currently being served by a society focused on consuming everything in sight. It also points towards solutions.

When it has been necessary to use the first person in this book, we have written from the point of view of the senior author. But that person is really an amalgam of both David Suzuki and Holly Dressel. It has been a joint effort all the way, a dual offering for the next millennium.

David Suzuki and Holly Dressel
Vancouver, British Columbia

CHAPTER 1
SHARING EACH OTHER'S SKIN

There is enough for everybody's need, but not enough for everybody's greed.

— MAHATMA GANDHI

When we began work on the first edition of this book at the very end of the twentieth century, it was becoming clear that the environmental goals put forth with so much hope at the focal point of decades of environmental activism, the Earth Summit in Rio de Janeiro back in 1992, were not only unable to keep up with increasing environmental degradation, they were under relentless assault by mainstream economic and political forces. "Globalization" was still a relatively new term that was being heralded as a social and economic salvation for the world, and most people were still unaware of the massive giveaway of national regulatory rights that had taken place under the world trade agreements signed the same year as the summit. Popular demonstrations and civil disobedience had not yet brought the actions of the World Trade Organization (WTO), the World Bank, the International Monetary Fund (IMF) and Chapter 11 of the North American Free Trade Agreement (NAFTA) to the attention of the average citizen, especially in North America.

Most people were also blissfully ignorant of the fact that genetically engineered foods had entered their diets and that the biotechnology industry was releasing commercial products that were beginning to have frightening impacts on the environment, along with a growing potential for harm to human health. Creating genetically altered organisms to serve our own whims and purposes also raises some of the most serious ethical and social concerns our species has ever had to face. All through the late 1990s these growing threats to global environmental stability were largely ignored as the world's media pursued stories of political sex and celebrity peccadilloes, dot-com proliferation and the exciting new Information Revolution. *From Naked Ape to Superspecies* was a book intended to address the serious threats our cultural obsessions posed to natural systems, and to remind readers that without clean water, air and viable soil, no cultural or economic life, even a virtual one, could exist for long on this planet.

Today, the continuing exponential growth in human numbers, consumptive demand, technological power and economic reach is putting increasingly unbearable pressures on the most basic commodities produced by the Earth. Global wars are being fought over oil, water is being rapidly privatized by multinational corporations all over the world, and there are so few intact natural systems left that entrepreneurs are now invading thousands of national parks as well as preserves set aside for indigenous peoples to dig for oil and gold, or to log and "develop" the area. These escalating activities have also placed many of the most basic, democratic rights that Westerners take for granted under serious threat.

Ecologists tell us that once the complex, interlocking relationships that make up a natural environment, like a forest, a fishery, good agricultural land or a watershed, are undermined beyond a certain critical threshold, it will collapse, usually quite suddenly. If recovery of a forest or a fishery is possible at all, it may take thousands of years. With so much at stake in terms of the air we all breathe, the food we eat and the water we drink, convincing people that we need to reassess the direction in which we are headed has become even more urgent. Put simply, we must learn to live in other species' skins, as well as in our own.

Cyber-Utopia

Transcendence is a wrong-headed concept. It means escape from the earth-bound and repetitive, climbing above the everyday. It means putting men on the moon before feeding and housing the world's poor.... The revolutionary step would be: to bring men down to earth.

— CYNTHIA COCKBURN, *MACHINERY OF DOMINANCE*

When I'm out on a lecture tour, I have tried as a scientist and journalist to remind people that we humans are embedded in and dependent on four simple things that we normally take completely for granted: clean air, water and soil and the fiery energy of sunlight. I tell my audiences that the fact that these four "elements" support all life on earth was once understood so well by nearly all human societies, that Earth, Air, Fire and Water were at some point in their history considered sacred. What we have only recently learned scientifically is the fact that all four of these elemental components are cleansed, replenished and, in fact, actually created by an intricate web of the very life forms their presence makes possible. This web of living things, which scientists call "biodiversity," but most people understand simply as "nature," is able, for example, to capture the energy of sunlight and then transform it into sugars, that is, storable chemical energy, for other living things to eat. They take up carbon dioxide and release oxygen in photosynthesis, making an atmosphere that other life forms can breathe. And they filter water and create soil, the other major prerequisites of life as we know it. I explain that over billions of years microorganisms, plants and animals, that is, life, and its interlocking mechanisms, like metabolism and filtration, have survived and flourished. Every different form has gradually developed the unique mechanisms necessary for its own survival, adapting to conditions from the Arctic tundra and the Gobi Desert to the rain forests, mangroves and peatbogs of the great wetlands. Because this is happening on a planet that has access to finite resources of water, air and so forth, each life form has had to find ways to constantly reuse and recycle the limited materials available to it.

I try to construct my speech in such a way that the audience understands the scientifically demonstrated fact that all this sharing of finite materials means that one organism's waste becomes another's opportunity for survival. Then I explain that the air we breathe contains molecules and atoms that have been exhaled by living organisms around the planet, and that these elements once resided in that life's various ancestors, back over millions of years. Water, an absolute prerequisite of life, for instance, is constantly circulated through the hydrologic cycle so that our morning coffee may carry molecules that once composed a dinosaur's tear, or dissolved the poison that killed Socrates. Therefore, as living organisms that also evolved upon this planet, we, too will leave marks on those molecules of air, water and soils, with repercussions for the rest of animal and plant life on land and in the oceans, which in turn will affect the atmosphere and the weather. That means, I tell the audience, that we have to step carefully to avoid depleting or contaminating these resources any faster than they can be replenished and cleansed. It means we have to live within limits set by nature, not politics or economics; it means we have to share — or else we are robbing our neighbours, ourselves in later life and, most important of all, our own children of a healthy and productive future. Once the implications of these statements sink in, most of the people who make up my audiences start asking questions designed to figure out *how* we can live in greater harmony with those things that make our lives possible. There are always a few people, however, who challenge the entire idea that humans have to operate within limits imposed by the finite nature of the planet.

In the late 1990s, information experts writing in *Wired* and *The Economist* were predicting a brave new world, where ideas spread electronically through cyberspace would enrich a humanity finally liberated from dealing with the physical world. They enthused that we were about to become a species unfettered by entropy and gravity, physical cause and effect, natural or workforce variables. We would be able to feed our families through the actions of our minds alone, and they talked seriously about extending our lifespans to 150 or even 250 years. I tried to point out that the mainframes and the motherboards of the machines we use to do all this disembodied thinking still have to come from the earth, and will be deposited back there when broken or outdated, where they're even now leaching shocking amounts of toxins into our water tables. I also said that if humankind aims to live forever, we'd have to stop reproducing entirely.

But rational arguments about limits often fail to persuade those who believe in the near-limitless potential of human creativity. They envision a new cyber life form that would somehow gradually merge with its own machines and move beyond eating, breathing and defecating, maybe even beyond dying. A few years ago these information technology (IT) experts were proclaiming that our economies were evolving in the same way, beyond consuming pork bellies and steel casings and extruding smoke and toxic wastes, towards dealing with pure ideas, in a cyberland that could ignore the dirt, the organic messes and the unpredictable chaos of the earth.

The Information Revolution that gave birth to these ideas was so important we devoted an entire chapter, "Sez Who?" to analyzing what media and cyberspace were doing to our culture and our relationship to the more tangible aspects of life on this planet. Communication technology breakthroughs have dazzled the public imagination and cornered economic investment, often to the exclusion of many more practical concerns. In fact, the repercussions of changes that personal home computers have brought to our daily lives, to say nothing of the computerization of all financial and business information worldwide, are hard to overstate. Yet that's just what the early theorists of IT did. Once they had grasped what computerization could do for the speed and proliferation of business information and in our personal lives, they projected their exuberance into the realm of science fiction. They predicted this materially based and materially convenient technology would actually transcend the limits of the material world, thereby freeing humans from the constraints and limits of both their bodies and the world's physical realities. In other words, the theorists believed that not only would IT help disseminate and promote traditional economic and cultural activities and communications, as indeed it has, they thought it would completely *replace* physical goods, creating a new, non-physical world. Since so much of the public was still barely grasping what these obvious geniuses were on about, many people fell for their vision of the future, and the dot-com bubble blew up to gigantic proportions. The sudden, spectacular economic collapse of the bubble followed only a couple of years later, but that event was considerably less analyzed and promoted in the media.

Western culture invested almost everything in IT in the last two decades of the twentieth century, diverting its attention from vital global issues of peace, ecological sustainability and human rights. Today, IT's more elaborate predictions of cultural revolution have disappeared, and its major proponents, as one ex-geek puts it, have been "reduced to bitter screeds about the good old days, posted on web sites that nobody reads." As Canadian journalist and ex-IT worker Ken Hechtman says, "Anybody without tangible, physical products got taken to the cleaners. People quickly figured out that if something isn't necessary to your personal or business survival, like food, fuel or the screws holding your printer together, they could either do without it or figure out some way to steal it."

One way the cyberworld was going to generate revenue and support all of us was through advertising. Our personal web sites, selling everything from fiction to ideas and products, would be self-supporting, if we just sold a little line space to corporations. But it soon became clear that people would not pay for things that were either intangible or easily available elsewhere. Hechtman uses the example of banner ads, those annoying graphics that jump uninvited onto our web pages and email. "As a web page owner, you used to get one cent an impression, that is, a penny every time someone opened your page and saw the ad. If they clicked on through to the advertiser's page, you, the host, got a nickel. But, of course, no one clicks on them now. Web designers even warn people who don't have ads on their home page not to put anything important in the banner-ad section, because people have trained their eyes to not even look there. So today most advertisers don't have to pay a single cent for an impression." Hechtman says the host is only paid a percentage of real sales of a real product actually generated through that host. Which isn't much, and quality has declined fast. "Banner ads used to be for Nike or Novartis, real companies selling real goods to normal people. Now they target only the really stupid, like the back pages of cheap comics. You see ads for weight loss, bladder control or singles clubs — sucker games, not real products. The only other way you can make a little money on the Net is through direct mail spam, and people are beginning to learn how to avoid that too."

The Fifth Estate

What finally provokes a journalist to resign in protest of bias? The answer is: when she begins to feel that that bias is doing her nation harm.

— PATRICIA PEARSON, JOURNALIST

If the ethereal nature of the IT world contributed to its demise as the major economic power of the future, other, more useful and less venal aspects of the cyberworld are still alive and well. Until a few years ago, fledgling journalists providing alternative and independent analyses were still sometimes able to string for the major newspapers or radio stations, learning their profession on the frontlines of world events while making some kind of a living. Today, with a very few exceptions, people like Hechtman, quoted above, an expert on the Taliban who was briefly kidnapped by them during the war in Afghanistan, are posting their stories on the Internet. The Guerrilla News Network, GlobalPublicMedia.com, Disinfo.com, Indymedia.org, TomPaine.com, Straightgoods.com, Undercurrents.org and many more do provide a real alternative to corporate-funded mass media, and are now keeping even average North American families apprised of the many underlying reasons for the war in Iraq, for example, or the latest machinations of the WTO. They deserve a good deal of the credit for stimulating the massive peace marches and the unprecedented resistance to the war in Iraq by most of the people on Earth. There is not much material support for the professional journalists doing this important work, however: they labour mostly for free.

It's just as difficult to get people to pay money for profound and vital ideas as for banal, commercial ones. There has been a virtual enclosure of mainstream news information — nearly all North American news is delivered by five mega-corporations, three of them, like Disney, General Electric and AOL Time Warner, being far more concerned with selling their entertainment and manufacturing products than with contributing towards an educated public. So today most of the really serious journalists who want to write about the vital issues necessary for informed social and political decisions, do so for the love of it; the people running the new alternative news sites survive hand-to-mouth.

Besides the obvious centralizing and entertainment compulsion of television news, today there are also fewer and fewer commercial opportunities for radio, magazine or newspaper journalists to dig beneath conventional, mainstream analyses and receive a salary for their work. There's *The Guardian* and *The Independent* in the United Kingdom, *Le Nouvel Observateur* in Paris, specialty magazines like *Mother Jones, The Nation* and *Harper's* in the United States — and almost no hard-copy alternatives in Canada except for *Adbusters,* which is largely geared to the U.S. market. This centralization of mainstream information means it's difficult for the few alternatives to reach large audiences, or for non-corporate journalists to be given significant access to breaking stories.

As we point out in chapter 4, "Sez Who?" healthy, independent media, the "fifth estate" standing apart from government, church, business and the citizenry, have always been considered an indispensable component of a democratic society. Journalism is supposed to be composed of independent professionals who aren't funded by anyone with a serious economic or political stake in the information produced. But what is now considered normal, mainstream journalism has become so compromised by the concentration of media ownership in commercial hands that on March 31, 2003, two major television networks, Global in Canada and ABC in the United States, separately ran two unannounced half-hour news specials about the war in Iraq that actually warned viewers away from their usual coverage.

In these programs, slipped in without fanfare on a Monday evening, mainstream television journalists criticized their own talking heads, empty visuals, and restricted access — even their own computer graphics. Over boxes of four different stations (including their own) each running an interview with a different expert, but showing the exact same "war" footage on one side, the voice-overs told their audiences to "notice that the visuals have absolutely no relation to what the talking head is saying." Over the animation of the war's progress, audiences were asked to "note that the trucks in the animation always move steadily ahead, and the only enemy they encounter is puffs of smoke . . . " Then actual footage of the war showed real trucks bogged down by snipers, lack of fuel and general chaos. Web site addresses and capsule reviews of some of the alternative media mentioned above were even offered, in order to give audiences a chance to get another viewpoint.

This is an interesting development; just before the Iraq conflict, during the rapidly eclipsed war in Afghanistan, journalists were not considered professional or serious unless they worked for big corporations like Disney, CNN or AOL Time Warner. By the Iraq war, however, it was becoming far sexier to get information from journalists who weren't "embedded" (the new term for news coverage that was "official" and also controlled by the U.S. military). Many networks, especially the CBC, used non-embedded stories preferentially. The salaried employees of the big media corporations, so unlike the classic hard-drinking, chain-smoking journalists of a generation or two before, are required not only to look good, but are much more carefully restrained from reporting stories that might embarrass or otherwise discomfit their distant, long-tentacled employers. Nevertheless, even some of these reporters still know what their job is supposed to be, and many within these corporate behemoths have gone public in recent years in order to expose what has been done to their profession. The people we chose five years ago to assess the media, like Leonard Grossman, formerly of NBC and PBS, have now been joined by many more.

In the story by former *National Post* columnist Patricia Pearson quoted at the beginning of this section, she explains that she left her job with a national newspaper because it became increasingly impossible for her to defend popular and even official Canadian positions on many issues because of interference and bias within the huge media consortium that employed her and vetted her stories. She describes the atmosphere of scorn and bias rampant at her former paper, the *National Post,* very succinctly. "When CanWest, controlled by the Asper family, acquired the paper from Conrad Black, I no longer dared to express sympathy for Palestinians," she wrote. "When my editor, of whom I am fond, revealed a deep suspicion of environmentalism, I self-censored in favour of con-viviality... [during the prelude to the war on Iraq], debate — so critical for Canadians at this juncture — was trounced at the *Post* by a sort of Shock and Awe campaign against any liberal position . . . " After weeks of this treatment, Ms. Pearson, granddaughter of Lester B. Pearson, the man many people feel is the most quintessentially Canadian of Canada's prime ministers, resigned. "I cannot sit back," she wrote, "and watch this nation attacked, relentlessly and viciously, by a newspaper that would trash so much of what we believe in, from tolerant social values to international law, belittling [Canadians] for having our beliefs, while turning around and saying that what makes America great is the Americans' ardour in defending their beliefs."

This corporate-concerned, centrally controlled view of the news is exactly what analysts such as Leonard Grossman and the late Neil Postman foretold five years ago when they discussed the direction mass media were headed. They also suggested antidotes. Grossman thought that academic and public libraries, many of which were threatened by the IT revolution, should learn to use Internet connections to their advantage, and this has come true to some extent. Libraries that have not been downsized or closed due to budget cuts have since often become hotbeds of technology that are now helping people research their interests more effectively. Grossman had hopes for universities as well, but they are steadily becoming ever more corporatized. In the sciences especially, universities increasingly serve the interests of the businesses that fund their research and academic departments, rather than devote themselves to the exploration of pure knowledge. So there are now fewer disinterested large institutions to which people can turn for unbiased scientific, political and social information. Again, the major source of non-corporate-influenced information lies in the chaos of data available on the Internet. Here, those people with the critical ability to tell bogus from real research can eventually access global sources, whether institutions or non-governmental organizations (NGOs), that still honour academic principles above economic ones.

Playing Genetic Roulette

> *According to sources, the [new GM] rice could be mixed with regular rice in the process of distribution, or its pollen might cross-fertilize with regular rice. In such cases, people could consume allergenic substances in rice unknowingly.*
> — "GOVERNMENT UNLIKELY TO OK ALLERGENIC GM RICE,"
> *YOMIURI SHIMBUN,* SEPTEMBER 8, 2003

When we first began researching the commercial applications of biotechnology for our radio series, that industry resembled the Information Revolution in terms of economic and philosophical excitement. It was, in fact, close to hysteria. The sky was literally the limit, and there were almost no downsides to the technology of splicing and recombining genes mentioned in the press — only a vision of cured diseases and fabulous new foods, products, and methods of manufacture. Food products were and still are looked upon by the public with some reserve, but who can resist the hope that terrible diseases, injuries and hereditary defects may be susceptible to revolutionary new cures? As a trained geneticist, I must admit I revelled vicariously in the astounding advances in technological dexterity that were being acquired. But I was even more acutely aware of how much we have yet to learn.

It's in the nature of revolutionary new areas of science that most of our current ideas will prove to be wrong or at least far off the mark. The way that scientists work when faced with a new set of observations, is to try to make sense of the data by creating a hypothesis that then can be tested experimentally. When the tests are carried out, chances are that the hypothesis will fail to be confirmed and will either be discarded or modified and tested again. This is how science progresses — by proving our current ideas are wrong or off the mark. Roger Perlmutter, executive vice-president of research and development for the biotech company Amgen, says: "Things we take as the absolute truth now are going to look pretty silly a few years from now." Craig Venter, the brash and controversial entrepreneur who was a key figure in accelerating the completion of the sequencing of the human genome, admitted in 2000: "We know far less than 1 percent of what will be known about biology, human physiology and medicine. My view of biology is, 'We don't know shit.' " But neither Venter nor Amgen have been reluctant to release commercial adaptations of our ignorance into the environment.

Back in the 1970s, when it was becoming clear new techniques for manipulating DNA were about to revolutionize the whole field of genetics, I expressed my concerns that in the intoxicating atmosphere of new discoveries, scientists would forget the tentative nature of their insights and hypotheses and begin to view them as established truth. Although work in biotechnology has become ever more impressive and exciting, even today, almost three decades later, we still have the barest understanding of how genes are regulated, how their products interact within cells, to say nothing of the complex interactions of whole suites of genes that are turned off and on during development and differentiation. These are extremely large and important blanks in our knowledge. They mean that *we still cannot reliably predict what will happen when we move genes from one organism to another*. In fact, one of the strongest warnings about the degree of our ignorance comes from biotech's greatest triumph to date — the elucidation of the sequence of all 3 billion letters in the human genome, the goal of the great Human Genome Project that was begun in 1999 and completed in 2002. Decoding the human genome represented a staggering technological feat. Twenty or thirty years earlier, no one but the most wildly imaginative scientist could have anticipated that such an enterprise was feasible, let alone that it would be completed so quickly.

Based on the amount of DNA within a human nucleus and our knowledge about the average size of genes, it was expected by every expert working on the project, as well as regular geneticists like myself, that the human genome would yield a motherlode of about 100,000 genes. This expectation was based on a lot of things, including comparisons to other, simpler organisms. The first complex organism (eukaryote) to have its entire genome sequenced was the single-celled yeast *Saccharomyces cervisiae*. Among its 12 million bases were 6,000 genes. In contrast, the simple but multicellular nematode *Caenorhabditis elegans,* which is made up of only 957 cells, was found to have 97 million bases that encode 18,000 genes (30 percent of which were related to human genes). Scientists focused on the fruit fly *Drosophila melanogaster,* perhaps the most popular organism for genetic study because of its rapid reproduction; it has more cells in one eye than are found in an entire *C. elegans.* To their astonishment, scientists found that even though it has a genome of 180 million bases, *Drosophila* had only 13,600 genes. That finding should have been a warning that surprises were in store; and, in fact, when it was announced in 2002 with great fanfare that the human genome had been essentially completely sequenced, the project heads admitted that instead of the expected 100,000 genes, humans appear to contain only a mere 30,000.

Scientists have long known that for most of the entire span that life has existed on Earth, bacteria ruled and in their world very simple, single-celled microorganisms developed nearly all the survival strategies that life uses today. Despite their small number of genes, these creatures established the ground rules of life — absorption, metabolism, excretion, growth, reproduction, photosynthesis, sensation — that have become part of the basic mechanisms for all other life forms. Even multicellularity and differentiation, when they came along, also depended on these fundamental processes, and do not require orders of magnitude of more genes. What the deficit in genes in our own genome should inform us is that despite the tremendous progress in our knowledge, we are still a long way from understanding the complexities of genetic structure, function and interaction. In other words, it's now even more obvious that genes are far from being the sole deciding factors when traits are expressed in an organism; much more complex relationships and processes are at play. And yet in most commercial genetic engineering as it is currently practised, splicing genes from one organism into another is considered to be all that's necessary to induce a desired trait, and unpredictable side effects are not supposed to happen.

The Human Genome Project was not just a triumph but also a very humbling insight. This latter aspect was quickly glossed over. Gene technologists as a whole do not seem to have heeded their fundamental erroneous assumption as a signal to exercise more caution. Rather, the sequencing of the human genome is being trumpeted as the beginning of a whole new universe of insights and techniques that will benefit humankind. In early 2003, for example, I received a self-congratulatory email from the International Human Genome Sequencing Consortium when the human genome was finally and officially decoded. There wasn't a single mention of the phenomenal surprise — that they had overestimated the number of genes by 300 percent. The press release devoted a lot of space to discussion of what was referred to as "grand challenges" for the future, that is, the various practical and especially commercial applications of this newfound information. But the small number of genes found didn't rate discussion, even though to scientists that shock should have been one of the most important "grand challenges" to emerge from the whole project.

If we continue to avoid serious discussion of the obvious inconsistencies and inadequacies of current notions, we are bound to miss real opportunities and to encounter such stunning surprises again and again. Of course, that announcement was a public-relations exercise meant to justify the hype and encourage continued financial speculation and support in the field. In fact, to this day, very few of the ballyhooed benefits of biotechnology have been realized, even in pharmaceuticals. Taken as a whole, genetically engineered crops have been proven to provide far more benefits to chemical and biotech companies than they do to consumers and farmers. Gene therapy has also not worked reliably; indeed, there is strong evidence to suggest that injecting DNA into patients is lethal. Cloning in mammals seems fraught with unexpected and negative results. And even genetically engineered drugs like insulin have been accompanied by severe side effects. (See references to chapters 5 and 6 at the end of this book.)

One way to tell where we are in this quest for knowledge is to look at the articles in the field. When scientists achieve the level of understanding and control they really need to apply their knowledge and techniques predictably, there won't be a lot of scientific papers on genetic engineering being published, because they won't be able to add anything to scientific understanding. At present, however, biotech articles fill the journals, and new journals and papers in the field are mushrooming — for the simple reason that we're still learning a lot we didn't know. As a geneticist myself, I have no doubt there will be enormous benefits that will develop from our insights into molecular biology. It is the haste to exploit incremental insights into some sort of practical application that is my major concern. When a specific sequence of DNA can be synthesized or isolated, inserted at a precise location within a genome, and the outcome derived exactly as predicted, then I would say the technology is mature enough to take to the next stage of growing in open fields or testing on humans. We're very far from that right now. The drive behind the commercialization of DNA technology is money, not the improvement of agriculture, the environment or human health. That drive is making us reckless in the claimed benefits and actual results.

As we will see in detail in chapters 5 and 6, we are conducting very crude and frightening experiments with our own bodies and the entire larger environment when we insert laboratory-created organisms into the viruses, bacteria, plants and animals that have until now been our partners and allies in keeping the processes of life going. Incidentally, the research institute responsible for creating the extremely dangerous allergenic rice mentioned in the quote preceding this section was indignant that the Japanese government might interfere with its work on an "edible vaccine" against hay fever. "It doesn't make sense," spokespeople for the institute said, "that the ministry has refused a new product under the existing standards."

Drugs as Food, Pathogens as Product

Got Pus?
— SLOGAN FOR ANTI-BGH CAMPAIGN, BASED ON THE INCREASED
RATE OF UDDER INFECTIONS SUFFERED BY BGH-TREATED COWS

This is one of the very few hypervirulent organisms ever created.
— LISA MORICI, LEAD AUTHOR, UNIVERSITY OF CALIFORNIA AT
BERKELEY STUDY ON TB

In chapter 6 of this book, we move from investigating the technological flaws and difficulties of the gene-splicing processes of biotechnology to the social implications of allowing companies and individuals to claim exclusive ownership over any form of life. The cascade effects that accompany relinquishing the idea that any given organism ought to have the first rights of ownership to its own cells become vivid when we remember that human cancer patients have discovered that their cells have been patented without their consent and sick children who can't afford the costs of their genetic diseases have found their genes being used in the marketplace to enrich doctors and companies they've never heard of. More important, the idea of life form ownership is affecting everything from the spread of dangerous pathogens to people's choice of food. Companies "owning" a gene in a living organism can forbid other people from using it; they can take it off the marketplace and hide it. So, if a company is trying to decode the genome of a pathogen, once it has a part of its methodology patented, it can prevent other researchers, including government and public-interest groups, from even looking in the same direction. This, as many scientists are beginning to notice, is becoming a serious impediment to scientific progress. It is also very much in the biotech companies' interest to destroy distribution, advertising and markets for *non*-genetically engineered products.

In New England, a battle has been raging for at least a decade over whether commercial dairies have the right to tell their customers that their milk products *don't* contain a genetically engineered substance. In this era of mad cow disease, E.coli and fears of virus-jumps from cattle, poultry and pork products, the core purpose of food labelling is at stake, though few people have realized it. And, of course, the whole idea of "pure food," that is, food unadulterated by man-made substances, is also in peril.

The notion of government authorities getting involved in maintaining pure food and drink can be traced back to the late Middle Ages when unscrupulous producers adulterated wine and liquor with chemicals and added lead or mercury to children's sweets in order to create tempting colours. By the turn of the twentieth century, with the advent of processed food and drink, the whole idea of national inspection agencies, local "seals of approval" and pure food and drug legislation developed rapidly in order to control new and potentially harmful products like Coca-Cola which, in the nineteenth century, not only had three times the caffeine as an espresso, but was well laced with cocaine, a new plant-based drug. Coca-Cola was marketed to children as young as three or four. Today, however, we're seeing a reversal of this protective trend. Because of the vast amounts of money potentially available to chemical companies in the form of biotech patents, food producers are being taken to court for labelling their products as pure and unadulterated.

Maine's Oakhurst Dairy is only one such company that has been taken to court in New York State, Vermont and elsewhere in the United States by Monsanto Corporation for labelling its milk as "BGH-Free." No other comments are on the label, just the fact that it doesn't contain a particular substance that doesn't occur naturally in milk. Having lost on the state level, Monsanto is now taking its suit claiming "deceptive and misleading advertising" to a federal court. Recombinant bovine growth hormone (rBGH) is the infamous, genetically engineered hormone added to U.S. milk for years now that has been banned in Canada and in all of Europe because of the stresses it puts on dairy cows. Because the hormone forces the bodies of cows to produce significantly more milk than they could normally, the immune systems of the animals are affected. The cows become more susceptible to illness, especially infections of the udder, which increases the amounts both of pus and of antibiotics in rBGH milk. On average, their life expectancy drops precipitously.

As a recent article in *Mother Jones* put it, "[Monsanto] has invested heavily in rBGH, spending $100 million on development and $150 million for a new production plant in Georgia. Prosilac, its brand name for the drug, is now injected into roughly a third of the U.S.'s 9 million dairy cows. It brings in more than $1 billion annually in sales." The lawsuits against the Maine dairy wouldn't exist if Monsanto hadn't been able to patent the growth hormone in the first place; it now sees lost dollar signs in every ounce of milk sold that *isn't* paying the company a premium.

While some would argue that it's debatable how useful most of the products of biotech are, since they all come with their own unique problems, it's not difficult to see what might happen to the abilities of public agencies attempting to protect the food and drug supply when there is now so much economic incentive to privatize all aspects of both. As the Pew Initiative, a think tank set up by the Pew Charitable Trusts to study genetic technologies states, "The history of biotechnology is that the regulatory system is always playing catch-up. The question here is whether the regulatory system can begin to think about who is now in charge."

In chapter 6, we explore these and many other social, legal and moral aspects of the patenting of life that go far beyond the dangers to health and the environment. For example, with so much money being dangled in front of them in the form of patents, researchers are juggling with DNA at a feverish rate that could prove catastrophic. In December 2003, scientists at Berkeley "accidentally created a highly virulent form of tuberculosis by trying to alter its genetic structure." The new mutant form is not only much more deadly, the opposite of its creators' intentions, it multiplies more quickly. "This breaks a longstanding assumption among scientists that disabling a potential virulence gene weakens a pathogen," said Lisa Morici, lead author of the study. And yet that very assumption is at the heart of every single claim that biotech is safe; the gene-splicing technique nearly *always* uses a supposedly disabled pathogen in the form of a vector to introduce its new genes into an organism in the first place. The scientists were so surprised they did the experiment twice. "Further investigations suggested that the genetic changes had the unexpected effect of undermining the host's own immune responses to TB." Despite such "surprises," it is very much in any lab's economic interest, under the present patenting regime, to go on experimenting in all directions in order to come up with some kind of salable genetic commodity.

The Global Casino

> *Today, globalization is being challenged around the world . . .*
> *and rightfully so . . . . For millions of people, globalization has*
> *not worked. Many have actually been made worse off, as they*
> *have seen their jobs destroyed and their lives become more*
> *insecure . . . . They have seen their democracies undermined,*
> *their cultures eroded.*
>
> — JOSEPH STIGLITZ, *GLOBALIZATION AND ITS DISCONTENTS*

The main reason biotechnology is creating products it doesn't understand and can't control is not because we can't live without herbicide-resistant soybeans or — this has literally been created, although not approved for the market — lawns that glow in the dark. The main reason our governments and industries push these technologies is in the service of commerce and trade. Our system of economics demands constant new sources of capital and new products for consumption; what better way to get them than by playing mix-and-match with DNA? Not only the make-up of our own cells, but even our most spiritual and altruistic ideals have been enfolded into the all-encompassing field of economics. And we are learning that this vision of the centrality of money and economics that has gone global brings a burden of social and ecological costs.

Since the late 1990s, even the international environmental community has been forced to downplay its more serious concern, the destruction of natural systems, in order to pay attention to economic interests. In 2002, my daughter, Severn, attended the tenth anniversary of the seminal global environmental meeting in Rio de Janeiro, the famous Earth Summit of 1992. By 2002, however, Earth Summit was considered too radical a term to describe the meeting's goals; it was instead dubbed the "World Summit for Sustainable Development." In 1992, we had raised the money to take five children in her organization, ECO (Environmental Children's Organization), to Rio. They went to remind adults that their decisions affect the kind of world children will inherit. As a twelve-year-old, Severn had made an electrifying speech at a plenary session. Ten years later, as a seasoned environmental campaigner of twenty-two, she was invited to make another such speech. However, in those crucial ten years, our national economies and interests had been exchanged for what was very aptly called a "New World Order" of international trade agreements and economic globalization, which has now assumed the forefront of every discussion about human and planetary needs.

Sev set off for the global reunion of environmentalists with very high hopes. Once she arrived at the lavish, heavily guarded venue, she found herself sitting through mind-numbing meetings where delegates debated endlessly, not about vital issues, but over tiny word changes to already watered-down resolutions (should it be "a" or "the"?). She ate lavish meals in the splendour of elegant conference centres, and rubbed shoulders with government officials and distinguished delegates from all over the world. She saw very little of the kind of urgent, committed work going on that she had dreamed about. Severn was so overwhelmed with a sense of impotence that she left the formal conference to visit Johannesburg's infamous townships and attend the "Alternative Summit." There, she took part in rallies with the poor and homeless. "I was electrified by the passion and sincerity of the many thousands of people who were not included in the formal discussions about the future of our planet," she remembers. She learned, like many before her, that reforming passion does not come from the wealthy, important politicians and world leaders; in fact, their summit seemed largely hypnotized by issues of economics and technology with, as she put it, "the health of real, suffering natural systems as . . . just an afterthought to concerns about how to maintain constant economic growth."

The kind of misunderstanding and even cynical manipulation of these vital issues by international economic interests that Sev witnessed in Johannesburg is summed up in our discussion in chapter 4, of how the media are manipulated to serve global economic interests, as well as in our two chapters, 7 and 8, on the economy. But that analysis didn't anticipate the outpouring of commitment and idealism over the past five years that has led to a true rebirth in citizens' activism across the world. New organizations, like the International Forum on Globalization that set up the Alternative Summit in South Africa, have also been helping people understand the thorny issues of economics and power that stand between them and more equitable, just and sustainable societies. I'm talking about the anti-globalization movement, surely the most hopeful development in world consciousness since the heady days of strong environmental legislation back in the 1980s.

The shutdown of the WTO ministerial meeting in Seattle happened right after the first edition of this book hit the shelves. We both attended, myself as a teach-in speaker and Holly as a reporter, and that gathering, as well as subsequent protests, were among the most heartening developments we have ever witnessed. This people's movement, as global in its embrace of complex issues as in its geography, sprang up with unprecedented spontaneity. It is not only bewilderingly diverse but completely decentralized. That is, there are no definable leaders, and not even a clearly definable theory that lies beneath it. It is an environmental movement, for example, but it's also much more. And although these massive protests take heroic efforts to organize, there seem to be no real leaders or central organization as such. The NGOs that have suddenly appeared to deal with these issues number literally in the thousands, and range from two-person cottage industries in southern India to well-funded, well-staffed offices in Washington, D.C., or Ottawa. The anti-globalization movement continues to be one of the most difficult to characterize in human history. And despite the unequalled power, wealth and ill will allied to crush it, it's also proving to be one of the hardest to combat.

What Is Government For?

> When Congress gives up the right to debate and amend trade agreements, they stop any meaningful participation by the American people in decisions that will affect every facet of their lives, as well as the lives of every person on this planet.
> — DENA HOFF, MONTANA FARMER AND CHAIR OF THE FREE TRADE TASK FORCE FOR THE NATIONAL FAMILY FARM COALITION

These days nearly all the anti-globalization protesters are pointing out that the nation-state, as we have known it, is ceasing to exist; its interests are being subsumed by those of private multinational corporations. Since we wrote about NAFTA in 1999, trade organizations have proliferated, and today, nearly all of the western hemisphere is about to join the Free Trade Area of the Americas (FTAA). This expansion and proliferation of centralized, global control is exactly what mainstream analysts like Maude Barlow, chair of the Council of Canadians, worry about. She sees that the values of husbanding resources as well as sharing and helping one another socially, which characterized the large and important nation of Canada, are now becoming tiny, local concerns dwarfed by the global needs of these powerful trade agreements. What is hardest to understand is why nations, and especially national politicians, have allied with economic interests to mastermind their own destruction. One explanation is that after the fall of communism, all of us, politicians included, fell under the spell of its assumed victor, free-market forces. Now corporate lobbyists and successful CEOs became the heroes we once sought in our kings, presidents and prime ministers. But anti-globalization protesters, as they grope towards a very different horizon, see both communism and capitalism as failed systems.

Since we first wrote this book, the possible hazards of the trade treaties against which the protesters were mounting their offensive have become increasingly destructive realities. The GATS, for example, the General Agreement on Trade in Services, is typical of trade agreements negotiated at the WTO in Seattle and elsewhere. Although it impacts directly on every type of service a government might wish to render its citizens, from law enforcement and education through health, automobile insurance and product regulations, it was worked out in secret by primarily economic experts, with little if any input from the social service, environment or health sectors of the countries involved. This is also true of all the North American trade treaties such as NAFTA and the forthcoming FTAA. In Europe, European Union members are elected by the populations of all the countries involved and have to answer for their policies to their electorates, who are kept informed about most of the details of agreements signed in their name. This may explain why all of Europe still has well-funded government medical and insurance services, solid farm subsidies and strict labelling laws for genetically engineered products, while the United States and Canada do not.

Most of the people who have negotiated NAFTA and are working on the FTAA aren't elected. One would think the details of these iron-clad treaties would become part of political candidates' election promises. However, in the United States, Canada and Mexico, the electorate typically hears little about the details of trade agreements and treaties that will radically affect their access to health services, education, jobs and their own country's resources, until the deals have been signed. Until, in fact, it is too late. And when these agreements start to alter our daily lives and enter the political discussion, you hardly ever hear restrictive trade provisions being blamed in Parliament. Furthermore, as we'll see, the framers of many trade agreements have managed to make it illegal to discuss their provisions in a public, political context. The worst part about all this is that in general, as in the United States' 1998 WTO decision aimed at forcing Europe to consume the hormones fed to American beef, or the 2003 U.S. challenge intended to make the European consumption of genetically engineered food unavoidable, these agreements are set up to favour the trade patterns most congenial to the activities of very large multinational corporations, rather than to raise health, safety, environmental or social standards of the member countries. As Elizabeth May points out later, world bodies that create rules to control trade could just as easily make trade healthy both for people and the planet.

The new Free Trade Area of the Americas is currently being forged to try to help the United States counterbalance an expanding EU and its tendency to protect its farmers and demand some economic accountability from corporations. The FTAA has been described by Maude Barlow as "NAFTA on steroids." She and Tony Clarke, in their very useful citizens' guide "Making the Links" (available through the Council of Canadians, www.canadians.org), point out that: "The U.S. has been using FTAA negotiations to lay the groundwork for challenging and compelling the EU to substantially reduce and eventually dismantle its export subsidy program for its farmers." Agriculture is only the tip of the iceberg. This same agreement is also clearly intended to force member countries to allow private corporations increasing access to goods, resources, communication tools in the form of newspapers and television channels, and virtually all public services, like schools, hospitals and prisons.

The notion that global trade organizations can scuttle popular domestic legislation, especially in terms of the services we expect from our governments, explains in many cases why, over the past few years, Canadian politicians such as Jean Chrétien and Bob Rae, who were elected on platforms of protecting their constituents' environment and social services, switch priorities once in office. For example, recently Canada, especially in Ontario and the Maritimes, has been jolted by skyrocketing car insurance rates. Citizens are threatening to unseat their representatives, and the governments in Nova Scotia and New Brunswick almost fell over the issue. To the public, it looks as if provincial leaders simply dithered about setting up simple and popular government-mediated car insurance plans like Quebec's, and they think that's led to soaring costs for auto insurance. However, the truth is that provinces that did not have public insurance in place prior to the signing of NAFTA, as Quebec, British Columbia, Saskatchewan and Manitoba did, have probably lost the power to enact such legislation, regardless of how much their electorates demand it.

Although it has hardly been foremost in the media discussion of the crisis, one compelling reason why the beleaguered Canadian premiers have not been able even to explain their position lies in the kind of trade agreement chill Elizabeth May worried would affect many kinds of local legislation following the disastrous MMT (a manganese-based gas additive) case in 1999, described in chapter 8. That is, if Canadian legislators publicly discuss their reasons for hesitating, the American insurance industry could sue them for restraint of trade. In May and June of 2003, CBC Radio stories at least mentioned the real problem behind the car insurance crisis, citing "an insurance industry ready to go after any government that adopts a public system." They pointed out that, "Some of the insurers doing business in Nova Scotia [and other Maritime provinces] are American firms. They argue they would be entitled to compensation under NAFTA if public plans take away business."

Maritime premiers banded together to see if Ottawa could advise them on how to live up to their mandate, but Maude Barlow says, given the language of Chapter 11 of NAFTA and the even more rigid regulations of the FTAA, it is "extremely unlikely" that our provincial governments will be permitted to enact this kind of legislation ever again. Argument enough, say the anti-globalization lobbyists, to regard the trade system in the Americas as anti-democratic and even anti-economic. It's like a blanket invitation for any important industry in any of the signatory countries to jack up its prices and then be rewarded with a legally captive market. And that danger to our pocketbooks says nothing about its effect on health care, education and the environment.

Also because of NAFTA, citizens in all three signatory countries face new threats of toxic poisoning because governments have signed away their power to protect us. Lindane is a powerful pesticide that has been linked to non-Hodgkin's lymphoma and seizures in children. It was one of the most hazardous toxins in the waste soup of Love Canal that brought about the Niagara Falls, New York, suburb's clean-up and the establishment of the U.S. Superfund program. Lindane has been banned in most developed countries, including Canada. But in 2001, the manufacturer, Connecticut-based Crompton Corporation, sought a reversal of the ban in Canada, as well as $150 million compensation for lost business. It is completely within its legal rights to do so under the terms of Chapter 11 of NAFTA and will likely win. This follows on the heels of a successful suit by S.D. Myers, an American PCB waste disposal company, seeking to force Canada to accept its poisonous exports. It won its lawsuit, including $75 million of Canadian taxpayers' money in "lost business," while the ban was in effect.

Even when it comes to drug regulations, governments have lost their ability to protect us. To offset criticism because of health effects of cigarettes on youth, the Canadian Commons Health Committee, on behalf of the country's federal government, "was poised to recommend the plain packaging of all cigarettes, to discourage their use," says Maude Barlow. But Julius Katz, former chief U.S. NAFTA negotiator, appeared before the committee on behalf of his new employers, the U.S. tobacco giants, and threatened to sue under Chapter 11, which prevents any tampering with precious trademarks. "The idea was quietly abandoned," Barlow reports.

We have recited just a few of numerous examples of the pernicious implications of trade agreements. Many people reading this may still be puzzled as to why so many young folks show up to shout slogans whenever a trade treaty is discussed. But the global resistance to the current doctrine of free trade is a direct response to such egregious intrusions into the democratic process. It has been born out of the recognition that it is the real reason for the deterioration in quality of life, from environmental and health protections to food security, education and the protection of culture. The movement to oppose globalization may be a positive development, but there is still much to be done, because trade agreements continue to proliferate, becoming ever more invasive and anti-democratic.

The Other World

> *Someone asked Gandhi, "What do you think of Western civilization?" He ... replied, "It would be a good idea."*
> — QUOTED BY SATISH KUMAR IN "GANDHI'S SWADESHI: THE ECONOMICS OF PERMANENCE, "*THE CASE AGAINST THE GLOBAL ECONOMY*

The discussion of the Third World that we undertook in chapter 9 is often a touchy issue, with constant changes of terminology that reflect a collective attempt to be polite or politically correct. Sometimes the industrially undeveloped, resource-rich countries on this planet are called the "Global South" or the "Developing Nations." But our friends, colleagues and correspondents from places like the Philippines, India, Malaysia, West Africa and Central America told us they prefer the term "Third World." "It implies," says one, "another way. Another world, that could be different from what the First World, the Developed World, the Global North, the West offers as the 'only way' to progress and happiness for people." So, in that chapter, we talk not only about the effects of our First World's global economic and environmental practices on the poorer countries on Earth, but the increasingly interesting ideas and values that are coming out of their cultures and could so enrich ours. Much of what has happened since September 11, 2001, especially in the developing countries of the Middle East and Africa, reveals the extent to which the men and women who planned and fostered the New World Order of globalization operate solely in a top-down manner, never considering that the "other world" has anything to offer them. Even when spreading from what could be construed as benign motives, rather than providing opportunities for dialogue, they have dictated all the terms of economic and political survival, as well as rules that alternately restrict or encourage education and social services to the entire world. A popular theory, voiced by many U.S. analysts following 9/11, blamed the attack on the World Trade Center on "jealousy of American riches and freedom." It illustrates the extent to which the United States, and the First World in general, assume that poor countries have absolutely nothing, beyond their resources, to offer the rich. Instead of possibly being enraged at our efforts to destroy their cultures, economies and beliefs, it's assumed they merely long to change places.

It is an axiom among the Native American elders of Canada that for every benefit of a new development there will also be a cost, a philosophical equivalent to the physicist's Third Law of Thermodynamics, "for every action there is an equal and opposite reaction." So when the First World steps in to change a regime, there can be very positive effects, like greater freedom for women. But there are likely to be equally serious reactions, as well. Cultural cohesion, a shared understanding of life that provides whatever social stability a country enjoys, once disrupted, is often replaced by a kind of shared hysteria that ends up swinging a society completely out of balance. In short, swift, radical changes have unpredictable effects. Most analysts agree the disruption of normal social life in Afghanistan by war caused its swing to radical Islam and extreme repression of women in the first place; while in Ethiopia, violent jumps between extremely centralized right- and left-wing regimes, as the country was used as a battleground by the United States and the Soviet Union in the Cold War, destroyed its ability to produce food for its people.

Social change, to be both benign and lasting, has to evolve gradually from within and under peaceful conditions, with give-and-take from both administrators and their populations, as Tewolde Berhan Gebra Egziather, head of the environmental protection ministry in Ethiopia, makes clear with some heartening examples in chapter 9. He also explains that when rapid policy change is tied in with outsiders vying for money and power, real horrors can be created, especially in terms of a major pre-occupation of the poor in the Third World: food security. The chief issue confronting the big trade negotiators at the failed WTO ministerial in Cancún in 2003 was the agricultural subsidies that are central to food production. Economist Lyle Bivens and Adam Hersh of the Economic Policy Institute, in a recent article called "A Rough Row" (posted on www.TomPaine.com on September 9, 2003), say that "Free-trade purists who believe in knocking down all trade barriers . . . [think that] once these protections are removed, the WTO will achieve the promise of economic development. This view is deeply mistaken." They point out that even if the politically impossible were to be achieved, that is, if *all* the agricultural subsidies enjoyed in the First World by powerful American agribusinesses and prosperous European farmers were actually removed, the GDP of developing nations would rise by only 0.14 percent over the next twenty years. In other words, they explain, "The annual per capita income of India, for example, would rise from $480 to $480.67 over the next seventeen years."

Not such a great deal for the poor. Bivens and Hersh argue that the free trade pundits present the current tariff inequalities as the major reason for the vast amounts of inequality between the First and Third Worlds. People, Bivens and Hersh say, should not allow this dialogue "to distract them from the more important problem inherent in the WTO — that the price of admission . . . is the surrender of economic autonomy, even in areas wholly unrelated to international trade . . . . A nation can do away with all barriers (tariff and other) . . . yet find itself in contravention of WTO rules by, for example, failing to devote sufficient resources to prosecuting street vendors selling pirated copies of *Star Wars*. Seriously." The point is, agricultural subsidies, like many government regulations, are neither all bad nor all good. Like other national laws, they depend on

the local circumstances. EU farmers are well-known for receiving generous agricultural subsidies, not only for export but also for domestic
production. These subsidies have enabled production of very high-
quality foodstuffs on restricted land bases while keeping those land bases
beautifully tended, and maintaining ancient villages, watercourses and
cultures. Government protection and incentives explain how French
farmers can continue to raise exotic specialty cheeses, wild strawberries,
rabbits fattened on chestnuts and lambs on ocean-salted pasture and still
maintain comfortable family incomes, rather than mass-producing
industrial food products, as in the United States and Canada. Would we
really want to eliminate such products completely? In Germany, Denmark, Holland and other northern countries, national policies also keep
food purity at their highest levels in the world. Government programs
and progressive subsidies make sure there is little or no tolerance for
toxic chemicals, hormones, pesticides or genetically engineered organisms, and are combining with increasing tax support to expand acreage
devoted to organic production.

In India and many other countries of the Third World, tariffs on fish,
for example, enable local fisherpeople, with their simple technologies, to
compete with international trawlers and find a market. There are loose
copyright laws so that entrepreneurs, one step above the homeless, can
pirate Hollywood movies by filming them in theatres with cheap video
cameras and showing them in tents to entertain their poor neighbours.
In India, there are also laws intended to keep vital medications as cheap
as possible, and to favour the production of local foods that have no international appeal. Books and all printed matter also have government support, which has made India one of the most literate of all the developing
nations. Without these trade barriers, all these marginal ways that millions of Indians have found to help support themselves and retain their
quality of life would vanish overnight.

In the United States and Canada, as in Europe, it can be argued,
nonetheless, that many government subsidies are ill-conceived and encourage overproduction and the misuse of natural systems. Worse, Bivens
and Hersh point out, when it comes to agricultural subsidies, "the lion's
share of their benefits is gobbled up by the giants of corporate agribusiness." If subsidies on chemical fertilizers, pesticides and heavy machinery,
for example, were removed, organic farming, with all its benefits to the

land and to consumers, would be far more competitive. On the other hand, if subsidies and marketing boards that currently help family farmers survive were cut, national food security would be destroyed.

The argument made for globalization is that completely unregulated trade in any imaginable good or service is the answer to any problem. It only needs some "fine-tuning," like getting rid of *all* the last trade barriers, however important they are to a country or its citizens' survival, a kind of "one size fits all" approach to the complexity of economic survival in many very different countries. The citizens of the Third World, who have made their desires known in vast popular demonstrations against their own governments as well as against bodies like the IMF and WTO, think no amount of fine-tuning will make up for the fundamental flaws in a system that is geared entirely to profit and is run by unaccountable entities. They know getting rid of tariffs is a very small thing compared to the real advantages rich nations possess in technology, education, transportation and health, that is, basic modern infrastructures that they do not possess and which directly or indirectly subsidize all First World production. Until all these things really are equal, they still need laws to help protect their production from us.

That's why Bivens and Hersh argue, "The prescription for improvement — mildly sweetening the very bad deal in place — is far too modest. Instead, the WTO needs to give economic autonomy back to the vast majority of its member nations." It may not have much choice in the matter. A WTO ministerial in Cancún in 2003 was aptly described by cover art on *The Economist,* showing a cactus in the shape of a fist with its third finger raised and the caption "The charming outcome of the Cancún trade talks." This important meeting, supposedly held on the key issues of agricultural trade, was cancelled spectacularly in mid-action without resolution of any major issue. For the first time, the domination of the United States and Europe over the supposedly globally inclusive WTO was broken by an alliance between the more numerous poor countries. As analyst Gwynne Dyer put it, the United States and the European Union were shocked that this new "Group of 21, led by China, Brazil and India," simply refused to discuss some of their "more brazen proposals." The big powers were trying to revive the Multilateral Agreement on Investment (now termed the "Singapore issues"), an iniquitous charter that was killed off by huge protests in 1998 because of the massive powers it awarded to private business corporations that would have gutted the autonomy of most national governments.

So at the moment, rich countries such as the United States, Canada, Japan, Australia and the nations of the European Union, who used to call all the shots, are threatening to abandon the same WTO they so staunchly defended from protesters when it was entirely under their control. In very menacing language, they have claimed they will go back to bilateral trade deals, which will be worse for the Third World. As Dyer wrote in a column, EU commissioner Pascal Lamy "damned the WTO as a medieval structure in urgent need of reform, and U.S. trade representative Robert Zoellick bluntly threatened to take his business elsewhere." However, most analysts agree that might not be so easy; having tested the fruits of alliance, the Group of 21 is more likely to begin to function like a new version of the Organization of Petroleum Exporting Countries (OPEC). After all, it's the poor countries that the richer nations need rather desperately if they are to continue to expand their economies. As Dyer says, "Just as trade unions were not really about overthrowing capitalism but about redistributing power within the system, so the G21 is not about destroying global trade but about redistributing its benefits."

This new group will have to be more vigilant than our own governments have been about the new agreements in the works, such as the FTAA, which are especially punitive to small countries, but which also gravely affect the national autonomy of rich ones. The fact is, most countries in the world, rich or poor, have fought costly and exhausting wars with invaders and have struggled with their own lack of internal unity in order to forge geographic areas where people see themselves as "the same" — as Senegalese, Indian, Kenyan, Malaysian, Dutch or Canadian, united by a shared history and certain shared values. In order to have stable national identities, countries also have to have some legal and economic autonomy. It is already difficult enough for the poor to try to be independent of the rich. But trade agreements that destroy fundamental national sovereignty over controlling imports and exports, protecting local farmers and producers, or even the right to make laws favourable to the growth and flowering of their particular cultures and beliefs, are having a profoundly destabilizing effect all over the planet. That's why it's good news that some national governments are still interested in protecting local interests, as they did in Cancún.

Fighting the (Aptly Named) Gross National Product

Money has not lived up to its potential as a liberator because it has been perverted by the monopolization of its creation and by politically manipulating its distribution — which makes it available to the favoured few, and scarce for everyone else.
— THOMAS GRECO, JR., *NEW MONEY FOR HEALTHY COMMUNITIES*

The most vital differences between ourselves and the people of "the other world" lie in values. Very often what is valued in places we dismiss as poor is not, in fact, our money and power, but their own communities and traditions, which they would like to keep. Of course, they would also like to acquire a better motor scooter or a more comfortable home. Even in the First World, if you landed from outer space and watched typical North American television, you might deduce that acquiring more money to buy stuff is the primary need and desire of all humanity. How is it that money and acquisitions have become such central, controlling forces over the lives of people in the developed world?

In the original edition of this book, in the chapter entitled "Follow the Money," we analyzed the difference between money and wealth. This discussion, as well as the way that we have been ceding power to private corporations rather than to accountable, elected governments, remains virtually unchanged, except for the amount of print devoted to the issue. Large numbers of people, even in the First World, are realizing that the value of democratic power, natural systems and cultural diversity really cannot be expressed in monetary terms. The takeover of public political systems by private financial ones has become so clear, even in the world's most prosperous democracies, that many thousands of NGOs and other organizations devoted to these issues have sprung up almost overnight. In fact, there has probably never been such an enormous global growth in organizations devoted to the support of civil society in all of human history. Among the most revolutionary and practical are the movements that are attempting to regain control of the democratic process by restricting the amounts of money that private corporations and the rich are allowed to contribute in each election. This concept has really caught fire with new organizations like MoveOn.org, even though there's a long way to go before citizens have wrested control from the moneymen. Nonetheless, significant national, state and provincial laws have been passed recently that provide very important precedents and a path to follow in the future.

In the United States, the McCain-Feingold Bill was passed in 2002. Its very progressive provisions would vastly limit the money that political candidates could accept when campaigning for office by virtually banning "soft" money. As Open Secrets, one of the many new NGOs concerned with the issue, puts it, unlimited sums of money "allow corporations, labour unions and wealthy individuals to wield tremendous influence over the political process — much more influence than the average voter." The passage of this bill by U.S. Congress was indeed too good to be true. Several groups from both the right and the left, such as the AFL-CIO, the American Civil Liberties Union, the Republican National Committee and the National Rifle Association, challenged the law under the U.S. Bill of Rights, claiming it infringed on their free speech and, in fact, in practice would have crippled unions more than corporations. A federal court ruled to permit soft money contributions again but restricted how they can be spent. That federal court ruling is now being appealed to the U.S. Supreme Court.

In Canada, Jean Chrétien's Bill C-24 was called "a good start towards limiting the political influence of corporations and wealthy individuals . . . that doesn't go far enough and contains too many loopholes to be truly effective" by Democracy Watch, one of the burgeoning number of Canadian groups concerned with this issue. The $10,000 annual limit is high (out of reach of most citizens), contributors don't have to name their employer — which would lead to obvious problems — and parties and candidates won't have to disclose their financial information until after elections. Still, campaign reform is now an issue that no European or North American politician, from presidents and prime ministers to local congresspersons or Members of Parliament, wants to be seen to be opposing. And that's a very big step.

Apart from chapter 7, the third chapter, "Bigfoot," which tackles the growing human population and its demands on the productivity of the natural environment, has perhaps changed the least since the first edition of this book. We are getting ever closer to restraining our numbers as more countries legalize birth control and realize the importance of trying to improve health facilities for children under five and young mothers. But we are also getting farther away from controlling our rampant appetite to discover and then consume every type of material good. That's one of the reasons why in 2002 we published our second book, *Good News for a Change,* in which, among other things, we analyze what kinds of consumption have the least or greatest effect on the environment and tackle, in a systemic way, what kind of lifestyle is actually sustainable. That could be very daunting news. But those who believe that a continuously expanding economy and constantly increasing consumption are the only ways to avoid economic depression and war would be surprised to find out that life in a regulated, steady-state economy can be very enjoyable.

The highly respected Wuppertal Institute in Bonn, Germany, worked out the figures. If everything we still have on this planet were shared out equally so that there were no starving babies in Africa, no sweatshops and misery in Asia, no utter dependence on volatile commodities like coffee in South America, then people all over the world would have roughly the same lifestyle level as the West Germans did in the early 1970s — with slightly fewer cars. That is, we'd all have centrally heated, comfortable, private homes or apartments, televisions, family vacations, good food, nice clothes, government-subsidized education, health and other benefits — every comfort the Germans had then. Obviously, achieving such an equitable distribution of wealth would be a daunting challenge, but it does show what could be achieved and, moreover, that there's not too much to lose for those who want to try.

Sharing One Skin

When we say the Okanagan word for ourselves, we are actually saying, "the ones who are dream and land together."
— JEANETTE ARMSTRONG, OKANAGAN ELDER

The last chapter in this book is about values — what really makes humans happy, what we really need to survive. Since we first wrote it, the public discussion of values and happiness, like the discussions about biotech and globalization, has widened and deepened exponentially, again at least partly because of the open exchanges on the Internet, which certainly was not the idea its commercial boosters had for it. By the same token, President George W. Bush's exhortation to go out and shop following 9/11 to "fight terrorism" also backfired. Many people decided instead to stay home with their families, to take long, solitary walks, to enjoy the simple things more. The aftershocks of Afghanistan, Iraq, wars and bombings in Israel and Palestine, to say nothing of the continuing assaults on the natural world, have not stimulated a raucous party atmosphere. These events cause us to think about mortality, and mortality makes us ponder what we've accomplished and what we really care about. That brings us to what's really valuable in life.

When we were making the radio series for which this book is named, I had the honour of interviewing a remarkable woman, an Okanagan elder from the interior of British Columbia named Jeanette Armstrong. In our interview and in her writing, Jeanette speaks a great deal about her upbringing out on the land with her father and grandmother in the 1940s. She remembers one day in particular, sitting in the sun on a peaceful, beautiful hillside surrounded by ripening berries, darting birds, wild sage and butterflies. "Down in the valley," she writes, "the heat waves danced, and dry dust rose in clouds from the dirt roads near town. Shafts of searing glitter reflected off hundreds of windows, while smoke and greyish haze hung over the town itself. The angry sounds of cars honking in a slow crawl along the black highway and the grind of large machinery from the sawmill next to the town rose in a steady buzzing overtone to the quiet of our hillside. Looking down to the valley, my grandmother said in Okanagan, 'The people down there are dangerous, they are all insane.' My father agreed, commenting, 'It's because they are wild and scatter anywhere.' "

Jeanette spends the rest of her article explaining what they meant by this. It's not as harsh as it sounds, and her father's statement in particular contains his pity and his worry that these other members of the human species are not connected to what really matters. She says: "One of the reasons I explain this is to try to bring our whole society closer to that kind of understanding, because without that deep connection to the environment, to the earth, to what we actually are, to what humanity is, we lose our place, and confusion and chaos enter. We then spend a lot of time dealing with that confusion."

Like so many other indigenous peoples, Jeanette Armstrong talks at length about values, and in her own way answers the challenge of globalization, that is, the notion that trees, rivers, animals, birds and cells are just machines, useful to humans for their economic production. "Land bonding," she says, "is not possible in the kind of economy surrounding us, because land must be seen as real estate to be 'used' and parted with if necessary. I see that separation is accelerated by the concept that 'wilderness' needs to be tamed by 'development' and that this is used to justify displacement of peoples and unwanted species. I know what it feels like to be an endangered species on my land, to see the land dying with us. It is my body that is being torn, deforested and poisoned by 'development.' Every fish, plant, insect, bird and animal that disappears is part of me dying. I know all their names, and I touch them with my spirit. I feel it every day, as my grandmother and my father did."

I was trained as a scientist, Holly Dressel as a historian. But it has become clear to us, as well as to so many others, that we cannot save our natural systems — and future generations — without a fundamental change in what it is that we value. If we really feel ourselves separate from nature, not only nature but we ourselves also suffer. The message of *From Naked Ape to Superspecies* was a documentation of our catastrophic assault on the planet that needs to be curbed immediately, so we went to work on *Good News for a Change*. In that book we tried to see how people might be able to turn the steamship of our modern economy away from its collision course with nature. Working to develop the skills, resources and technologies to live in balance with the biosphere, people from all walks of life — architects, farmers, businesswomen, factory workers, activists, Native people, teachers and researchers — see themselves and nature as inseparable, sharing the same skin. They share the same approach, the same hesitant, humble, gentle methods of working with the natural world instead of against it or bludgeoning it into submission. These people lie ahead in this book as well, in the thoughts and methodologies of so many of the contacts we met who helped us analyze and understand the problems we are facing.

CHAPTER 2
BUGS Я US

*We don't know nearly enough to manage the ecosystems on
our own. If we think that we can eliminate those natural
ecosystems and substitute prosthetic devices, like creating clean
air or water with fusion energy or sustaining the stability of
cropland — in fact, [if we think we can] keep the planet in
that delicately balanced, highly peculiar state on which
humanity depends for its continued existence — then we
are kidding ourselves.*

— E.O. WILSON, ECOLOGIST

Back in the fall of 1991, there was a great deal of hoopla surrounding a
$200 million scientific experiment getting under way in Arizona. Eight
people were sealed in a giant bubble called Biosphere II in the middle of
the desert. It was meant to be a miniature version of Earth's natural systems. The project was undertaken not only in anticipation of long flights
in space but also to demonstrate how well we understand the world
around us.

Like Biosphere I — the Earth — Biosphere II contained a variety of
ecosystems and a sampling of plants and animals that were supposed to
provide the same "ecosystem services" that nature does; it would purify
and produce water, recycle wastes, provide oxygen and absorb carbon dioxide, use photosynthesis to capture sunlight, and produce plants and food.
The bubble covered several hectares and housed a variety of ecosystems,
including a desert, a miniature ocean, a grassland and a tropical forest.
Nearly 4,000 different plant and animal species were deliberately assembled
to populate this airtight system. It was a kind of ark of technology that was

supposed to provide a self-sustaining environment that would support the "bionauts" for two years. Biosphere II was intended to provide proof that we humans have learned enough about the world to re-create it at will, wherever we choose, and there was enormous media excitement the day the crew was sealed in.

I'm a scientist, and I've spent a great deal of my career studying ecosystems. I remember, as a boy, watching a sealed flask containing water, a plant and a fish. The fish and plant lived for weeks, giving the illusion that each was providing what the other needed and that they therefore had created a harmonious balance. Eventually, of course, microorganisms took over and the fish and the plant died. But the illusion is striking. I remember thinking at the time of Biosphere II that this was just a bigger version of the sealed flask, and that it was doomed to fail. We simply haven't yet reached the stage where we understand the workings of things like the water cycle and air purification enough to duplicate them. Yet the media and many experts were treating the event as if it were already an accomplished triumph, a portent of things to come as we rocket off to the stars.

Once the bionauts got inside, however, things started to go wrong pretty quickly. Oxygen levels plummeted, primarily because the mix of soil organisms did not produce the proper proportion of gases. Oxygen concentration dropped to a level found above 5,182 metres. As a result, the bionauts suffered some of the alarming problems associated with low oxygen intake. Then nitrogen levels skyrocketed, creating a risk of brain damage. Most of the insects that had been carefully selected to pollinate the plants died off, dooming most of the bionauts' intended sources of food and air and water purification. And many other species careered out of control — cockroaches, katydids and a species aptly named crazy ants swarmed over everything. Some plants and vines also proliferated wildly. The bionauts spent increasing amounts of their time killing insects and hacking back vines. Eventually, the defeated pioneers, malnourished and sick, gave up and came out. Had they stayed much longer, they might have died. There was very little hoopla surrounding this failure; in fact, most people still don't know what a resounding defeat it was.

Yet I couldn't help feeling that for all its disasters, Biosphere II was a successful lesson — it provided powerful evidence of how little we understand the natural systems that sustain us, that deliver what are termed the "ecosystem services" of replenishing air, water and soil and removing waste without our ever having to think about them.

Gretchen Daily, an author and ecologist at Stanford University who investigates the extinction of species, says efforts like Biosphere II are humbling. "It shows some of the good sides of humanity: our curiosity, our determination. And in many ways those are noble qualities. But at the same time, we're like little kids. We've amassed this power, and we've deceived ourselves somewhat. In modern urban societies today, and in the richest and most powerful parts of the world, people are so removed from these ecosystem services that they've all but faded from view."

The failure of Biosphere II is a direct result of a revolutionary change in the human perception of how the natural world works that began more than 300 years ago. In the seventeenth century, the French philosopher René Descartes postulated that by stepping outside the world and becoming observers of nature, rather than being part of it, we would be able to see how it really worked. Later in the same century, Sir Isaac Newton used a scientific approach to discover laws that were universal and timeless. He viewed the world as an immense clockwork mechanism that could be taken apart and its components studied. This methodology is called "scientific reductionism." The scientific method it gave rise to works by pulling things apart and reducing them to their smallest possible components so they can be analyzed under controlled conditions. By measuring input, output and changes within isolated fragments, scientists assume they can determine how each part functions and, by extension, how the whole of nature works. Newton hypothesized that once we knew these secrets, we would be able to use our intellect to harness those functions for our own benefit.

This scientific view of the world sees our species as being at the pinnacle of a steady evolutionary march from simple to complex, from instinctive to intellectual, from a state of nature to civilization. Ever since Descartes, we have considered human intellect — the activity of the mind — to be separate from and superior to the body. As we've become increasingly separated from nature in urban settings, we have come to believe that our intellect, manifested through science, empowers us with the capacity to understand and control the world around us.

As scientific reductionism was applied, our inventive capacity blossomed. Within a hundred years, steam- and coal-powered engines were replacing manual labour and amplifying muscle power. Europeans were able to travel farther and faster than ever before, and to bring back new foods, luxuries and medicines from around the globe — to say nothing of conquering other lands and their inhabitants with powerful weapons. People began to understand how many common diseases spread and how to control them, primarily through better sanitation. We learned to analyze soil, human nutrition and animal breeding, and to adapt them to our own uses.

The technological revolution that scientific reductionism led to has been like one long party. It has brought power and glory. It has enabled us to extend our dominion over the globe and over nature. But as failed experiments like Biosphere II are showing, reductionism provides only part of the whole picture. When we broke aspects of nature down into their component parts — when we took a mosquito out of its surroundings and labelled it a pest, for example — we lost sight of the patterns of nature and the symbiosis within ecosystems, the way every part has a purpose. We lost sight of the patterns the mosquito was part of — its relationships to the fish and the birds it fed, the microorganisms it lived with, the ecosystem in which it played out its life cycle.

We've been trained to look at big creatures as the most noble, the most important, the most pivotal and valuable, with ourselves as the biggest creature of all. We've ignored the relationships all large creatures must have with small ones, like ants, bacteria and fungi, which provide the real fundamentals of survival. But science is always changing. There is a new way of thinking, a fresh view that is beginning to focus on relationships and patterns — and on the role of the most minute, as well as the greatest, creatures in the world around us.

The Meek Inherit the Earth

So far as we know, there is not a single other planet in this universe that has even rudimentary protoplasm. And here we have millions, maybe tens of millions, of unique forms of life that manifest themselves as tigers and polar bears,

thistles, birds and snakes. And we don't know if it's 10
million or 100 million kinds. It's crazy. We don't even have
that basic grasp of what makes our planet tick — let alone
how to keep it ticking.

— NORMAN MYERS, ECOLOGIST

A major reason we're still so far from being able to construct a viable human habitat like Biosphere II, despite all our apparent technological wizardry, is that we really haven't devoted enough attention to the details of the natural world. After all, if we were put in charge of managing a shoe factory, we would at the very least need two important bits of information to do a good job. First, we would require an inventory of every item in the factory. Then we'd have to understand how each of the pieces interact and fit together: if you put soles on before heels, whether you glue or nail first. We need the same kind of information if we intend to interfere with any part of nature's productivity.

So what is the state of our Earth inventory? Although estimates of the number of species range from 2 to 100 million, the consensus among biologists seems to be that there are around 10 million. Globally, the total number of species identified so far is between 1.4 and 1.6 million. The reason for the disparity is that different scientists often mistakenly identify the same species independently. That means that we have names for only 20 percent of all the non-bacterial life forms on Earth. When it comes to bacteria and other microorganisms, we have no idea at all.

How much do we know about how the pieces work together? Edward O. Wilson, a professor of ecology at Harvard, is one of the world's pre-eminent experts on biodiversity, the variety of life on Earth. He uses an ironic example to illustrate how this interdependence works, and how the big and the small organisms typically fit into it: "How can I put this without sounding callous? If all humanity disappeared, the rest of life, except for domestic animals and plants, which represent only a minute fraction of the plants and animals of the world, would benefit enormously." The forests, Wilson says, would gradually grow back, and relative stability would return to the ecosystem services that control global temperature and atmosphere. The fish in the oceans would recover, and most endangered species would slowly come back. Of course, there would be no humans around to enjoy this, but as far as the survival of numbers of species goes, the planet would be better off.

On the other hand, says Wilson, if all members of one of the groups of smaller creatures, such as ants, were to vanish, the results would be close to catastrophic. Ants turn and aerate a very large part of the Earth's soils. They're major predators of other insects, and they're the chief scavengers of small animals, removing and breaking up more than 90 percent of any small, dead creatures as part of the soil-nutrient cycle. They even pollinate many plants. Wilson, who has studied ants all of his professional life, says, "If they were to disappear, there would be major extinctions of other species and probably partial collapse of some eco-systems." The Earth as a whole would suffer. If we go further and imagine the impact of the extinction of a microorganism like the common gut bacterium *Eschercia coli,* or mycorrhysae, the invisible soil fungi that facilitate the uptake of nutrients by plants, we see that the catastrophe would be even worse. The functions of the creatures living in the air we breathe, beneath our feet or in our own bodies are exotic and unimaginable; they are working in concert to keep us alive. As E.O. Wilson says, "We are living on a relatively unexplored planet. And that's why the amazement and the enthusiasm that are invariably aroused when the subject of life on Mars comes up [are] just out of pro-portion to reality.... We are standing on an unexplored world, for the most part, and there are many amazing things still to discover here."

The Whole Is Greater Than the Sum of Its Parts

Newtonian physics was a liberation [at first], but it became a straitjacket by the early twentieth century. Now we have a different kind of liberation. We have the liberation of quantum physics, the liberation of systems theory and the liberation of the Gaia hypothesis, which all help free us from the Newtonian straitjacket.

— FRITJOF CAPRA, PHYSICIST

In the early part of the twentieth century, scientists learned to divide an atom into its components — protons, neutrons and electrons. They cre-ated a model of the atom that showed it as small, solid objects circling a core, rather like a miniature solar system. Naturally, the scientists believed they were elucidating one of the greatest mysteries of the universe — the nature of energy and matter.

But quantum mechanics reveals that protons, neutrons and electrons do not conform to the solar-system model. As we zero in on such particles, they become fuzzier, or less distinct. The whereabouts of electrons, for example, can be indicated only as a calculated probability. The renowned nuclear physicist Fritjof Capra says, "There are no particles in atoms, such as the balls in the atomic model. There are no hard and solid spheres or objects like grains of sand. A sub-atomic particle is essentially a set of relationships reaching outward to other things, and the other things are again relationships to yet other things."

As the twentieth century progressed, physicists began to make a dramatic shift that reflected this reality, focusing less on discrete objects and more on relationships. They began to realize that the relationships and interactions among organisms, cells and photons were as important to an understanding of the whole they formed as is an analysis of their parts. Systems studies, the new science that has grown up around quantum physics, states it clearly: "The whole is greater than the sum of its parts." Capra, one of this discipline's most eloquent spokesmen, says that the traditional or reductionist view of evolution has been that rocks, oceans and the air are the stage on which life plays out its drama. "By an act of creation," he says, "a fluke of chance or some other mystery," life somehow rose up on this stage, evolved from the first cell and has been evolving to "higher forms" ever since. What we are now learning about evolution, however, shows that life is created and shaped by other life, in continuous, interlocking patterns; and as the failure of the Biosphere II project suggests, these patterns aren't easily controlled or knowable. An imbalance in the mixture of soil organisms — microscopic life forms — was enough to topple an entire ecosystem.

A lesson to be drawn from the Biosphere project is that the small, usually overlooked aspects of our ecosystem — the insects and the microorganisms — are in many ways the most important. Insects, for example, make up the most numerous, diverse and successful group of animals on Earth, and only a minute percentage is harmful to or competitive with us. When scientific technology developed powerful weapons like chemical pesticides to use against them, we ended up killing thousands of species just to get at the less than 1 percent that we find troublesome. We are now beginning to see that insects and microorganisms are vital parts of ecosystems: they form a major portion of the diets of birds, fish, mammals and amphibians, and as predators are still the main force keeping other insects under control.

Similarly, most bacteria are not agents of disease. The vast majority of microorganisms and insects, both invisible and visible, are benign to humans. We're only just learning that if we didn't have them in a natural balance, there wouldn't be many of the other things we like — including ourselves. If biology follows the lead of a hard science like physics, its future will lie in revealing not just bits and pieces of nature but the complex interrelationships that actually make organisms and ecosystems work. The only question is whether we will make this jump into a true analysis of the world around us before the outdated notions of reductionism leave us with few intact systems to study.

Bugs Really Я Us

> *The basic pattern of life is a network. Whenever you see life, you see networks. The whole planet, what we can term "Gaia," is a network of processes involving feedback tubes. And the world of bacteria is critical to the details of these feedback processes, because bacteria play a crucial role in the regulation of the whole Gaian system.*
>
> — FRITJOF CAPRA

The clockwork approach of scientific reductionism viewed nature as something to be controlled and kept at bay. Nature's most alien forms — the insects and bacteria, long-time intimate guests on the human body — were increasingly seen as being extremely far away from humankind, the apex of creation. When we discovered that some of them carried or caused disease, they were banished, along with nature itself, as something extremely different from civilized humans. Virtually any means to destroy them was considered appropriate, and these attitudes have remained with us in the twenty-first century. We crush insects; we develop appallingly dangerous poisons to eradicate them; we

invent antiseptics, vaccines and antibiotics to protect "us" from "them."
In fact, the difference between what we consider civilized society and
degraded or desperate surroundings is sometimes simply the number of
visible or invisible bugs.

But for the first 2 billion years of the more than 3.5 billion years that
life has existed, bacteria and other microorganisms were the only living
things on Earth. Through that long period, *they invented* all of the fundamental processes of life, from replication to responses to gravity, light
and temperature, from fermentation to photosynthesis and mobility. And
even today, the total biomass — the weight of living things — of all
microorganisms is greater than the biomass of all other creatures put
together, including trees and whales and people. It is still a bacterial
world. By the time aggregates of cells — multi-celled creatures —
appeared, they had no need to reinvent all these basic life properties.
Instead, they incorporated bacteria into their own protoplasm to live
symbiotically and continue to perform these functions. As Fritjof Capra
says, "We owe practically all life to bacteria."

The cells of multi-cellular organisms contain structures called
organelles that perform highly specialized roles such as the generation of
energy, photosynthesis and chromosome movement. These organelles
are evolutionary remnants of bacteria. The mutually beneficial collabora-
tion between organelles and their single-celled hosts is now so profound
that host and guest can no longer survive without each other. Each of us
is an aggregate of some 60 trillion cells. Within each cell, there may be
hundreds of organelles, the once-independent bacteria. So when you
look in the mirror, remember that you are, in fact, gazing at an immense
aggregate of trillions of cells and quadrillions of bacteria. A human being
— or a dog or a potted plant, for that matter — is not just an individual.
Each one of us is actually a community of organisms.

And just as we are an aggregate of a number of organisms, each
exquisitely balanced with the others and made up of a diversity of specialized cells, all life forms within the biosphere can be thought of as making
up a kind of super-organism. The immensely varied organisms within different ecosystems are held together by the global matrix of air, water, soil
and energy from the sun. Collectively, the individuals within a species, as
well as the numerous species and the many different ecosystems on the
Earth itself, have the properties of a self-regulating entity. The engineer

James Lovelock called this interlocking system of all life on Earth "Gaia," after the Greek goddess of Earth.

Remarkably, all of these levels of variability, from genes to species to ecosystems, have somehow worked in concert to create vast changes in the planet's biophysical makeup and to make the Earth more and more liveable for creatures like us. As Capra says, "The environment of life, which we thought of as independent, is itself shaped by life. Not only that, it is shaped in a certain way. The whole system regulates itself and maintains conditions conducive to life . . . . Gaia, like us, maintains her body temperature over the millennia, and not only overall temperature, but also the composition of gases like oxygen and CO_2, the salinity of the oceans — all these processes are maintained at levels that are conducive to life. How does Gaia do this? Through multiple feedback tubes, a network of processes."

The Fish Need the Forest, the Forest Needs the Fish

When the forest began to disappear . . . we had flooding, we had erosion . . . and the role of science was misapplied. The Army Corps of Engineers looked at the floods and said, "The answer is to build dams," . . . so now we have 114 dams on the Columbia River. And the relationship between those dams and the salmon is that the temperature of the Columbia and the Snake rivers has permanently risen. It is quite common, in the summer months, to find the Columbia and Snake rivers exceeding 65°F. At 65°F those waters are lethal to salmon.

— TED STRONG, YAKIMA TRIBE MEMBER

One example of the complex network of relationships within ecosystems — between their largest and smallest members — can be seen in a controversy raging right now in the Pacific Northwest. Because of the rapid decline of Pacific salmon, oceanographers, botanists, ichthyologists and other scientists have devoted a good deal of time in the past few years to try to understand more about their life cycle. Up and down the Pacific coast of North America, five species of salmon have spawned in rivers and streams for millennia. But early in the twentieth century, human activity, such as the construction of dams, mining, logging, urban development and agriculture, negatively affected salmon habitat and life cycles. Populations began to crash, and human beings intervened in order to restore the salmon stocks.

This work was commenced in a typically reductionist way. It was quickly found that mature adults could be milked of eggs and sperm, and that eggs could be fertilized in containers. The fertilized eggs could then be incubated to hatching. Thus fish hatcheries could be used to vastly increase populations of young salmon. The rather simple assumption was that if more young could be pumped into the system, then somehow, after years at sea, more mature adults would return. When the fry couldn't readily reach the sea from the hatcheries, they were loaded into tankers and trucked downriver.

Lack of knowledge led to incredible blunders. Take as an example what happened to the sockeye salmon, a prolific and highly prized food fish. Sockeye fingerlings were reared and released in rivers close to the ocean on the assumption that this would help them survive by shortening the treacherous journey to the sea. Years later it was discovered that sockeye fry never go directly to the ocean. Instead, they first spend a year in a freshwater lake. The human attempts to help the sockeye simply ensured that millions perished.

Salmon-breeding programs have cost billions of dollars, but in terms of saving wild races and keeping their numbers high, the investment has not been very successful. In 1994, hundreds of millions of dollars were spent to help salmon in the Snake River in Washington State, for example, but only 800 chinook returned in the fall. The same year, only one adult sockeye made it. Forty-nine wild varieties of salmon in the state have been extirpated. Because the wild populations are dying out, experts have tried to create new genetic strains or bypass the wild part of the animals' life cycle by growing them in net pens. Despite numerous lessons on the hazards of introducing exotic species into new habitats, fish farmers are rearing millions of Atlantic salmon from the East Coast in the Pacific. And despite assurances by government bureaucrats that they will not reproduce, the escapees are not only moving into rivers and spawning but their offspring are surviving.

These catastrophic developments have come about because of our truncated and fragmented way of looking at the resource. Management efforts focus on different "resources": terrestrial game, marine life, forests, rivers and estuaries, which are often administered separately by state, federal, international and local bodies. Ted Strong is a member of the Yakima band and is a former head of the Columbia Inter-Tribal Fish Commission, an organization trying to hammer out a management strategy for the Columbia River basin. He says there's another way to look at the river: "Tribal peoples are often said to be backwards; we look to the past. But if we wanted to impose a value system and order to salmon restoration, we would ask that the scientific community look more to the past than to the future. A recently concluded study by an independent science group on the Columbia River stated that what the salmon needed was 'a complex, interconnected array of wetlands, in-stream flows, gravel beds and so on that are created, manipulated and managed by natural forces.' That's a very fancy way of saying, 'They need a natural river.' And that's what the Indian people propose in our own salmon-restoration plan."

Restoring a dammed, farmed and logged river basin is a pretty tall order. And in any case, many people see nothing wrong with the status quo. Why not try to domesticate salmon? What does it matter if there are fewer species? Pinks and chums never fetched a good price anyway. And, they figure, maybe the Atlantic salmon will replace the native species that have been lost. If species and their environment really were unrelated, if the world existed primarily to fulfill the immediate needs of humans, and if we were able to survive indefinitely in industrialized isolation the way we thought we could back in the 1950s, those might be viable arguments. But when a new breed of scientist decided to look at the totality of the salmon and their habitat, they discovered an amazing relationship between fish and their surroundings that means our reductionist method of trying to help fails to take into account the true nature of a species.

Salmon spend their adult lives at sea, thousands of kilometres from their birthplace in forest and mountain streams. Marine carbon and nitrogen isotopes leave a unique "signature" that scientists can detect and, of course, after years in the ocean, salmon carcasses are full of these two important atomic nutrients. After the adult salmon make the arduous journey from the ocean back to their natal streams, they spawn and die. Various predators — bears, eagles, wolves — catch them and leave their remains on land, where the carcasses are quickly covered with fly eggs, which hatch into maggots that consume the flesh and then pupate. When the adult flies emerge, they nourish fish, mammals and birds throughout the forest. In the water, the dead salmon are digested by fungus, which in turn nourishes bacteria that sustain insects, copepods and other invertebrates in the stream. And like fallen logs in a forest, their decaying flesh provides nourishment for their own offspring. When the fry emerge from the gravel, a banquet awaits: 25 to 40 percent of the carbon and nitrogen in juvenile salmon comes from the remains of their parents. Isotope studies show that 30 percent of the vital nitrogen and carbon in aquatic algae and insects, and 18 percent in vegetation along the river, comes directly from salmon. So what may seem to be a waste is, in fact, an evolutionary mechanism used to nourish a new generation and to keep the stream itself healthy.

We know that salmon have to have clean, cool, fast-running streams for their habitat. When forests surrounding rivers and streams are clearcut, salmon populations plummet and often disappear. Salmon need the forest to shade the rivers and keep the water cool, to provide nourishment and to prevent soil from silting the gravel of spawning beds. But we are just learning that this relationship is reciprocal — forests need the salmon, too, to remain healthy. University of Victoria biologist Tom Reimchen has shown that in one season a single bear will carry about 700 partially consumed salmon carcasses into the forest. He has found their remains as far as 200 metres from the river, spanning valley bottoms where the biggest trees grow. After consuming salmon, bears, eagles, wolves and ravens defecate and spread salmon remnants through the forest, providing the largest pulse of nitrogen fertilizer the trees get all year. In fact, you can correlate the amount of growth by the width of tree rings with the amount of marine carbon and nitrogen, which reflects the size of the salmon runs! Grizzly bears were extinguished in Oregon by 1931. Preserved hides of those animals reveal that up to 90 percent of the nitrogen and carbon in their bodies was of marine origin. So the forest and the salmon, the insects and the bears, all need one another — and all are linked through the water and the salmon to the ocean, thousands of kilometres away. When forest, fish, bears and birds are seen this way — as a single living entity, each part of which supports the other — it is clear that our attempts to manage these resources as separate parts are bound to fail.

From Top Down to Bottom Up

The soil is virtually a living organism. It's not just a bunch of grains with bugs walking through them. It is a mass of organic, living material in an inorganic matrix. It's dynamic. It's full of life. And it does not produce anything for human beings unless it's sustained in that living condition.

— E.O. WILSON

One biologist who studies the interrelationships in ecosystems is Orrie Loucks. Field biologists like Loucks are becoming increasingly rare, but it's only through fieldwork that we can access the facts about complex ecosystems. This research simply cannot be done in a laboratory, where scientists can work with only tiny parts of the overall puzzle. "Twenty-five or thirty years ago," Loucks says, "much of the research in ecology was focused on the productivity of ecosystems. We looked at photosynthesis, for example." This approach made sense. After all, in human society, productivity is directly related to the available supply of energy, and the planet's primary source of energy is the sun. So it seemed logical to assume that sunlight and its capture would be limiting factors for the rest of life as well.

But scientists like Loucks have found something quite surprising: energy does influence productivity, but the ultimate regulator of ecosystems is decay. Processes that release nutrients back into the ecosystem for future production are what enable those ecosystems to function. "The breakdown of organic matter," Loucks says, "and the release of the nutrients that are in that organic matter are systems that are vastly more complex than photosynthesis." Decay is more complex than photosynthesis because it involves the co-operation of thousands of species of living organisms, from insects to worms, fungi and bacteria. These species are very sensitive to factors like toxic chemicals, small changes in soil moisture and temperature, and any change in the mix of species that co-operate to break down the organic matter. If anything happens to disrupt the process of decay, then nutrient supplies are curtailed or unavailable to the next season's crop of plants.

Biologists have long studied big things like bears, trees and fish because they are impressive or important to us. The small creatures like insects and bacteria seemed insignificant alongside these lords of the Earth. We could see with our own eyes how wolves, elephants and humans regulated the numbers of the herbivores, plants and so on. For example, when sea otters on the west coast of North Africa were exterminated by fur hunters, their favourite food, sea urchins, exploded in number and drastically diminished the supply of kelp, which is the urchins' main food. Kelp forests provide food and shelter for a bewildering array of organisms, including many species of fish. So as kelp disappeared, so did many fish. When sea otters were reintroduced in California, British Columbia and Alaska, they reinstated "top-down" regulation, reducing sea urchins and enabling kelp to return, along with fish.

While we attempt to manage resources by top-down regulation, it has had varying degrees of success. We've devoted far less attention to bottom-up regulation. It wasn't until the early 1980s, for example, that scientists began to realize that soil organisms and their interactions control the extent to which plants grow. Elaine Ingham, a soil pathologist at Oregon State University, describes it best: "Soil is alive. Sand, silt, clay . . . are the mineral fraction of the soil, like the building blocks of a house. The bacteria that grow in that sand, silt and clay, along with some organic material as food, form the tiny bricks . . . . Fungi tie these bricks together into walls, floors and ceilings." These creatures, which Ingham calls the architects, design housing for all the larger organisms, tiny animals and plants, such as microarthropods, earthworms, rotifers, protozoa and nematodes.

These are just a few of the creatures that live in soils. We really don't have a clue how many there are, what each one does or how they work together. In one of the few such descriptive studies ever carried out, scientists in Denmark scooped up a cubic metre of earth in a beech forest and took it into their lab. With the naked eye alone, they found more than 50,000 small earthworms and their relatives, 50,000 different kinds of insects and mites, and 12 million roundworms and their relatives. When they took just a gram from the same soil and put it under a microscope, they found 30,000 protozoa, 50,000 algae, 400,000 fungi and billions of individual bacteria of unknown numbers of species. They identified about 4,000 species of bacteria, almost all completely new to science. When the scientists then took another gram of soil from an estuary not far from the beech forest, they found another 4,000 species of bacteria, almost all different from those in the beech forest and also all new to science!

When such soil organisms are said to be "identified," that simply means they have been given a name. It doesn't mean that anything is known about what they eat or how long they live, what they all do in concert with one another, what makes one of them flourish in one soil but not in another, or how one soil type becomes home to a copse of oak while another supports white pine. One of the very few things we do know about these unexplored worlds is that without the spaces created by bacteria and microorganisms in the "bricks" of sand and clay, there would be no place for plant roots to move into the soil. They'd have to expend too much energy, as anyone knows who's tried to grow plants in baked clay or the hardpan residue of chemically treated fields. Furthermore, without these spaces, air and water don't infiltrate. Not only does the soil remain barren, but its ability to act as a sponge and watershed purifier is gradually destroyed. Such soil may no longer be conducive to growing the pollinators that enable the area to produce a certain kind of flower or tree, or a species may suffer from huge pest invasions because the nematodes or other tiny predators that kept the pests down are no longer there.

Back in the 1940s and 1950s, in true reductionist fashion, scientists teased soil apart into its mineral, animal and fungal components. They saw soil primarily as an inert matrix that held plants up. To improve productivity for human agriculture, they added water and food in the form of chemical fertilizers, and dealt with pests by blanketing them with poisons. Laboratory soil analysis told them what minerals to add and in what quantities — how much nitrogen, potassium and calcium mix, for example, would give lush dahlias or fields of beans.

The initial results of this approach were spectacular. But now, at the beginning of a new millennium, we're starting to get some very clear signals that something is wrong with the scientific analysis. In recent years, yields have fallen and researchers are scrambling to invent stronger fertilizers and more high-tech seeds. One of the reasons we're having trouble is that we've converted many agricultural soils into hardpan, leaving none of the spaces that animal and fungal architects kept loose and open. We did that by adding chemicals that are too concentrated and don't contain the complex mixture of leaf litter, dead animals, manure and organic material that provides nutrition for the community of creatures that makes soil so rich for plant crops. Moreover, our fields are soaked in toxic poisons designed to kill whatever we've identified as pests, whether these are plants like dandelions or milkweeds that compete with our crops, insects like corn borers or Colorado potato beetles, or diseases like rust or blight. Unfortunately, these poisons also kill many of the creatures that make the system function in the first place.

Recent evidence shows that all the money spent on these chemical cocktails may be wasted. Despite our arsenal of powerful toxins, most agricultural pests are still controlled by natural enemies like spiders, wasps, birds and other predators that live adjacent to or in the fields. Moreover, in forty years of pesticide use, we have not managed to eradicate a single pest, and we actually have more problems with them than when we started. Even the heavy machinery needed to work large tracts of industrial farms has a negative impact on living, productive soils. As Elaine Ingham says, "You can't keep running your tractors across the surface of the soil, squishing it, without destroying all that space, the living room and the dining room and the kitchen, that these organisms need to grow in."

Industrial farming methods — growing single-crop monocultures in huge fields aided by heavy machinery, chemical fertilizers and pesticides — produced double and sometimes triple yields at first, which seemed to prove they were the right thing to do. But as Ingham says, the reason that these heroic methods worked for a generation or so is that nature had built up a reserve in soil: "When you take soil out of a native system, it has accumulated a lot of organic matter. The past lives of all these organisms working on all that organic stuff have built a kind of savings account of nutrients in the soil. For the first forty to sixty years, you're slowly but surely mining out that organic matter, pulling out the material that's in your savings account."

Ingham's economic metaphor is apt. If we use the organic material in soil at the same rate as it is replenished, we can farm indefinitely. If you have a bank account and each year you draw only the interest, you can do that year after year. But if you dip into the capital, as she notes, "You eventually hit bottom when there's nothing left in your savings account and you're in serious trouble. And the chemicals won't work, because you don't have the organic matter in the soil; you've mined it out. You don't have any more organisms because there's no food for them." Not only does the soil on hillsides begin to wash away, but fertilizers and toxins run through the soil and end up in surface water and groundwater.

Elaine Ingham and other soil experts who understand that the richness of soil is its biodiversity maintain that the bank account can be filled again. She has had remarkable success overcoming serious hardpan problems by drilling holes in which to insert organic material, compost and manure, and by advocating no-till methods, heavy mulching and careful crop rotation. She even has a business that helps farmers get off the chemical fix. This, she says, is the direction in which we should be moving if we want a real future for soils and crops. Nurturing rotting compost, worms, beetles, slime moulds and funny smells is at the heart of nature's production. It's not clean and tidy and orderly, like petrochemicals in a lab. But we're finally learning that if we don't pay attention to nature's smallest creatures, to bottom-up intervention and to the role of the tiniest living things in providing for the rest of us, we may find ourselves in a truly frightening place.

Death Takes a Holiday

> *Eleven percent of all bird species are on the verge of extinction as a consequence of what we've done. A report of the world's plants suggests that 12 percent... are on the brink of extinction, and I suspect that's an underestimate, because there are many of the world's plants that we don't know. It's not just in land and soils; it's in oceans and rivers. A huge fraction of our freshwater clams and mussels are on the verge of extinction. The pattern is ubiquitous. Whenever we have sufficient data, we recognize somewhere between 10 and 40 percent of all the world's species are already on the verge of extinction.*
> — STUART PIMM, CONSERVATION BIOLOGIST

Damage to soil doesn't stop with modern agricultural practices. Industrial toxins, the acidic fallout from burning sulphurous fuels like coal, nitrogen oxides from car exhaust, and PCBs and dioxins all end up in forests and streams, as well as on cropland. More than a decade and a half ago, scientists like Orrie Loucks began getting calls from farmers and other people in rural areas around the Ohio River valley who were troubled by the number of dying trees. Loucks and his colleagues set up two test plots. The soils in both plots were carefully matched for mineral type and forest variety; both had the same species of white and black oak and hickory. One site was in southern Ohio, where wind drift and proximity to industry caused very high doses of pollution and a pH level of about 4.1, which indicates very acidic soils. The other site was in southern Illinois, and had low doses of pollution and an average pH of 5, a level that enabled the soil to absorb acids and toxins without negative effects. In the southern Illinois soils, the researchers found an average of thirty earthworms per cubic metre. As Loucks says, "That's rich. You can get all the bait you need to go fishing in fifteen minutes." In contrast, the polluted soils of Ohio yielded only one earthworm per cubic metre, and those were of a species that lives in dead wood. No other species were found.

Loucks's group began to track 150 other invertebrates in the two plots of soil. At the polluted Ohio site, the species that still existed in relative abundance were present in populations about 60 to 70 percent of the size of the Illinois populations. And between 40 and 50 percent of the species found at the less polluted site weren't present at all in the Ohio soil. In other words, half of the life forms that sustained the Ohio oak-forest ecosystem had disappeared, and those still remaining were severely depleted. If worm populations reflect the decline of soil microorganisms, they mirror the toll on larger organisms, too, because the trees in the Ohio plots, reports Loucks, are also dying. Leaf litter under the trees, which normally decomposes between fall and spring every year, is just lying there. In the low-pollution site the litter is about half a centimetre thick, but in the polluted sites it is almost three centimetres. That means that the calcium, nitrogen and all the trace elements that the creatures in the soil — and, ultimately, the trees — need to survive is tied up in the undecayed leaf litter. And as the soil studies on the site have shown, the organisms responsible for decomposing the leaf litter and releasing its nutrients have disappeared, or have become depleted to the point where they can no longer perform their function.

Orrie Loucks's team has developed a keen admiration for the way battalions of different species transform cellulose and putrefying bodies into plant food. Because bacteria cannot easily decompose large pieces of leaves, for example, it is absolutely essential that larger creatures, like earthworms and insects, initially shred the leaves into small pieces. If the sugars in these leaves were just left to the fungi and bacteria, they would devour it all themselves. However, mites eat the bacteria and fungi, like miniature livestock grazing on grass in a pasture, and that keeps the bacteria and fungi under control. The system is held in a dynamic equilibrium, so one part doesn't grow and reproduce too quickly. The raw litter is turned into nutrients at a rate that leaves an optimum amount available to the following year's plants.

As Loucks says, "It's a kind of orchestra, a symphony of different species complementing each other . . . to achieve balance in the breakdown of organic matter." He goes on to ask an important question: "At what point in species impoverishment did ecosystem function become so impaired that we destroyed the capacity of the trees to grow? . . . Every aspect of that system is now dysfunctional. Did the dysfunction start at 20 percent loss of biodiversity? At 10 percent?" Unfortunately, Loucks's team has been unable to get the research funding to find out why. He says, "I can't imagine any question that's more important. If the Earth is threatened, as it is now, with the loss of 30 to 50 percent of its biodiversity, shouldn't we immediately try to understand what has been put at risk?"

Scientific funding these days, however, is aimed at discovering new technological fixes, not at teasing out the intricacies of how nature works or what has gone wrong when productivity drops. Once again, we have to look at our cultural and historical traditions to discover why we have such huge scientific blind spots. Loucks believes that "human observation tends to be directed at events that are big and dramatic, or events that go on over the course of a year, like a major drought. But events that induce profound changes for the Earth and for human prospects, and that stretch over periods of two to three decades, are judged as speculative. And so changes that take place among little organisms over long periods have been just generally discounted. You know, it's only a relatively small audience out there that is able to say, 'Earthworms are king.' "

Destroying the Engines of Life

*It takes a long time — millions of years — to create species as
fully developed as the ones around us . . . . We are destroying
species a hundred times faster than they could be created, even
if we left the natural environment alone. We are doing the
equivalent of drawing down hard on our bank account; and
you can't be drawing down on your bank account at a
hundred times the rate you're putting new money in without
going broke very fast.*

— E.O. WILSON

Today, we've quickened the pace of extinction to a level never seen
before. In tropical rain forests alone, for example, we're eliminating, at a
very conservative estimate, about 25,000 entire species every year.
Biologists are alarmed by the human-caused acceleration of extinction.
We've learned from the past that when major extinction events have
occurred, new niches may be opened up for opportunistic species to
move in, but it takes long periods of time for the variety of species to
be restored.

Norman Myers is a well-known author and lecturer on biodiversity,
affiliated with Green College at Oxford University. He says that the great
powerhouses of evolution are in the tropics — rain forests, coral reefs and
wetlands. They weren't destroyed in the previous great extinctions of
species on this planet. And because they were still intact, it took "only"
about 5 million years for evolution to refurnish the planet with life forms
as complex and diverse as those it had lost. But now, he says, "We're
knocking out not only large numbers of species, but also the capacity of
evolution to come up with replacement species as readily as it has done in
the past . . . . This time, the bounce-back period won't be just 5 million
years. It'll be more like 25 million years."

Most of us lack the perspective that would make the extinction crisis
seem real. To try to make it clear, let's first note that all organisms of
Earth need energy to carry out every life process, from movement and
metabolism to growth and reproduction. And that energy comes from
the sun. Photosynthesis enables plants to capture photons — packets of
energy in sunlight — and convert them into storable chemical energy.

And all the rest of life, from microorganisms to animals, needs the sunlight captured by photosynthetic activity to flourish and multiply.

A few years ago, when there were only 5 billion of us, Stuart Pimm, a prominent conservation biologist and professor of ecology at the University of Tennessee in Knoxville, calculated the fraction of all terrestrial photosynthetic activity that human beings take for their own use. We use about 15 million square kilometres for agriculture — most of that is land taken from places that were once forest. We also consume large parts of forests for our fuel and building materials. We graze very large portions of the planet, roughly 60 million square kilometres. About half of the non-frozen land area of the world is some sort of pastureland. Added up, this means that we are already exploiting about 42 percent of the planet's land mass. If you adjust the figures to reflect the fact that there are now more than 6 billion of us, they will show that we are fully exploiting about half of everything the planet produces annually.

To many people, that may not sound so bad. After all, about half is still left. Indeed, the United Nations Commission on the Environment and Development recommended in 1987 that all countries protect 12 percent of their land base. That target seems to accept that 88 percent of the land is ours to use as we wish. But we are merely one species among perhaps 10 million, and we are still completely dependent on the services performed by all of that biodiversity. By co-opting such a large part of the planet's photosynthetic activity to fuel our energy needs, we deprive many of the other species that would have used it, and thereby drive them to extinction. Because our population has been growing exponentially, we know that the human population could double again, to 10 to 12 billion, in only thirty or forty more years. Will we then take over all available land spaces to grow food for ourselves or our domesticated animals? And Pimm adds a fact that makes the outlook even grimmer. "We, humanity, have highlighted the planet. We have taken the best bits. The forests of Europe, Japan and North America are the bits that are just right. Not too hot, not too cold, not too wet, not too dry. Now, as our population expands, it's moving into areas that are not as good, parts of the planet that are basically indigestible. As a consequence, that other half, that 50 percent of the primary production that is still forests and tundra and grasslands, is not going to support us anywhere as well as the parts we've already used up do."

Our effect on rivers is just as overwhelming. Riparian habitat, the zone along riverbanks, has long been recognized as one of the richest areas of diversity of life on the planet. In harsh desert or tundra regions like Israel or northern Canada, riverbanks shelter almost all the life that exists. That's why dams or river diversions there wreak such environmental havoc. So much water is taken out of the Colorado, the river whose force dug the Grand Canyon, that by the time it reaches the ocean there's none left. The Tennessee is the fifth-largest river in North America, and there isn't one inch of it that hasn't been dammed or channelled. These rivers function as ecological circulatory systems that conduct life's fluid to the land, yet we've altered them so extensively that water shortages are a looming fear for the entire planet in the next few decades. We're only just beginning to realize what our industrial and engineering activity has done to the biological productivity of habitats and nurseries of biodiversity like wetlands and estuaries.

The story of human disruption of complex terrestrial ecosystems is being repeated in the oceans. We're seeing fisheries crash all around the globe as they're overharvested or destroyed in other ways. Shrimp farms and agricultural runoff devastate the mangrove swamps and coral reefs that serve as nurseries and habitats for most forms of ocean life. Recent calculations show that we already consume about a third of the biological production of the ocean, using practices that have almost all of our fisheries on the brink of collapse. As Pimm says, "We're skimming the top level of life out of the ocean — the sharks, sailfish, bluefin tuna — all the top predators, which are the equivalent of the lions, tigers and bears of the terrestrial environment."

Eighty percent of the species for which we have names come from tropical rain forests. We all know what's happening to them. Stuart Pimm says, "If you fly over Haiti, you'll see what looks like an ecosystem haemorrhaging a flow of red blood out into the blue sea." When people cut down a natural forest for fuel or farmland, particularly in a hilly country, soil once held in place by trees is exposed to wind and rain and simply washes out to sea. And in the ocean, soil sediment clogs the rich coral reefs that are incubators for much of the fish population. Forests and reefs made tropical island paradises of ease and plenty. They were the wellsprings of life — all life, including our own — and we're destroying them not just for farms, but also so we can build fancy hotels behind concrete walls along every shoreline. Our culture does not see baby fish and soils and clean water as riches, only the discos and bars of development.

As we said in the introduction to this book, our culture has never before been in the position of having to see the world — the air, the seas, the rivers, our soils — as finite, as something that we could ever use up. For at least a thousand years, Western culture, *our* culture, has believed that when we run out of things in one place, we can just go to another place, we can substitute something else or we can use our inventiveness to create new forms of riches. But we've hit the wall in terms of real riches, the ones you can eat, the ones that sustain lives. Now that we're finally sitting down to take inventory, we're realizing that everything from lions and tigers to earthworms and microbes is disappearing. We can finally compare our numbers and our yearly needs with how much mathematically measurable energy the Earth is capable of producing. The figures are clear.

We have to realize that our culture's old method of dealing with the world — taking all we can and moving on — is not going to work anymore. But that's just *our* culture. All over the world there are remnants of other human cultures that in effect hit the wall of limited resources and saturated populations many, many years ago. Some, though by no means all, aboriginal cultures have remained the same for thousands of years because they were composed of the people who didn't leave when the resources got scarce. The adventurers among them went off to colonize and exploit new lands. But on every continent, a small core group of people stayed home. They were able to stay because they had figured out the rules of sustaining a finite resource base. They kept their numbers in check, and they kept their exploitation of the resources in check as well. They co-operated in this difficult endeavour by evolving religious beliefs that helped them conserve their own particular piece of the planet, which was to them the whole world.

Only a small fraction of the more than 6 billion of us who now exist will ever explore anything "out there" again. Of course, there's always the chance that some day, at least a generation from now, hundreds or even thousands of adventurers might try their luck in space. But most of us aren't going anywhere, for the simple reason that there isn't anywhere to go. It may be time, then, to ask a few of the aboriginal people who are still around to give us some advice about how a finite world and a controlled human population would work. And in the process, we may learn something about all of physical reality.

We Are the Land and the Land Is Us

> *One of the things that we as Okanagan people know is that our very flesh is our land. Our very breath that we take is our land. Everything that is about us is our land . . . . And so when we call on our spirits in the spirit world, that's how we talk to them. When we refer to everything that is, we have only one word meaning the all, everything, the land, the water, the birds, the insects — everything, including ourselves. We say* tamihu. *And that means everything, including us, has that life force in it.*
>
> — JEANETTE ARMSTRONG, OKANAGAN ELDER

What is the life force that Jeanette Armstrong, the elder of the Okanagan people of British Columbia we met in chapter 1, speaks of? It's the same thing that we've been talking about in a scientific way throughout this whole chapter. It's the fundamental interrelatedness of all life on Earth, which we are discovering on the molecular, microscopic and visible-ecosystem levels. She is using only slightly different words to describe the same reality that researchers and scientists like Orrie Loucks, E.O. Wilson and Stuart Pimm have been discovering and have called "the biosphere."

Jeanette Armstrong, besides being a writer and a community leader on the Okanagan Reserve in Penticton, British Columbia, is also a critic of modern culture. Her essays have appeared in books such as the Sierra Club's *The Case Against the Global Economy.* A grandmother herself, she remembers going out on the land with her own aunts and uncles and grandparents. She says, "Going out on the land was like being with members of the family."

Her family taught her about the interrelatedness of life, not through scientific studies, but by speech and example. She says, "When my grandmother would say 'Grandmother' and name a lake, or 'Grandfather' and name a mountain, or refer to 'these grandfather trees' and talk to the wildlife that way, I never used to see the differentiation between human beings and the life forms around us. And I think that perspective, in terms of forming relationships to the land, being in love with the land in the same way that you're in love with people, is an essential part of the human being."

Armstrong lives in British Columbia, where the ancient forests are melting like snow under the advance of modern exploitation. She tries to explain how her people still identify with the land and all the creatures around them. She says, "When those grandmother or grandfather mountains or lakes or rivers or streams are destroyed, it's like your grandparents being hurt . . . . For our people, it's no different than a family member leaving. It's that connection, at that level. And it's more than just an emotional regard for the land, or a love of the land because of its pictorial quality or its beauty. It's an actual relationship with our land in that way. I would hear my grandmother talk about a certain place, and then when we would be there, she would speak and sing to that place. We would share our food with it, and it would share its food with us . . . . In that kind of principle, with that kind of attitude, you can't go out and destroy that land, unless you're insane, no more than you could destroy your own grandparents."

Like Armstrong, people in the vast majority of traditional and indigenous groups believe that the Earth is alive, that it is an organism. And they believe that human beings are as much a part of the natural world as insects or whales or clouds. Most of them believe that humans have a responsibility to take care of the other creatures around them, that calamity will result if we are greedy, wasteful or destructive. They cement this understanding of the physical world not with scientific data, but with emotion and experience.

Brian Swimme is a cosmologist and mathematician teaching at the California Institute for Integral Studies. He's a quantum-mechanics physicist at the cutting edge of modern science. One of the most exciting things we're learning about how the world works, he thinks, is how closely it resembles the pictures painted by traditional peoples and ancient religions. The empirical, scientific way of knowing how the universe works "is simply a new way, one that is powerful and wonderful, but that corroborates much of traditional understanding in often surprising ways." One example Swimme cites is what he calls "the third great discovery of modern science: the realization that the Earth's system is itself self-organizing or self-regulating. And one easy way to capture this is to simply say the Earth is alive. Of course, it's not alive the way a salamander is. It doesn't give birth to baby Earths. But it's alive in the sense that it actually organizes itself so that the complexity of its life forms might continue."

We have scientific data, as opposed to traditional intuition, to back this up. Swimme says, "One easy way of seeing this is that life forms require a particular spectrum of temperatures for their molecules to operate. And the Earth maintains itself in that range. It enables life to come forth. And the reason we know this is that the temperature of the sun has changed drastically over billions of years. It was 25 percent less intense in the beginnings of life. So life itself, the bacteria alive at the time, had to alter clouds to trap heat, to warm the place up for more life. It's a stunning realization, because you can't perceive the Earth as self-regulating in our own local frame of reference."

Ancient and traditional cultures have long taught that the Earth is alive. Swimme says that although we just got these data together, "intuitively, this was something understood by indigenous peoples for about the last 50,000 years. Four hundred years of science have enabled us to begin to approach this intuitive truth from an outer, or empirical, way. Data don't make a truth superior, but they do make it different. Taken together with the old form of knowledge, we have the opportunity for a stronger understanding of the self-regulatory or living nature of Earth."

Swimme believes that this fundamental change in how science views the world will have far-reaching effects on Western culture. He says, "Now we begin to see ourselves not just as individuals on planet Earth, but rather as sensing creatures that live within the organized life of Earth. We're part of a whole. And this perspective is, I think, very different from what we've been used to in modern industrial culture." It is scientific reductionism itself that is helping us to see the limitations of our old cultural methods of dealing with the world. If we are to conserve the things we care about, like salmon, like clean air or water or soils full of earthworms, we can no longer see the world as a simple clockwork we can easily take apart with no damage done. We need to come to an understanding of how amazingly complex and delicate the connections that keep life going really are.

We are still enormously ignorant of the complexity and interconnectedness of the world around us. As projects like Biosphere II have reminded us, we have no ability to replicate, let alone improve upon, the natural world. But we do know that the marvellous productivity and regeneration of nature flows from the rich diversity of living things. With extinction cutting out species right and left, we are tearing at the fabric of biodiversity on which we absolutely depend. Modern science is now revealing the depth of our embeddedness in and dependence on nature, as well as the catastrophic changes we are causing. We have to understand all these basic truths. One such truth is that not all bugs are dangerous or dirty; they are, in fact, part of our family, part of our life history and an integral part of the web of life. We need to respect and accommodate them, and all other life forms, if we are all to go on helping one another to survive.

CHAPTER 3
BIGFOOT

In 1992, in Rio, all the 100-plus nations said, "Okay, we have to reduce our impact on the planet." But now, more than ten years later, we have more people, more cars, more pollution, more resource consumption, fewer species, less water, less topsoil. So we are farther away from sustainability. But how far? We have to know. If we can't measure that, we will never be able to move to sustainability. How much nature do we use? And how much nature do we have left?

— MATHIS WACKERNAGEL, CO-DEVELOPER OF THE ECOLOGICAL FOOTPRINT MEASUREMENT

With stunning speed, humanity has laid claim to every bit of land on the planet and a wide zone of the oceans along our coastlines. We've even planted flags on the moon and sent our machines to Mars. The fate of every ecosystem on the planet is now determined by human activity. We exploit potentially renewable resources like fish, trees, soil and fresh water far beyond their rates of replacement. We poison air, water and soil with toxic effluents. We transform the landscape with cities and farms, dams and roads. And as we exert our will on the terrain around us, wild things are pushed out of the way — and out of existence.

We don't seem to worry about this as much as we should, I think largely because we have an unwarranted faith in the ability of science and technology to pull us out of the mess that our technological prowess has created. In my opinion, no more destructive belief exists than the idea that we have escaped the constraints imposed by nature on all other species. We assume that by enabling us to exploit and alter our surroundings, our intellect has freed us from dependence on specific habitats. We believe we are no longer part of nature, because we have acquired the ability to control and manage the forces impinging on us.

This illusion of escape from nature has been reinforced by our extraordinary transformation in the twentieth century from country dwellers to city dwellers. In an urban setting, we live in a human-created environment, surrounded by other people plus a few domesticated plants and animals, as well as the pests that have overcome our defences. Living among such a dearth of other species, we no longer recognize our dependence on the rest of life for our well-being and our very survival. It is simpler to assume that the economy delivers our food, clean air, water and energy, and takes away our sewage and waste. We forget that the Earth itself provides all these services, and so makes economists and the economy possible. We are biological beings, as dependent on the biosphere as any other life form, and we forget our animal nature at our peril. As we undermine clean air, water, soil and energy; as we burn fossil fuels beyond the capacity of the Earth to reabsorb the greenhouse molecules thus created; as we use our surroundings as dumping grounds for toxic effluents; and as we degrade pristine areas once teeming with other life and resources, I believe we're embarking on a suicidal path.

Those are pretty strong words, and many people will take issue with them. But having interviewed hundreds of experts and scientists in many parts of the world, I have become convinced that they are true. There are real, measurable limits on the world around us. The resources of our planet are finite; a limited amount of air and water can support a limited number of living creatures, which we need for food and shelter. So what are the limits? Is there any way to calculate how many of us is too many and how much consumption or waste disposal is too much? Once these questions seemed too vast to figure out reliably. But with computers, we have the technical means to find answers.

Mathis Wackernagel was a graduate student of Bill Rees, a population ecologist at the University of British Columbia. Together they developed one of the first reliable ways to measure how much of the planet's productivity we're already using. They call it our "ecological footprint."

We can already measure the number of organisms other than humans that a given ecosystem can support indefinitely. Ecologists call this the carrying capacity of the ecosystem. Obviously, if there are more individuals feeding on coral or trees than the reef or forest can support, then the habitat will be destroyed over time. Take, as an example, deer in a northwestern forest. As long as there's plenty of food around — plenty of plant biomass — the deer eat and their numbers increase rapidly. "But as they become too numerous," Rees explains, "they overgraze and destroy the very basis of their own existence. And then they crash down to a level that is back in balance with what the now-altered environment can sustain over time."

Throughout the realm of all living organisms, one of the strategies of evolution is reproduction in excess of the number needed to perpetuate a species indefinitely. Other species are able to live off this excess without jeopardizing the prey's long-term survival. Thus numerous species — from puffins to salmon, seals and humans — can take from the lavish abundance of herring in the Pacific Ocean without endangering the herring's survival. But there are still limits that set the carrying capacity of each species.

Rees notes that many economists reject the idea of applying the notion of carrying capacity to human beings. They argue that the planet can sustain our present numbers and more because we aren't like other animals. We're intelligent and inventive, they say, and these qualities have enabled us to go beyond our biological roots. Unlike the deer, we don't have to satisfy our needs from our immediate surroundings. Unlike bears and coral reefs, we can engage in trade. Therefore our population capacity in a given region needn't be limited by the resources in that region. "The idea here," Rees says, "is that almost any place on the planet, any significant region, will have certain things in surplus and other things that may be scarce. But if we're creative, we can trade our surpluses for those things that are locally scarce. And that enables both us and other regions to overcome our local carrying capacity. The population then can rise. The consumption of goods and services can rise far beyond the capacity of the local area to support that population."

Even then, surely, there is a limit; if the numbers everywhere get large enough, eventually the problem will resurface. But those who believe in infinite growth throw in another critical factor: technology. Rees says, "Many people believe that as technology improves, we'll be able to find a substitute for any resource that gets scarce, even scarce everywhere. In other words, human ingenuity can [find a] substitute for any good or service provided by nature . . . . So when you put the whole issue of trade and technological advance in the basket, it seems that carrying capacity is an irrelevant concept for human beings."

If economists don't believe that carrying capacity limits human beings, what is our wealth (i.e., natural wealth, like food and water) based on? Economists believe that human inventiveness can create even natural wealth and will satisfy our every need, including basic ones like food, clean air and shelter. The late Julian Simon, an economist, once told me that the more people there are, the better off we'll all be, because there will be more geniuses like Albert Einstein to come up with a never-ending supply of new ideas and inventions. As an ecologist, Bill Rees doesn't think this idea makes much sense. He feels it violates something intrinsic about human beings: "For all our technological wizardry, we're still real animals; we're still organisms dependent on physical processes and physical materials."

Rees and Wackernagel decided to turn the concept of carrying capacity on its head to help apply it to humans. They decided that instead of asking how many people a region could support, they would ask how much land it would take to support the physical needs of each human being. Within any system of trading, regardless of the level of technological development, each person still requires a certain productive area on the surface of the Earth to maintain his or her physical needs and absorb his or her wastes. Rees says, "So all we've done is to invert the traditional carrying-capacity ratio. Instead of asking how many organisms per unit area, it's how much area per organism? Or, if you like, how much area is needed to support a defined population, like a city or a country?" The answer gives you that human's, or that population's, ecological footprint.

Canadians are among the wealthiest people in the world. We have a lavish national lifestyle. We drive big cars. We can eat fresh strawberries year-round. We have big houses full of lots of stuff. It takes a great deal of land to provide this lifestyle, to grow fibres for our clothes and provide space and material for our houses, as well as to grow food to feed us and pump water to sustain it all. We also need land to absorb the wastes we create. The carbon dioxide from our cars and the wastes we flush down the toilet or bury in landfills all require a certain amount of water, soil and time before they can be absorbed and made non-toxic. Therefore, as a country, we have a very big ecological footprint. Rees has figured out just how big: "We estimate that the ecological footprint of an average Canadian, taking into account food requirements, fibre requirements, wood requirements, paper requirements and so on, plus just the assimilation of carbon dioxide, would bring us up to about seven hectares per capita. There is a marine component there as well — that includes something on the order of 0.7 hectares of productive ocean surface. So . . . the average Canadian needs seven hectares of biologically productive land of many, many different categories all over the surface of the Earth, dedicated exclusively to [his or her] support."

We have a very big country, so seven hectares each may not sound too bad. We Canadians do not, in fact, use the entire ecological capacity of our country. But Canada isn't alone on the planet, and we do sell our trees, fish and crops. Because of this trade, our forests are in serious decline, our fisheries are collapsing and our agricultural soils are eroding at an extreme rate. As Rees says, "The collapse of our fish stocks is not the result of domestic consumption, but rather the result of our exports . . . . The loss of old-growth [forests] again is not the result of domestic consumption . . . . We export most of our forest products. Most of our grain is exported from Canada, so erosion and other forms of dissipation of our prairie soils have to do with consumption taking place in other countries. Furthermore, such things as atmospheric contamination with, among other things, carbon dioxide . . . is a global problem. It doesn't really matter what we do in this country. If the rest of the world is on an unsustainable path that destroys the atmosphere or upsets climate conditions or makes a hole in the ozone layer, then we all suffer the consequences."

Fair Is Fair

The time has come,
To say fair's fair,
To pay the rent,
To pay our share.

— MIDNIGHT OIL, "BEDS ARE BURNING"

If we Canadians each need seven hectares of productive land mass to be supported in the style to which we've become accustomed, what, in fact, is the amount of land available to each existing human on this planet, if it were divided up equally? The answer to that question is simple, says Rees. "Anybody can do the math. The reality is that there are . . . more than 6 billion people on the Earth right now. There are only about 8.9 billion hectares [of agricultural land]. So if you divided the total amount of ecologically productive landscape by the human population, . . . we'd each get less than a hectare and a half of land." That's 170 metres by 100 metres — equivalent to about a city block.

If you added in everybody's fair share of the productive parts of the ocean, each person would get about two hectares of land and water. "That figure gives us, right away, a measure of how far Canadians would have to move in terms of reducing our consumption to reach our fair Earth-share," Rees says. "Or, conversely, it tells us how much we would have to improve our technological efficiency so that we could continue the same level of material consumption. In fact, in order to bring everyone on the planet to the same general level of consumption and well-being of the average Canadian, we would need four or five more Earths — right now! These figures are frightening, but they're real, and they give us a clear idea of how far we have to slim down consumption, population and wastes, as well as readapt technologies, if we are to have a decent future for the majority of humanity. At this point, scientists tell us this is still not an impossible goal."

The ecological footprint measure also gives us some idea of what is possible. We can see just how lavish and excessive Canadian consumption is when we compare it with the per capita consumption of poorer countries like Bangladesh. There, they consume ten to forty times less than we do per capita. The Chinese consume about a seventh of what we do. It follows that if we reduced our consumption, even a little, people in developing countries would have a much better chance of increasing theirs. It also means that any idea of raising the masses of the Third World "up to our level" is impossible, given the scientific and mathematical measurements of the planet's productive capacity. The claim made by economists and corporate leaders that more consumption around the world will benefit the poor is simply untrue. In fact, the opposite is true. *All the expansion approach does is use up the future faster, primarily for the benefit of the already excessively rich.* According to the United Nations Educational, Scientific, and Cultural Organization (UNESCO), if we divide the world's population into fifths, "in 1990, the richest group received fifty-nine times more than the poorest. Despite growth of almost 3 percent per annum worldwide of per capita GNP over the past three decades, this is almost double the ratio of 30:1 in 1960. In that year, the richest 20 percent already controlled 70.2 percent of global GNP. In 1990, this had climbed to 82.7 percent. In 1960, the poorest 20 percent had to get by with only 2.3 percent, which by 1990 had fallen to 1.4 percent."

Mathis Wackernagel, the co-developer of the ecological footprint concept, suggests that we are looking in the wrong direction for solutions. "We have been very good at looking at atoms, at plants, animals, even looking at poor people. Because we are rich, we tell the poor people what to do. But we have been very bad at looking at ourselves . . . . We, the world's middle class, are the wealthiest 20 percent of the world. We have the biggest ecological footprint. We have to learn to look at ourselves and see what share of the global pie we occupy."

We've gotten used to consuming so much of the planet partly because we don't see the connection between our way of life and other parts of the world. The bananas and coffee we have for breakfast weren't grown on our own land base. They came from thousands of kilometres away, but we don't think about that while we're grabbing a bite and reading the morning newspaper. It's as if we're provided with goods and services by servants about whom we never have to think or even notice. Pretending that limitless growth is possible, and that it will benefit the servants because they'll get more, too, is just a way of avoiding the reality that we're taking more than our fair share.

Economists often argue that countries with few natural resources and large populations, like Germany or the Netherlands, can nonetheless achieve prosperity because of their ingenuity. The Netherlands is often cited as a model of modern economic development. Despite its proportionately huge population, roughly 434 people per square kilometre, its total land base and natural resources are very limited. Yet Holland has a positive balance of trade and one of the highest per capita gross domestic products (GDPs) on the planet. Instead of natural capital like trees, land and fish, it is argued that Holland's riches are generated through human ingenuity that leads to the production of value-added goods like fancy cheeses and fine electronics.

According to Bill Rees, "Many economists marvel at the fact that sometimes the Netherlands has an export in agricultural produce. But that's a money measure.... Dutch agriculture [actually] requires a land area outside the country about five times larger, some say as much as seven times larger, than the agricultural land within the country." That's because in order to grow their produce, the Dutch have to import petrochemical fertilizers, pesticides, machines and seeds from other countries. But those costs are simply not subtracted from the revenue generated when their flowers and vegetables are sold. Rees says, "The point is, Holland and countries like it, most of the developed nations, for that matter, are often used as models for the Third World to follow. But ... it's not possible for the Third World to follow these models, because in many respects the Third World is providing the surpluses that these countries exploit in order to have their extremely high standards of living. So for every country that has an ecological deficit, there has to be another part of the Earth that has an ecological surplus. If every country runs an ecological deficit, then we are quite literally consuming the Earth. And, in fact ... that is exactly what we are doing."

This imbalance has begun only very recently. Wackernagel says, "We now live in an ecologically full world. And that's new. After the Second World War, the ecological footprint, even of rich people, was smaller than the Earth's ecological capacity available on a per capita basis. So expansion was easy. We weren't competing directly with each other. But now the world is overly full, and our research has shown that ... the ecological footprint of humanity is 35 percent larger than the ecological capacity of the world."

Thirty-five percent larger? How can we be using up more than there is? Wackernagel means that we are now using up natural capital, not just the interest; we are using more than the production that nature provides annually above and beyond what it needs to sustain itself. "It's like an account," he says. "You can use just the interest, or you can use a little more. And what happens? Natural capital disappears. That's what we see. Topsoils vanishing. Deforestation. Water tables going down. Biodiversity loss. So basically, we are depleting our natural capital."

Baby Blues

We certainly have predispositions to do many things, like have [a lot of children]. But one of the most fundamental is the predisposition to learn, and to learn all our lives. The most distinctive human capacity is to modify our behaviour depending on what we see around us. And we can see around us the running out of resources and the risk that our numbers are posing to society. We can learn from that.

— JARED DIAMOND,
EVOLUTIONARY BIOLOGIST

Like any other species, human beings have always had an ecological footprint. But for 99 percent of our existence, our needs were simple and nature was vast and endlessly self-renewing. That relationship has changed with sudden speed. When I was born in 1936, there were 2 billion people on Earth. There are now more than 6 billion of us. We have exploded into the age of space travel, computers and global telecommunications. And we have a global economy that is predicated on the need for constant growth.

The problem is that we live in a finite world. Endless growth is an impossibility. Anything that grows steadily over time, whether it's the diameter of a tree, the amount of water we use, the rise in population or the increase in GDP, grows exponentially. Anything that grows exponentially will double in a predictable period of time. Suppose the size of a city grows at a rate of 1 percent a year. In seventy years, it will have doubled. Growth at 2 percent a year will result in a doubling in thirty-five years; at 3 percent, in twenty-three years; at 4 percent, in seventeen and a half years; and so on.

The University of Colorado physicist Arthur Bartlett once gave me a graphic illustration of the effects of exponential growth. Imagine a test tube full of food for bacteria. Now imagine that we introduce a single bacterium that will proceed to divide every minute. So at the beginning, there's one cell. A minute later, there are two; in two minutes, four; and in three minutes, eight. That's exponential growth. At sixty minutes in this example, the test tube is full of bacteria and all the food is gone. Then when was the test tube half full? The answer, of course, is at fifty-nine minutes. So at fifty-eight minutes, it was 25 percent full; at fifty-seven minutes, 12.5 percent full. At fifty-five minutes, the test tube was only 3 percent full.

If, at fifty-five minutes, some bacterial genius spoke up and said, "I think we have a population problem," the less astute majority would probably retort, "What are you talking about? Ninety-seven percent of the test tube is empty, and we've been around for fifty-five minutes!" But if the doubling continued, then at fifty-nine minutes most bacteria would probably realize they were in trouble. Suppose they threw money at their scientists and begged for a solution. And suppose that, in less than a minute, those bacterial scientists created three test tubes full of food! Everyone would be saved, right? Well, no. At sixty minutes, the first test tube would be full. At sixty-one minutes, the second would be full, and at sixty-two minutes all four would be full. By quadrupling the amount of food and space, they gain only two extra minutes.

Most biologists believe we are long past the fifty-ninth minute with respect to our use of the planet's life-support systems. Fortunately, we aren't living in a test tube. We're living on a planet that can renew itself — but only if we give it a chance. In 1968, Paul Ehrlich, the famous population ecologist from Stanford University, published a blockbuster, a book called *The Population Bomb*. He averred that our numbers were on a collision course with our resources, and predicted massive famines. Many people now say that book exaggerated the issue and attacked the solution from the wrong angle. After all, we're still managing; our problems didn't get as bad as Ehrlich said they would in the ensuing years. Many credit the global economy with the failure of Ehrlich's predictions. However, part of the reason why we haven't come to ruin is that a lot of people did pay attention to Ehrlich's warning.

Over the past year or two, we have, for the first time, seen a drop in the rate of global population growth. This phenomenon has to do with more enlightened birth-control policies and populations that are beginning to see for themselves the effects of overcrowding. Moreover, many agencies now working on population issues didn't exist in the 1970s. The World Fertility Survey, for example, polled some 500 million women in poor countries around the world. The women told the agency that they didn't want all the children they had, and that they also wanted to space their pregnancies more carefully, but that they lacked the means to do anything about it. If these women were able to have the number of children they really wanted, the Earth's population would stabilize at somewhere around 8 billion, rather than 12 billion, over the next thirty or forty years.

Werner Fornos is a population expert, a former university professor, and, for the past twenty years, president of the Population Institute in Washington, D.C. He says that when he got into population studies in 1972, only 5 percent of couples in the world used modern methods of contraception. That figure is now up to 50 percent. Mexico, once full of huge families, has rapidly declining fertility figures. Brazil has achieved a 60 percent use of contraceptives. Italy, the home of the Vatican, has the lowest fertility rate in the world. And there are many other success stories. But the large cohort of people now passing through their reproductive years means that even in a country like China, which has adopted draconian measures in an effort to stem the tide of babies, the population is still growing at a rate of 21 million people per year. China will soon be surpassed in total population by India, which has reached a billion people and is growing at a rate of 2 million every month. If we do nothing to encourage slower population growth, it will mean, as Fornos says, "the difference between . . . a twenty-first century where we have a hope for a better quality of life, or [one] where the Four Horsemen of the Apocalypse will return, only this time they'll be Global Warming, Ozone Depletion, Famine and Over-Population."

We could help countries like Nicaragua, Honduras and Haiti with simple contraceptive aid so that women can have the number of children they really want. Fornos says that in the United States, "we're currently spending [the equivalent of] roughly a cup of coffee and a donut a year, per person, on international population assistance. I think we can afford to spend half a six-pack of whatever your favourite beverage is." But international population assistance is not just a matter of providing cheap pills or condoms. It's something much more fundamental, as has been proven through a great number of university and United Nations studies over the past ten years. The United Nations–sponsored Conference on Population, held in Cairo in 1993, agreed unanimously on a course of actions to reduce populations.

The first of these is to raise female literacy rates. There is a direct and almost unbreakable connection between the ability of women to read and a lower birth rate, a fact that first came to the attention of researchers through studies unconnected to population. It holds true virtually throughout the world.

The second is to provide more employment opportunities for women, and to give them access to credit, banking and inheritance rights, as well as the kind of training that makes them useful in their societies for reasons other than breeding. An elevation in the social status of women is crucial. Generally, the higher the status of women in society, the lower the birth rate, and vice versa.

The third action suggested at the Cairo Conference is one of the most effective and rewarding ways to reduce the birth rate — lower infant mortality. This can be accomplished by providing clinics for children under the age of five, simple and cheap rehydration powders to save babies from dying of diarrhoea, and hygiene courses for mothers. All these save existing lives and prevent many new births. But some people claim it's counterproductive to save infant lives, because it eliminates natural controls and thereby increases the population. Werner Fornos answers, "I'm involved in population because I love children. And I think we each have an obligation to take care of the children we bring into this world. But also, by reducing infant mortality to its lowest level, you get rid of the compensatory factor." Experience indicates that when women see babies dying, they continue to have them in order to ensure the number of children they really want. When they have some confidence that their first children will live, they immediately cut back on the number of children they bring into the world.

The fourth action to take is to provide universal access to means of fertility control, to prevent unwanted or unintended pregnancies. The main thing to remember about this aspect of a population-control program, says Fornos, is that "universal access to family planning does not mean abortion. I don't understand when our politicians are going to learn that abortion, like war, is a failure of society to come to grips with a much more fundamental problem. And in this case, the fundamental problem is the prevention of unintended pregnancy."

These four steps, according to Fornos, could lead to a future with a stable human population, where our major focus would be to curb consumption enough that it would be possible for the world to share resources more equitably while protecting the pristine wilderness areas left. There are many promising ways to reduce consumption. Already, Germany requires all manufacturers to take back packaging, obsolete cars, used refrigerators and old computers, and to assume responsibility for the products they make from beginning to end. In Sweden and Denmark, there are industries that recycle their wastes so efficiently that they produce no emissions; every bit of waste is used or returned into natural systems. We could use our vaunted creative energy and ingenuity in these directions, instead of using them to expand markets for clever new gadgets or marketing gimmicks to keep consumption at a fever pitch.

When the population discussion first came up, the perception was that the real problem lay in the Third World, and in general with the poor, who had social customs that demanded and rewarded large families. But we have to remember that while there are far more *people* in poor countries like India, China, Kenya or the Philippines, more than 80 percent of the planet's *resources* are being consumed by countries like the United States, Japan, Germany and Canada. If you're a Canadian or an American with only one child, that child will consume more than forty times what two little Bangladeshis will. The problem of overpopulation is not just numbers. It's a factor of both population and per capita consumption.

Way Down Below the Ocean

> *Worldwide longline fisheries are now extirpating all the top trophic levels of the ocean: the sharks, the sailfish, the bluefish tuna, all these top predators. And as a consequence, the very nature of our fisheries is changing. We now feed from the ocean at one trophic level lower than we did thirty years ago. At this current rate, you can forget swordfish, you can forget tuna. A generation from now, you'll all be eating anchovies, and I hope you like them on your pizza.*
>
> — STUART PIMM

Our basic environmental problems are overpopulation and overconsumption. We have the ecological footprint to help us measure our individual and collective impact. But how do we know whether the situation is getting critical? That is, how radically do we have to change and how fast? Despite 400 years of science, we're still far from having baseline data on natural systems. Still, and as always in science, we have to go with the best data we have now. We have to depend on the few experts who actually have seen what is going on around us.

Seventy-one percent of this planet's surface is covered by oceans. But for most of human existence, they've been as mysterious as the moon. The marine biologist Sylvia Earle is the Explorer in Residence for the National Geographic Society, and she has spent much of the past twenty-five years underwater, seeing the oceans firsthand. We forget that they are the basis for all life on Earth. As Earle says, "This planet is governed by the oceans' climate. It's also home for most of life on Earth. Most of our oxygen comes from the sea. Something like 70 percent of the oxygen is generated by organisms living in the ocean. Much of the carbon dioxide is absorbed there . . . . If we dried up the oceans, we would do away with the cornerstone of our life support."

We are encroaching on this system of life support. The extent of this devastation was estimated by a marine ecosystem expert, Elliot Norse, who calculated that the area of ocean floor destroyed annually by draggers, enormous nets that scrape everything off the ocean bottom in huge swaths, is 150 times the total amount of land cleared by logging in the world's forests! Right off the coast of Canada and the northern United States, says Earle, underwater robots have provided us with a view of George's Bank and the Grand Banks, "and where people have been with scallop dredges, it looks like a superhighway. It's like a bulldozer has just gone by. If you go to places where they haven't been scraping the sea floor, it looks hauntingly like a coral reef. There are sponges. There are . . . soft corals. They are beautiful with diverse life — crabs, starfish and various kinds of fish."

One of the worst aspects of this type of "fishing" is the waste. Only a few targeted species — scallops, for example — are kept. The others are thrown away. "It's like going through the streets of Toronto and knocking down the buildings and trees and picking out a few pedestrians that you might like to munch on," says Earle. The scale of fishing that is now permitted with new technologies only increases such practices. Fishing boats are like immense factories equipped with processing facilities and freezers, satellite positioning and sonar locators. They can stay at sea for weeks and pinpoint their prey with deadly accuracy. The weapons they deploy — longlines, dredges, nets — are massive. "Some nets are [big] enough to engulf something as large as a dozen giant jets in a single swoop through the ocean," says Earle. "In one bite, you could take the equivalent of twelve Boeing 747 aircraft, for example. These are being applied in the ocean in a way that takes everything. I liken it to using bulldozers to catch songbirds and squirrels."

Again, there are many things that can be done; limits on fisheries and protection of coral reefs are already legislated in many parts of the world. As Earle says, "We are in a peculiar position right now, in our lifetime. We're at a turning point. We have the power to protect the sea, and to change, in ways that will benefit not just the fish, the whales, the phytoplankton and all the rest of ocean life, but that will directly have a positive benefit on our future: our health, our wealth, our survival."

No one sets out deliberately to trash the very things we need for our survival. But our ability to focus on a challenge — especially one that brings money and prosperity, whether it's catching fish or farming the land — can blind us to the links that connect all parts of the planet. Oceans are connected to air and land, and the things we do to each of them have repercussions for the others. Our ecological footprint is now so massive that even the vast expanses of the oceans are showing it. Earle observes, "Taking care of the ocean starts at the tops of mountains. What we do upstream affects everything downstream . . . . What we put on our backyards, on our farms or what we allow to go into the atmosphere flows onto the land and finally into the sea. Ultimately, the sea is the sewer for the whole of civilization. Wherever we live on the planet, we are affecting the oceans of the world."

Losing Our Skin

We've created the situation with these huge fires or floods or insect epidemics [that] are overtaking us. We've been trying to industrialize forests. We've been trying to make them meet human dreams and expectations of completely efficient, productive, simplified landscapes. And what we need to do right now is just give that up completely. Let's recognize that there are real constraints on how much wood or grass or water we can take out.

— NANCY LANGSTON, FORESTER

We've become aware only gradually of what's happening to another cornerstone of our life-support system — forests. Alan Durning, a colleague and an environmentalist known for an important book, *How Much Is Enough?* has devised a visual image to help illustrate both the speed and the severity of human pressures on Earth, as well as their astonishing escalation in the past few decades. He asks us to imagine a time-lapse film of Earth taken from space. "Play back the last 10,000 years sped up, so that a millennium passes by every minute. For more than seven minutes, the screen displays what looks like a still photograph — the blue planet Earth, its lands swathed in a mantle of trees. After seven and a half minutes, there's a tiny clearing of forest around Athens. This is the flowering of classical Greece. Little else changes. At nine minutes — 1,000 years ago — the forest gets thinner in parts of Europe, Central America, China and India. Twelve seconds from the end, two centuries ago, the thinning spreads a little farther in Europe and China. Six seconds from the end, eastern North America is deforested. This is the Industrial Revolution. Little else has changed.

"In the final three seconds, after 1950, the change accelerates explosively. Vast tracts of forest vanish from Japan, the Philippines, the mainland of Southeast Asia, most of Central America, the Horn of Africa, western North America and eastern South America, the Indian subcontinent and sub-Saharan Africa. Fires rage in the Amazon basin, where they never have before. Central Europe's forests die, poisoned by the air and the rain. Southeast Asia looks like a dog with mange. Malaysian Borneo is shaved. In the final fractions of a second, the clearings spread to Siberia and the Canadian North. Forest disappears so quickly from so many places that it looks like a plague of locusts has landed on the Earth."

The forest industry says that there are more trees growing in the United States now than in 1920. That may be true, since the twenties saw the end of a timber rampage that denuded the entire eastern seaboard. But they are also counting as trees the millions of sickly, chemically supported seedlings (also known as plantation forests) planted around North America by timber and paper companies. These plants are part of a massive monoculture, and bear little resemblance to the functioning, diverse-species ecosystems that deliver all the services we expect from a forest. Our original natural forest cover is being either destroyed or replaced by these simplified systems, which are merely uniform stands of trees of the same age and species; other plants attempting to grow in such a system are generally cleared or killed with herbicides. This is very different from the complex interaction of natural species in a real forest.

The World Resources Institute (WRI) in Washington, D.C., a research group trying to promote sustainable development, publishes regular groundbreaking reports on the current state of the Earth's forests. Nigel Sizer, the head of the team that created these surveys, says, "Most of the forests that you would drive through in the United States, and even in a significant part of Canada, are no longer natural, fully functioning forests. A lot of them are plantation forests . . . very highly managed and manipulated by man. They've lost significant components of their biological diversity." Sizer says that not only are forests degraded, but, significantly, they are highly fragmented. "They are no longer able to withstand catastrophic disturbances . . . like hurricanes or massive flooding episodes. To a large extent, they're dependent now upon human intervention and human management to maintain them as functioning ecosystems."

How much of what we used to have is left? Sizer and his team estimate that four-fifths of our great natural forests are gone. And the 20 percent that is left is concentrated in three countries: Brazil, Canada and Russia. Forest disappearance is a new phenomenon, and it is directly related to our ecological footprint. We've logged forests for pulp, fibre and lumber, and to make room for agriculture and support our ever-increasing material demands. Now there are so few untouched forests left that there's a kind of gold rush for the remaining trees.

We think that Canada is a modern, industrialized nation, yet in terms of the way we deal with our forests we behave like a developing country. For example, during the early 1990s, the Government of Alberta's premier, Donald Getty, sought to diversify the province's economy. The government saw the province's boreal forest as an under-utilized resource and invited major transnational logging companies to exploit it. Like a typical developing country, Alberta offered massive financial incentives to foreign companies. The Getty government underwrote the cost of immense pulp mills, guaranteed access to vast tracts of land for next to no stumpage costs, failed to respond satisfactorily to the land claims of the First Nations of the territory and seemed to pay no attention to the ecological consequences of the massive clearcuts that would occur.

Most Canadians are surprised to find that their country is the target of this Third World type of resource exploitation. They shouldn't be. Canada is one of the few countries in the world with much of its original forest cover left. The boreal forest lies just south of the Arctic tundra. It's not as lush or diverse as a tropical rain forest, but it's a vast system that is just as important for removing carbon dioxide from the atmosphere and pumping out oxygen. Sizer says the Canadian government and its bureaucracy are no different from those he encountered in Brazil or Africa. "We see a very tight connection between the political system, politicians and campaign financing, and the operations interests of the large pulp-and-paper and timber companies in both countries. We also see very limited public access to objective information about the state of Canada's forests."

Furthermore, Sizer says, "Canada is one of the countries with the poorest national statistics on the state of its forests, which is very striking when you consider that the forest industry is one of the largest industries in Canada, and that this is a country with the expertise, the ability and the resources to collect better statistics than just about anywhere else. This is a striking situation, and I don't think it's a coincidence. In our work mapping out global forest conditions, the country where we received the most conflicting information and the most concerted pressure to ensure that we reflected this or that point of view, was Canada — more than Brazil, more than Indonesia, more than Russia. I think that that says a lot."

Of course, it's not just Canadians who have to worry about how we treat our forests. One of the special characteristics of the boreal forests that you find in Canada and Russia is that they store vast amounts of carbon in the soil below the surface of the ground. And with forest clearance, that carbon is released into the atmosphere, adding to the gases that cause global warming. With more warming, the northern forests will degenerate simply because of rising temperatures. This could begin what's called a positive feedback loop: forests cut or dying will release more carbon from their soils; more warming will cause still more carbon emissions; and the situation could spiral out of control.

Can anything be done to protect the remaining forests? Despite the lure of billions of dollars and the unprotected status of most of the remaining forests in the world, there is some hope for protection and even reconstruction. Sizer has some hopeful stories even from a poor country like Surinam. After the WRI released a report about rapacious Asian loggers who were about to move into that country, the people of Surinam objected to the selling off of their forests, and the international community offered assistance to Surinamese citizens' groups. And, says Sizer, "The government of Surinam did not sign the massive and highly lucrative contracts that the Asians were offering. In fact, the previous government lost the elections partly because they were proposing to sign those contracts. The people of Surinam objected strongly, especially the indigenous people in the hinterland, who do have a vote in that country."

We could take a lesson from the Surinamese. For large multinational corporations, the ultimate goal is to maximize profit in the shortest possible time and give investors the highest possible return on investment. These companies do not — indeed, their shareholders will not allow them to — consider the well-being of local ecosystems or communities. To gain access to forests, companies will do whatever they can, and are not above bribery, deceit or threats. In an industrialized area like British Columbia, companies routinely threaten to move their plants, lay off employees or pull out of the province in order to win concessions. Through the 1970s and 1980s, as the volume of wood being cut climbed steadily, the number of jobs in the forest industry dropped steeply. The reason, of course, was greater mechanization and automation that replaced employees, yet the industry and many workers blamed environmentalists. In a recent book, the forest expert Patricia Marchak has indicated that the province's system of stewardship over the forests must undergo a radical shift away from large companies and towards local community-controlled forests. The annual volume of wood being cut in British Columbia's forests exceeds by a large margin the amount that can be removed sustainably while still protecting the diverse values of the forest.

In seventeenth-century Europe, trees were considered unsightly, gloomy weeds, obstructions to healthful fresh air, agriculture and human activities. Trees weren't allowed in gardens or cities. But at that time, there were still enormous, untouched forests spreading out over the rest of the planet. Today we know how vital trees are and how badly we need them, yet we still treat them as if they're obstructions to human progress or, at best, a source of fuel or cash. Today, our ecological footprint hovers over every tree on Earth. Without making major changes in the way we live, we'll all come crashing down together.

Atmospheric Pressures

> *What this climate problem has become is much less of a science issue than an ideological battleground between those people who believe in protecting the commons and those people who believe that anything that interferes with private entrepreneurial rights is worse than whatever might happen at the collective level. This is all unconscious; nobody ever says it outright.*
>
> — STEPHEN SCHNEIDER, CLIMATOLOGIST

The atmosphere is a very complex mixture of gases that acts like an immense blanket that envelops the world. This blanket maintains a temperature conducive to life; it also gives us our weather and climate. Trees and air are inextricably linked: what we do to one affects the other. Among other things, forests act as a "carbon sink" for the atmosphere, absorbing carbon dioxide. But in addition to losing this carbon sink because of the destruction of forests, the atmosphere is sustaining other assaults. Stephen Schneider, a climatologist who spent twenty years at the National Center for Atmospheric Research in Boulder, Colorado, and is currently a professor at Stanford University, says there is absolutely no doubt that the world is warming up. But if you read the mainstream papers or watch the news on television, especially in the United States, you might be under the impression that climate change through global warming is still controversial and needs verification.

Schneider says, "The words 'global warming' have become almost a battle cry for various factions in the world who 'believe' or 'don't believe,' who think it's serious or not serious. Yet despite a number of uncertainties surrounding the climate-change debate, the actual existence of global warming is a fact established virtually beyond doubt. If we look at the thermometers of the world and see what they've done over the last century, we see that the world is about a half-degree Celsius or so warmer than it was a century ago. Mountain glaciers have receded virtually all around the world, and many are disappearing at a rapid rate. And finally, the sea levels have risen something on the order of ten to twenty centimetres. Those are all consistent with warming. In fact, looks at historical records convince us that the last two decades were the warmest in the past 1,000 years."

Schneider says the real issue about which scientists disagree is what's causing the warming. "The debate is whether that warming is just a per-verse act of nature or whether this is something that humans have done through increasing numbers of people demanding higher standards of living and using technologies that change the land surface and dump pol-lutants in the atmosphere. What created it? That's the controversy among the knowledgeable scientific community, not whether the warming itself exists, which is a virtual fact."

Schneider spends a lot of time advising the Intergovernmental Panel on Climate Change. That panel was set up by the World Meteorological Organization and the United Nations Environment Program to establish a global consensus about what is well-known and what is speculative about all aspects of climate change. More than 2,000 scientists contribute to this panel. After reviewing more than 20,000 scientific papers on weather and climate, it has now concluded that there is overwhelming evidence that human activity is a major cause of global warming.

Bill McKibben's famous environmental best-seller *The End of Nature* is about global warming. He says, "This is no longer a scientific issue that's in dispute. That doesn't mean that you can't find a few peo-ple who will say otherwise. You could also find people who don't believe in evolution. And you can find people who don't believe in the Holocaust, and you can find all sorts of things. But this is as solid a physi-cal fact as we have around us . . . . Tens of thousands of scientists have pro-duced airplane hangars full of reports and studies and graphs and charts. We now know, beyond any reasonable doubt, what is taking place and what's likely to take place in the next 100 years. We don't know exactly what's going to happen down to the last tenth of a degree or down to the last inch of sea level or something like that. But we do know that we've raised the temperature of the planet. Every year is warmer than the last, with nine of the ten warmest years on record within the last decade."

It doesn't take a rocket scientist — or a panel of experts — to see the links between the warming of the planet, the thinning of the ozone layer and human activity. So-called greenhouse gases are produced by natural processes of metabolism, decay, fire, transpiration and evaporation. We're increasing their production by chopping down trees and disturbing phytoplankton in the oceans. We know the fuels we've been burning since the Industrial Revolution — oil, coal, gas and wood — release carbon dioxide in the atmosphere, and there is now 25 to 30 percent more carbon dioxide in the atmosphere than there was before the Industrial Revolution. We're also adding to the mix human-made greenhouse gases like CFCs, which have been used in refrigerator coolants and aerosols. Besides reflecting heat, CFCs also deplete the ozone layer, which protects us from dangerous ultraviolet rays. The thinning of the ozone layer has been implicated as the cause of increases in skin cancers and deformities in sensitive creatures like frogs. Clearly, the twentieth century bore discernible evidence of human activity on the composition of the atmosphere. And now the vast majority of climate experts believes this activity has had a significant effect on the temperature of the Earth.

Of course, human beings have always made changes to the world. We've had to to survive. We've built houses, ploughed fields, dammed rivers, cut trees. But as Bill McKibben says, there was always a border, an edge where our disruption ended. There were always mountaintops, seabeds, the Antarctic, places where there was no human footprint. McKibben was motivated to write *The End of Nature* because of the terrifying implications of human intervention on the composition of our global atmosphere: "It occurred to me as I began to read the early science about global warming, about climate change, that finally we were affecting everything. That one species on this planet now had its thumb utterly on the scale, and was changing everything around us."

What does the future hold? Scientists really don't know how much the temperature will rise. Steve Schneider says the best estimates are that we can expect somewhere between one and four degrees of warming in the next century. He says, "By and large, most of us can adapt to one degree. But four degrees is virtually the difference between an ice age and a warm epoch like we're in now. It takes nature 10,000 years to make those kinds of changes, and we're talking about changes like that on the order of a century. There isn't an ecologist anywhere who thinks that we can adapt to that without dramatic dislocation to the species in the world, and to agriculture and other patterns of living that depend on the climate."

These changes go beyond a simple rise in temperature. These perturbations to the entire envelope of gases that sustains life have all kinds of other effects, from strange weather to localized crop failures or ecosystem collapses. "Then you have the speculative stuff," Schneider says. "What happens to hurricanes? What surprises might be triggered with ocean currents? What happens to ecosystems? Where will species flourish and where will species go extinct?" We're now experiencing an increase in the number of record-breaking hot days. We're not getting an overall increase in rainfall, but we are getting more of a phenomenon called rain dumps, when the rain falls all at once and does real damage. Insurance payments for catastrophic weather events in the 1990s exceeded all payments made during the previous three decades. All these things have driven the insurance industry to support action against fossil fuels and CFCs.

And there's more to worry about. "Could what we're doing now cause something irreversible, something we can't say, 'Oops, let's back out of that'?" asks Schneider. "Well, unfortunately, the answer is very definitely yes. Climatologists call these surprise scenarios." There are plenty of candidates for such surprises. The Gulf Stream could change direction or stop, so that while the world was warming, Europe would freeze. The destruction of the West Antarctic ice sheet could raise sea levels by many metres, flooding coastlines and inundating island nations. It might take a few generations for that scenario to play out, but once it starts we won't be able to stop it.

Many respond to experts like Schneider by saying that one of the most unpredictable processes in our world is climate, and that there have been many extreme events in the past that we and our ecosystems have all survived. But, Bill McKibben points out, "climate changes all the time, but it changes slowly. We're doing it at an enormous rate of speed — by most estimates, something between ten and sixty times faster than it changes normally. That has real consequences . . . . Natural systems can't adapt to that sort of speed of change." And he adds, "With the ability to change the climate, you change everything. You change the flora and the fauna that live at a particular place. You change the rate at which the rain falls and at which the rain evaporates. You change the speed of the wind. You may change the very ocean currents . . . . Nothing that we've ever done as a species is as large in its effect as this. An all-out nuclear war . . . would have been a consequence of the same magnitude. But happily, we stepped back from that brink. Unhappily, we're not stepping back from this one."

How can we step back? Well, everyone agrees that we must reduce the burning of fossil fuels as quickly as possible. This is entirely feasible with the technology we have now — solar energy, wind harnessed by windmills, fuels like alcohol and methane from fermentation, even used vegetable oils. Indeed, Denmark already generates 7 percent of its energy by wind and has plans to make that 50 percent by the year 2010. Globally, wind and solar energy are the fastest-growing sectors in the energy industry. We also need to reinvent trains and other comfortable forms of mass transit to lure people out of their private cars.

Even big oil companies like British Petroleum and Shell have seen the writing on the wall. They are hedging their bets with major investments in alternative energy. Greenpeace International has developed a "green" refrigerator that is extremely energy efficient and doesn't use ozone-depleting CFCs. It is available all over the world, except in North America, where inefficient ozone-damaging fridges still rule the roost.

When there is a conflict between an available clean technology and an entrenched dirty one, the challenge is politics and the need for legislative action, not technology. Governments in Canada and the United States continue to subsidize fossil-fuel energy to the tune of billions of dollars annually and then plead a lack of money to support alternative energy or public transit. We can do it; we just have to want to.

Poison in the Well

In 1976, somebody made a decision at Love Canal when they did a cost/benefit analysis. They put a dollar on my head. They put a dollar on the potential for my children. They put a dollar on my husband's head, based on what he was making, which at the time was $10,000 a year. And they decided that we were not worthy of a $20 million clean-up. They thought, for $20 million, that it was perfectly okay to kill not only

*my children, but the 900 families that were living at Love
Canal....And that's wrong.*
— LOIS GIBBS, FOUNDER OF THE CENTER FOR HEALTH,
ENVIRONMENT AND JUSTICE

One of my biggest heroes is Lois Gibbs. She's an environmental activist
who emerged from the grass-roots and founded the Center for Health,
Environment and Justice, one of the most effective environmental
groups in North America. She has been called the Mother of the
Superfund because her efforts helped create a mechanism in the United
States for licensing, evaluating and cleaning up toxic-waste dumps.
Her story is a graphic and frightening illustration of the effect that toxic
waste has on our ecological footprint. It is also an inspiring story of
how ordinary citizens can work together for change.

When Lois Gibbs was a young housewife back in the 1970s, she
lived with her husband and two young children in a suburb called Love
Canal, near Niagara Falls in Upstate New York. The new subdivision
had been built over a former industrial site and landfill, and 90 percent
of the residents worked at nearby chemical companies that manufac-
tured pesticides and solvents. Nobody thought anything about it until a
lot of the residents, especially the children, began to experience health
problems. Gibbs had two small children at the time. Her son, Michael,
was a perfectly healthy one-year-old when the family moved to Love
Canal. After living there for four years, he developed asthma, epilepsy, a
urinary-tract problem that required two operations to correct and an
immune-system disorder very similar to AIDS. Gibbs's daughter,
Melissa, who was conceived and carried at Love Canal, developed a rare
blood disease that caused internal haemorrhaging and left her with a
very high risk of developing leukemia.

Very frightened, and at a loss to understand her children's health
problems, Gibbs stumbled upon a newspaper story that described workplace diseases and symptoms resulting from exposure to the same chemicals that were in the landfill at Love Canal. Gibbs says, "I'm reading this
newspaper and I see epilepsy, asthma, depressed immune system, blood
disorders, leukemia. And I'm checking off my kids' problems, and it was
like, 'Whoa. This is really scary.' "

It wasn't in Gibbs's nature to make a fuss. Like most of us, she assumed someone in charge would know what to do. She figured that if she was scared, then so were other people. So she waited for somebody smarter and more skilled to show up and tell her what to do. "But nobody came knocking on my door. And so Michael got very sick, and he was in the hospital. And that was sort of the turning point for me." Gibbs remembers thinking, " 'You know, the reason he's here sick is not just Love Canal. It's not the city's and the government's fault. It's *my* fault. I made a decision that I would sit back and wait for somebody else to come and do this. And when they didn't come, I didn't do anything. What kind of responsible mother could I possibly be if I sat back, knowing my children were being poisoned?' And that's when I got up enough courage to go to the very first door and begin to talk to people."

Her neighbours, equally scared, were also waiting. When Gibbs made contact with them, they began to compare notes and to organize. They found that their homes and playgrounds had been built over 20,000 tons of mixed chemicals, mostly solvents and pesticides that had been buried there by Hooker Chemical Corporation, a division of Occidental Petroleum. Doing their own research, they discovered that among the 240 chemicals in their environment was the highly toxic lindane, which has been banned in the United States and Canada for decades. This poison, 100 percent pure, was lying on the surface of the ground, a ubiquitous yellow dust no one had really noticed before.

Gibbs determined that her own problems were not the worst. "We had statistics that showed that 56 percent of the children . . . were born with birth defects ranging from three ears to extra rows of teeth, extra fingers, extra toes or mental retardation. We found that the women at Love Canal gave birth at statistically higher rates to low-birthweight babies than [did women] in other communities. We had more miscarriages and stillborn children than other communities. And central nervous system problems and so on."

Gibbs was transformed into an environmental activist. She says the collective effort of 500 of the community's families changed her life. "Some people were just brilliant, smart people who could pronounce all the silly chemicals, could look at the technical reports. Some were just so creative that they came up with these ideas that were fun, media-worthy and focused on politicians. Other people were great at things like typing and cutting newspaper clippings. That's what makes organizations run. That's what creates power. And that's how we won at Love Canal." With Gibbs as a major spokesperson, they protested the lack of responsible dumping and demanded that the government do something. After years of fighting, they won. There was an investigation, followed by a clean-up. And all 900 families in the community received enough money to move away from Love Canal.

After all that, Gibbs launched the Citizens' Clearinghouse on Hazardous Waste, whose name was later changed to the Center for Health, Environment and Justice. The organization has been successful in stopping every single proposal for a commercial hazardous-waste landfill in the United States since 1980! Not one has been sited in the United States since Gibbs got to work. The disposal sites are legal to build — you need double-liners, leak-detection systems, monitoring programs, and then you can build them — but people have said no. "When they go out and they try to site these things . . . people literally come out and link their arms with their kids, with their husbands, with their cows, with their tractors, and they just stand up and refuse it. And corporations cannot fight that."

When we interviewed Gibbs, she had just returned from Rochester, New York, where Kodak's headquarters are located. She told us about the alarming number of families living near the Kodak facility whose children have been diagnosed with brain cancer. The parents and other concerned community residents allege that the number of local cases of nervous system cancers in children is unusually high, and together constitute a "cancer cluster" that they fear is linked to the emissions from the Kodak facility. Kodak denies any connection between its incinerator and the cancers. New York State agencies have so far not been able to show a link; a new federal study is trying to assess the situation, but it has always been very difficult to prove the exact connections between toxic emissions and human disease.

Things aren't much different in Canada. Elizabeth May of the Sierra Club of Canada has been helping out the community of Sydney, Nova Scotia; its inhabitants are living adjacent to the largest toxic-waste site in Canada and possibly all of North America. She says, "Within a community that includes tens of thousands of people, there are families living along a street who have toxic waste in a ditch running through their backyards ... people with arsenic pooling on their basement floors. They're sick every day, and the government says, 'We don't think there's any risk to your health. The kinds of illnesses you're having are completely consistent with the bacteriological, viral infections such as Canadians get every winter.' This is outrageous."

The source of the contamination was a provincially and federally operated steel mill and coking plant. While the most polluting elements of these operations have closed down for economic reasons, the legacy is an enormous amount of toxic wastes — 634,900 tonnes of sludge contaminated with polyaromatic hydrocarbons (PAHs), of which 40,815 tonnes are PCB-contaminated and located in an estuary now known as the Tar Ponds. That's more than *twenty-three times* the amount of toxins buried at Love Canal. There are also more than 50.6 hectares of land, the former site of the coke oven, that are saturated to depths of twenty-four metres with PAHs such as naphthalene, benzene, toluene and benzopyrene, as well as arsenic, cyanide and other heavy metals, all of which are incredibly dangerous carcinogens and poisons. As May says, "Cancer and birth-defect rates are far above the national average in Sydney. The site looks like something out of Eastern Europe. You have to understand, this is far worse than comparable steel sites in the rest of North America. People's backyards have eighteen times the Canadian Council of Ministers of the Environment guideline for arsenic for residential and parkland areas. Eighteen times! And the government is saying, 'Well, you're not going to run into trouble unless you go back out there and eat the soil.'" May acknowledges that the government is finally throwing money at the Tar Ponds. Early in his tenure as prime minister, Paul Martin promised $3.2 billion for such clean-ups. But money has been promised before, and so far, May says, it's been spent on "more studies, not to move people away." The only clean-up that's started on the Tar Ponds is a scheme that's trucking this dangerous material all the way to Quebec, where it's being incinerated upwind of Montreal.

What happened in Sydney is classic and tends to be repeated whenever citizens realize they are being exposed to toxic industrial wastes. In both Canada and the United States, when communities ask for a simple health study to establish facts about their illnesses, they are told the same thing: "The health problems are just a random clustering." Lois Gibbs heard it twenty years ago, and it's the same line that governments use today. When our ecological footprint gets so huge that our babies contract leukemia and are born without brains or eyes, when we've poisoned our wells and crops and animals, it ought to be time to ask what it costs Earth to have photographs and pesticides and car tires and the myriad other products that are so very toxic. And for people who say that we have come to need these products, and that we also need the jobs they create, Lois Gibbs has a very clear answer: we can have almost all of our products, and our health — if we want to.

One thing we can do is to make the cheap disposal of hazardous wastes more difficult. We also can force businesses to take responsibility for their products through legislation. Many countries have done it already. If waste-disposal costs were part of production costs, they would have to be added into the costs each company would have to deduct before it realized its profits, just as is done in Sweden and Germany. If the prices of products reflected their real costs to society, and if manufacturers couldn't just offload their garbage onto the community or the government, these companies would soon find ways to dispose of or substitute for the really hazardous substances.

Lois Gibbs's husband worked in a chemical plant back in the 1970s. She says, "He was a chemical operator. He made it, they dumped it, I yelled about it. Our dinner table conversation was very interesting." But workers often know how to make better products. Gibb says, "My husband could have gone to his bosses ... and made six suggestions about what they could do to keep, for example, vinyl chloride from going out the stack into the community, to keep the workers safe. It's really not always an either/or situation. It's a matter of whether the corporations are made to invest in protecting their workers and their communities or not." And if proper waste disposal cuts too deep into a product's costs for that product to remain viable, well, it's supposed to be a free-market system, without subsidies given to losing operations.

Ironically, toxics are one of our easiest environmental problems. We have the technology to eliminate, replace or contain them, and we even have an economic system that can guide our decisions about which substances simply aren't worth having around, like DDT, lindane or dioxins. But the best news of all is that the clean-ups work. Since Lois Gibbs won her fight with the government, her children have largely recovered.

Melissa's blood count is within normal limits, and all of Michael's problems are gone except for his asthma. Gibbs says that although the government has never bothered to run any follow-up studies, many Love Canal families report the same thing. Michael's urinary-tract problem was supposed to be lifelong, but Gibbs says it disappeared a year after they got away from Love Canal. At first, she thought it might have been a coincidence. "But when you looked at the hematology clinic results for Melissa's blood, you actually could see the levels of platelets increasing over time. And if you looked at the brainwave test from Michael's epilepsy, you could see his epilepsy actually disappearing. It isn't coincidence. The body is an amazing instrument. And if we can remove ourselves from these chemicals, we can heal."

Lois Gibbs runs a true NIMBY organization. That's the acronym coined by corporations and their PR firms to belittle environmentalists for their "not in my backyard" attitudes. The idea is that toxic waste, even if it robs babies of eyes, has to go somewhere, and that it's selfish to want to partake of the benefits of modern life, like steel mills, computer chips and weedkillers, if we are unwilling to accept the consequences. Gibbs says she wears the acronym as a badge of honour. "It means that people are unwilling to accept somebody else's poison in their backyard for somebody else's profit." She says corporate spokesmen often argue, " 'It's got to go somewhere, Ms. Gibbs. If we don't put it in your backyard, then whose backyard do we put it in?' And our answer is: 'It's not an either/or. It's not either your backyard or my backyard.' In today's technology, we know how to take care of hazardous waste. We know how to take care of garbage. We know how to take care of most of the stuff that we produce. What we are unwilling to do, what corporations are unwilling to do, is pay the cost to deal with their waste at the front end."

So we have to make them. And when we do, everybody's footprint will be a lot smaller.

CHAPTER 4
SEZ WHO?

We, the undersigned, senior members of the world's scientific
community, hereby warn all humanity of what lies ahead. A
great change in our stewardship of the Earth and the life on it
is required if vast human misery is to be avoided and our
global home on this planet is not to be irretrievably mutilated.
— "WORLD SCIENTISTS' WARNING TO HUMANITY,"
NOVEMBER 1992

From Tuktoyuktuk to St. John's, from Seoul to Rio, Shanghai and
Bombay, people have been told that the revolution in information that
has occurred during the past fifty years has aided the spread of democracy
and will continue to empower people everywhere. That's the reason we
turn to the purveyors of information, the media, whenever we feel that
an issue must be brought to public attention. That's why people like me,
who are concerned with specific issues, like the environment, depend on
the media to spread the word.

The media have indeed informed the public about threats to our air,
water and food. Ever since 1962, when Rachel Carson published *Silent
Spring*, more and more information has been made available. And the
public has responded. About fifteen years ago, public interest in the environment reached its height. In 1988, George Bush Senior promised that,
if elected, he would be an environmental president. In the same year,
Canadian Prime Minister Brian Mulroney was re-elected, and to indicate
his ecological concern he moved the minister of the environment into
the inner Cabinet. Newly created environment departments around the
world were poised to cut back on fossil-fuel use, monitor the effects of

acid rain and other pollutants, clean up toxic wastes, and protect plant and animal species. Information about our troubled environment had reached a large number of people, and that information, as expected, led to civic and political action. In 1992, it all reached its apex as the largest-ever gathering of heads of state in human history met at the Earth Summit in Rio de Janeiro. "Sustainable development" was the rallying cry, and politicians and business leaders promised to take a new path. Henceforth, they said, the environment would be weighed in every political, social and economic decision.

Yet only two weeks after all the fine statements of purpose and government commitments were signed in Rio, the Group of Seven industrialized nations met in Munich and not a word was mentioned about the environment. The main topic was the global economy. The environment, it was said, had fallen off the list of public concerns, and environmentalism had been relegated to the status of a transitory fad.

Something even more portentous happened a few months after the Earth Summit. A document called "World Scientists' Warning to Humanity" was released. It was signed by more than 1,600 senior scientists from all over the world, including more than half of all living Nobel Prize winners. Here's part of what it said:

> Human beings and the natural world are on a collision course . . . .
> Many of our current practices put at serious risk the future for
> human society . . . and may so alter the living world that it will be
> unable to sustain life in the manner that we know. Fundamental
> changes are urgent . . . . No more than one or a few years remain
> before the chance to avert the threats we now confront will be
> lost and the prospects for humanity immeasurably diminished.

Since scientists are usually extremely cautious about making public statements, it was remarkable that so many eminent scientists would sign such a strongly worded document. It was therefore a real shock to me that no one — not Canada's national newspaper, the *Globe and Mail*, or its national television network, the CBC, or any major American television network — bothered to report the "Warning to Humanity." Two of the most prestigious North American newspapers, the *New York Times* and the *Washington Post*, pronounced the warning "not newsworthy."

What happened to move environmental stories from the front page to the waste bin? Could our informed demand for action have moved us all so far along a road to recovery that the need to heed the "Warning to Humanity" is no longer urgent? Or could it be that having access to enormous amounts of information doesn't always mean that we're really informed? In fact, most of us are beginning to realize that even with all the wondrous new ways to access information, we're not always able to learn about and analyze the issues that are most crucial to our lives, especially through mainstream media like television and daily newspapers. Although we live in the Age of Information, we are increasingly aware that we receive insufficient, misleading or even *no* information at all about issues that ought to be of extreme importance to every one of us.

Finger-Lickin' Good

> *In North America, some of the protections, like the First Amendment for a free press, are being used by commercial interests to inhibit speech or to protect commercial speech. Here it's a matter of catching up with the techniques and providing the laws and regulations that will make sure that people are not intimidated, and cannot be shut down because they can't afford to defend themselves.*
>
> — LAWRENCE GROSSMAN, FORMER PRESIDENT OF
> NBC NEWS AND PBS

If you live within reach of any form of North American media, you probably remember this story: in 1996, Oprah Winfrey, the famous talk-show host, was taken to court by some angry cattlemen who accused her of defaming their product, U.S. beef. This story was played out in the centre ring of a media circus for weeks. Winfrey even took her show down to Texas, where she was being tried. Crowds of emotional supporters, as well as the usual mobs of journalists, gathered on the courthouse steps every day until she was acquitted. What the audience to this spectacle took away was some vague idea of the trials and tribulations of the incredibly famous. But unless you saw the particular *Oprah* episode on which the lawsuit was based, you might not have known that the court case was actually about something of far more direct concern to each one of us.

Howard Lyman, a former Montana cattle rancher and then activist with the Society for the Prevention of Cruelty to Animals, was the subject of that *Oprah* episode. He told millions of viewers simple and verifiable facts about modern meat production. As the mad cow crisis of 2003 has finally informed mass audiences, Lyman back then explained that herbivorous food animals like cows and sheep had been turned into carnivores and cannibals by being fed the ground-up and cooked-down flesh and blood of their own and other animal species. He says, "We started doing this after the Second World War. Remember, about half of the volume of . . . beef animals is not sellable to humans. You either have to pay to put it in a landfill, or find a way to turn it into a product that is sellable. The solution that they've come up with is to render it. Grind it. Cook it. Turn it into meal that's high in protein. And feed it back to other animals."

We're putting this grey, sugary-looking, rendered protein into our pets' food, and ultimately into our own bodies as well because we feed it to animals we eat — cows, sheep, pigs, chickens, turkeys. "Old spent laying hens have very little value," says Lyman. "To take those old hens and put them into a cage and load them on a truck is to get almost no economic return from them. It is much easier to take them and suck them out with a big vacuum cleaner into a grinder. We're grinding up live chickens and feeding them back to other chickens. We collect the manure and feed it back to them too. It's a whole new meaning to [the phrase] 'finger-licking good.'" Such practices are not done everywhere, but they're considered normal by the industry and are widespread in North America.

Howard Lyman's specific attacks on beef centred around the fear of bovine spongiform encephalopathy (BSE), or mad cow disease. During the British epidemic of this disease, it was found that feeding rendered sheep to cows can pass on a terrible brain infection called scrapie. And when humans eat the infected cows, they can get it, too, dying horribly of Creutzfeldt-Jakob disease (CJD) as their brains gradually fill with holes, like sponges. In the 1980s and 1990s, the same type of feed was being used in the United States. This feed was also implicated in the famous case of a single cow that died of mad cow disease and decimated the Canadian beef industry in 2003.

After Oprah Winfrey heard how industrial beef was fed, she announced, "Stop me cold from eatin' another burger!" It was riveting television, and the people who saw it demanded changes, which they got within a year. Lyman says, "The [public] response . . . was overwhelming." As a result, in August 1997 the Food and Drug Administration and United States Department of Agriculture passed regulations against feeding ruminant animals to ruminants. It was a baby step, since it did not stop cows from being fed to pigs or pigs to cows, but as Lyman says, "At least we ended up with a lot more people understanding the issue, responding, educating themselves . . . finding out what the facts are."

It looks like a happy ending after all. But the satisfying story of how important information was brought to the public so that citizens could take steps to protect themselves is not what made that *Oprah* episode a big media event. Something else happened. The National Cattlemen's Association sued both Oprah Winfrey and Howard Lyman, claiming $20 million under a new state law prohibiting anyone from making "disparaging remarks about food." In the end, Winfrey won. But because she did, the law was never challenged in a higher court. It's still on the books, and is still being used, actively chilling the hearts of any of us who might want to discuss publicly our concerns about the industrial food chain. Despite the media circus surrounding Winfrey's trial, there was almost no mention of the implications of the Food Disparagement Act, and what it means to people who might fall victims to bad beef. Perhaps reporters were afraid of being sued.

The question remains whether the public would ever have heard about this particular story unless people like Oprah Winfrey and Howard Lyman were willing to accept the risk of an enormous lawsuit. In all likelihood, the United States would still be feeding sheep to cows. As it is, the FDA is warning that there are probably hundreds of more cases of BSE in the country. The reporters who congregated on the courthouse steps didn't talk about the service the show had performed for the public, or about the truth or falseness of the accusations. And they certainly didn't go into the social implications of this kind of legal chill. In fact, they were at the court only because a big celebrity was on trial.

Oprah Winfrey can afford to defend herself against a $20 million lawsuit. But Howard Lyman, a typical environmental activist, is not a rich man. He incurred thousands of dollars in legal fees, as well as the loss of his time, simply for trying to inform the public about how the food industry works. And he's become a pariah with the media. He told us he's no longer invited to appear on high-profile talk shows, nor is he interviewed by big-circulation magazines. As a result, his message is being suppressed by media self-censorship. It's called libel chill.

So the basic concept — that information will help people in a democracy protect themselves — proved partly true. But we heard about the story mainly because of its entertainment value. Most reports focused on the scandal — a big celebrity gets sued — and on the sensational ending of a big celebrity triumphing. Following 9/11 and the second Gulf war, there has again been something of a resurgence in so-called "serious news," largely because Americans, in particular, are dying in battle in numbers considerably higher than media estimates. But whenever things are relatively quiet in Iraq or Palestine, the media slip back easily into celebrity mode, jumping on the latest scandals and local murders instead of providing anything like an analysis of the real challenges facing citizens even in peacetime. So the question is, how did the information media lose sight of their major reason for existing?

Bread and Circuses

Let's face it, television news is increasingly becoming entertainment. There's a complete blurring between what's significant as a global event and what's basically salacious gossip.
— ELIZABETH MAY, EXECUTIVE DIRECTOR,
SIERRA CLUB OF CANADA

It's important to know what's going on around us. Such knowledge allows us to form opinions about issues and then to act on them. In fact, it's axiomatic that a democratic government is impossible to sustain without a free press. And these days it has become the accepted wisdom that the more kinds of media we have, the better it will be for society. We now have the Internet and the 500-channel universe. It is hoped that with access to all this information, we'll have more power to control our lives.

Lawrence Grossman, past president of both NBC News and PBS, has thought deeply about the nature of information and public discourse. He says: "Most people get most of their information now in the United States from local television newscasts. And local television newscasts, as every study has shown, are filled with nothing but crime and happy talk and sports and weather and very little else. So while there is a great opportunity for an informed public to act in [its] own interest, there are questions as to whether we're really getting an informed public. And while everybody talks about this great new age as the Information Age, what we're really saying is that it's not the Information Age coming upon us as much as the Entertainment Age."

When there are no immediate military or terrorist attacks, that is, when the corporate media are left to their own devices, news shows on television and the radio are more about distraction and amusement than they are about discourse on serious issues. People don't want serious issues — or so we are told by those who run the media — they want entertainment, celebrities, sports, spectacles and local crime. And since mainstream media organizations thrive on ratings, they claim they are only giving their audiences what they want. "There's a great tendency to blame the people: why aren't they interested in things that affect their own lives?" says Grossman. "But at the same time we aren't doing anything about encouraging that interest, stimulating that interest, providing what is necessary to make sure that they get well informed. And the one thing we know about democracy is that we cannot just leave it to chance. And it's certainly unwise to leave it to a medium that is dominated by entertainment, diversion and the need for profits."

With the glaring exception of the United States, virtually every democracy in the world has publicly funded television and radio channels. But the very idea of subsidizing information this way is under profound attack, especially here in North America. The one U.S. public broadcaster, PBS, has been stripped of nearly all the public funding it once enjoyed. It now runs what amount to commercials for corporate sponsors between desperate pleas for viewer funding. In Canada, the CBC has been subjected to cut after cut, not to eliminate fat, as is claimed, but to make the organization less of a public broadcaster and more of a profit-centred corporation. At the beginning of the new millennium, it seems it's no longer appropriate to spend tax money on information. It's supposed to fund itself.

Lawrence Grossman ran both NBC and PBS, one commercially funded and the other supported with public money. He says that in our new era of telecommunications, with the convergence of satellites, computers and television, the race to broadcast is being "entirely driven by and won by commercial interests — understandably, because there's a huge amount of money to be made." We entrust these public airwaves to the free market because it is widely believed that money-making corporations are efficient and will do the best job. But Grossman says, "The marketplace fails in certain areas — not in entertainment, in sports, in commercial data distribution, but in areas like providing fair-minded, intelligent, full civic information and education." He says the marketplace also fails when it comes to providing education for "an increasingly older and maturing society that requires . . . job retraining, continuing education. The marketplace fails when it comes to providing culture and arts, particularly for minority audiences. The marketplace fails when it comes to quality children's programming. And so, if society values those, something has to be done to make sure that they are provided."

But who will provide them? Certainly not the current system of private networks. When it comes to providing objective economic analysis or civic information on justice issues and environmental problems, commercial television pays little serious attention. Most commercial broadcasters don't produce or broadcast the serious documentary programming that deals with such issues. The late Neil Postman, who was a professor at New York University, a famous media analyst and the author of best-sellers like *Amusing Ourselves to Death* and *The Disappearance of Childhood,* called television the "command centre of our culture" because it is our primary source of news and information. The problem, Postman said, is that "as a visual medium, television is good for fast-moving, exciting, exotic images. It is not good for serious public discourse," and that's a major reason why there's so little of it on the air.

Postman said a television news director knows what the medium is good for. So when he or she has to choose between footage of a man committing suicide in some small American city and a discussion of the country's annual budget, guess which one gets the most air time? Matters that require talk, complex concepts and language, detailed explanation and historical perspective just don't work as well as those exciting visuals.

Use What You Got

Electronic media can be used to exploit people, to exploit issues and to put the wrong focus on priorities. But so can words in

*a book. And yes, it's more of an emotional than a rational
medium. But the important thing is not to wring your hands
over whether people are smart enough to be able to see through
it. The important thing is to have a set of policies that enables
these media to be used for civic and democratic purposes, and
not just for superficial, bad or money-making purposes.*

— LAWRENCE GROSSMAN

Lawrence Grossman says we have alternatives. He says that the kind
of intelligent, fair-minded commentary we need is not the responsibility of the commercial broadcasters. Their responsibility is "to their
owners and stockholders, to maximize profit. And that's a perfectly
legitimate and reasonable objective. But it has *nothing* to do with the
public interest. Nor should it be confused with the public interest."

So what can we do? We can increase support for public broadcasters
while we still have them. And we can even let in some new players.
Grossman points out that we already have great respect for certain institutions in our society. "The universities, the museums, the libraries have
all been totally bypassed by the television and radio era. We should make
sure that they are not bypassed by the telecommunications era. They have
a lot to offer. It would be relatively inexpensive for them to provide the
content. Then people will have a choice . . . . When it's important to
them, and when there are issues of real national or even personal and
local interest, they will have a place to turn to get the legitimate facts and
information that they need."

In a sense, with access to the Internet, a good deal of this is already
happening. But real broadcasting needs funding, and Grossman says
that's not an insurmountable obstacle. "In my judgement, the money
should come from the commercial enterprises that are exploiting the
publicly owned spectrum. The airwaves are incredibly valuable, and commercial enterprises pay nothing for them. Some of their huge profits, a
thin slice, should be taken off the top to provide a public freeway that
serves people, connects to every home, every hospital, every school,
every library, every museum, every prison. This could provide the kinds
of material that society values and needs for its survival, and that democracy in particular requires, but that the marketplace, understandably and
for very good reasons, is not providing."

This plan is far from being hopelessly idealistic. A hundred years ago, there were no public libraries in most of North America, but a combination of public will, local funding and staffing, and philanthropy — such as the millions endowed by industrialist Andrew Carnegie — made free access to books a reality across the continent. We simply have to apply what we have done before to new methods of delivering information. Grossman says, "Never underestimate . . . people's ability to figure out their own self-interest. We have a lot of problems . . . with public participation in civic activities. One major reason is there's much more entertainment and diversion and amusement than there used to be. It's a lot easier to spend your time on that. But when people really feel that something is important to them, whether it's jobs or social security or pension protection or health care, they'll get on the case." Indeed, both the early campaign success of presidential candidate Howard Dean and the proliferation of politically active web sites like MoveOn.org or TomPaine.com illustrate how people are beginning to inform themselves.

To bring about change, loud, explicit demand is essential. We cannot leave these matters, as Grossman says, "to serendipity or accident, or to corporations that have . . . a bottom-line interest. We've got to make provision for a public freeway in which the major issues can be discussed, through programs with documentaries, through printouts, web sites or courses, where you can have town meetings and public discussion of issues . . . . You cannot force people to attend or to listen or participate, but you sure can take the steps required to make this stuff available, so that people will have nobody to blame but themselves if they make the wrong decision."

Real Time vs. Fast Forward

When you watch a subject like climate change in the news, it's never treated as climate change. It's floods in China, or fires burning out of control wherever, or people dying of a massive heat wave in France or the Midwest. But generally speaking, those short, fast stories aren't presented as what they are: one long, slow, compelling story of how human behaviour is changing the climate of the world we live in.

— ELIZABETH MAY

In 1989, my family and I had the wonderful privilege of spending ten days in the Kayapo village of Aucre, deep in the Amazon rain forest, some fourteen days by dugout canoe from the nearest town. It was a humbling experience, as it made me realize how limited my academic knowledge was for day-to-day survival. Walking into the forest each day to gather food, I often paused, overwhelmed by the complexity of the surroundings, the immensity of my ignorance and the vast store of information my Kayapo hosts possess for their survival.

The community of creatures making up that forest was also constantly receiving information from the world around it, information that flowed through the air, water and soil. The most obvious information was the day and night cycle, which signalled the nocturnal and diurnal responses of activity or rest. Even on the equator, plants know when to germinate and flower, and animals regulate their behaviour in tune with the vegetation. The flashes of colour in birds offered information, both to their own kind and to other species. The colour patterns of flowers were informative to potential pollinators, and the different shades, patterns and shapes of butterflies and caterpillars gave signals to both mates and potential predators.

I was aware of the cacophony of sound, but I couldn't decipher the information contained within all that insect and bird noise. Indeed, much of the information coursing through the forest — such as minute quantities of molecules functioning as pheromones, territorial markers or indicators of food or predators — was beyond my sensory capacity. I couldn't hear the sounds of insects and bats that lay beyond my aural detection. But even if I couldn't detect or decipher it all, I knew that the forest community was awash with information vital to life.

As I was led through the forest and along the rivers, I watched the way my Kayapo hosts "read" the information contained in their surroundings. Animal tracks, specific plants, the type of soil — all told the Kayapo of hazards or sources of food or useful materials. These forest people possessed knowledge that was generations old, and that had been painstakingly acquired through acute observation and trial-and-error experience. That vast quantity of information formed a knowledge base that not only helped the Kayapo survive, but also informed them of who they are as a people, what their history is, and their very reason for being.

The media expert Neil Postman said that as biological beings, we evolved sense organs — eyes, tongue, ears, skin, nose — that inform us about the physical world around us. What we see, taste, hear, touch or smell is immensely important to our survival. He said that, as is the case in the Kayapo world, "there was a time when the purpose of information was to solve specific problems, in either the physical or symbolic environment. People sought information because they needed it to do something in the world." But beginning in the 1840s, according to Postman, with the invention of telegraphy and photography, this began to change. By the time we got radio and television, to say nothing of the Internet, we had become inundated with information from all over the world. The information that comes to our senses today is fundamentally different from that received when we had to pay attention to nature. Rather than impelling us to action — as the smell of smoke, the crack of a twig or the stripes of a tiger used to do — information today pacifies us. Now, although there's a huge quantity of sensory input, there's little demand for *output*. As Postman said, "There's nothing we can do. I mean, we now can watch a murder, or a story about a murder, on television, or something just as lurid and awful, and continue to eat our chicken sandwich. We don't even get upset by it. We're not invited to do anything in any case . . . . This [passive observation] may have created the sense of impotence that people feel. After all, if you're constantly getting information about which you can do nothing, or very little, it tends to suggest that you're a rather impotent person. I mean, I have no plans to do anything about the troubles in the Middle East."

This sense of impotence may explain why the public often seems unable to act in response to those stories that seem so central to our well-being, to our very survival. "Maybe this accounts in some part for the surprising indifference that many people have toward toxicity in the physical environment," Postman said. "One may wonder, wouldn't people be interested in global warming, or in what's happening to the rain forest? I mean, what could be more interesting? But it's possible that we've simply been conditioned to think that information has nothing to do with us. It's just a kind of commodity."

Moreover, Postman added, information often comes to us in a form that is composed mostly of "ands," as in "This happened, and then this happened, and then this happened. You have a news show, and you tell people the budget was passed today. A man committed suicide. There's an earthquake in Chile. No one would even ask, 'Did the earthquake in Chile happen because this man in Oklahoma City committed suicide? Did this man commit suicide because the budget was passed?' There is no cause and effect. This is a significant difference between a visual medium, like television, and language."

There's something else about the way we get information today that's very different from the way we got it in the past. Telecommunications have become so sophisticated that the images created are difficult to distinguish from real-life pictures. We can speed up time and destroy space. We can even ignore our place in reality by entering the virtual world of cyberspace, where we can have the kinkiest sex without worrying about AIDS or getting caught; we can have the spine-tingling thrill of a car race, crash into a wall and still walk away; we can take part in a gunfight, lose and live to fight again. Virtual reality seems better than the messiness and complexity of real life.

As a journalist, I had always thought that my own programs allowed the viewers to experience nature in a way that would inspire them to love it. But now I realize that my programs are also a creation, not a reflection, of reality. To make a film about the Arctic or the Amazon rain forest, a photographer spends many lonely months in a blind or laboriously lugging gear in search of exquisite shots. Back in the editing room, hours of this hard-earned film are boiled down to sequences of sensational shot after sensational shot. As a result, the film is filled with colour and activity that will thrill and amaze the audience. But anyone who has ever travelled to the Arctic or to a tropical rain forest knows that the constant flurry of activity is not what the wilderness is like at all. Even in a species-rich rain forest, most of the activity goes on at night or high above ground in the forest canopy. You can wait a long time and see very few animals.

The difference between the film and reality isn't truth; the animals are really there, and in a responsible documentary they're behaving naturally. What's missing in the filmed version of nature is *time*. Nature must have time, but television cannot tolerate it. So we create a virtual reality, a collage of images that conveys a distorted sense of what a real wilderness is like. The destructive aspect of what we create is the implicit message that nature functions at a faster pace than it really does. And the further we humans get from the natural pace of nature, the less we realize what Earth time is. In cities, surrounded by technologies like cellphones and fast-paced videos, working in offices to serve an economy that demands constant, rapid growth, it's easy to assume that nature can deliver the things we want at a similar pace. Fish, trees or soil microorganisms don't grow fast enough for our speedy time-frame. But if the programs we create give an impression of a hopped-up nature, we might expect it to be able to meet our ever-faster modern needs.

Time is the one ingredient that is absolutely vital for nature. It is the vast sweep of evolutionary time that has allowed life to flourish and huge changes to occur. In the 4 billion years that life has existed, the sun has increased in intensity by 25 percent, magnetic poles have switched and reversed back, continents have smashed into each other and then pulled apart, ice ages and warm periods have come and gone, and the atmosphere has been transformed from a non-oxygen to an oxygen-rich one. Yet life has persisted, simply because of the immense periods of time it has had to make adjustments.

Today, the rate at which we are extracting trees, fish, topsoil and clean water, as well as creating pollutants and greenhouse gases, may match the speed of information technology and the economy, but it is not in synch with the reproductive rates of natural systems. More and more, our sources of information are no longer connected to the natural world and its limits. Politics, civic action and participatory democracy need time too. Democratic groups like PTAs and other voluntary human institutions take time to do their work.

Until we slow down the rate of growth in information and technology and learn to pay attention to the true pace of the non-technological planet, we'll keep making unrealistic demands that can't be fulfilled. At the very least, we need to understand that our accelerated rates of production and use of human-made technical information function at a completely different pace from that of the natural rates of information exchange, like those I experienced in the Brazilian rain forest.

Life in the Fast Lane

> *Computers . . . change the way children's minds process*
> *information and affect not only what they know but what they*

*are capable of knowing — that is, computers alter the pathways
of children's cognition. Newly immersed in data-based forms of
knowledge and limited to information transmissible in digital
form, our culture is sacrificing the subtle, contextual and
memory-based knowledge gleaned from living in a nature-
based culture, meaningful interactive learning with other
humans, and an ecologically based value system.*

— C.A. BOWERS, AUTHOR OF *EDUCATION, CULTURAL MYTHS,
AND THE ECOLOGICAL CRISIS*

David Shenk is a young computer enthusiast who revelled in the birth of
the so-called Information Revolution. He's still enthusiastic, but already
in the late 1990s, after only a few years of surfing the Net, he started to
have second thoughts. He wrote about them in a best-seller called *Data
Smog.* "Information is a resource," he says. "It is something that we all
need. But there is a point beyond which the speed of information, and
even the amount of information, starts to diminish the quality of our
lives. We always want to get more information, and to learn more and
more. But we need to distinguish between information and knowledge. I
think we live in an age now where it's very easy to get caught up in the
thrill of getting the information, and forget that it's all about taking that
information and turning it into something interesting inside your brain."

Although he's still under forty, Shenk has realized that we have
to "take the time and make sure that all the other parts of our lives are
in place, so we can turn information into knowledge. We need to communicate that knowledge and wisdom to people we know, love, respect
and spend time with. We need to live richer lives as a result of that
information — not just amass it and bombard ourselves with it."

To make matters even worse, the clamour for our attention grows
steadily more strident and outrageous. Shenk says, "There's kind of a
vicious spiral that results from this, in that the more people who are competing for our attention, the more outrageous things they have to do to
get our attention . . . . And so the neon signs become bigger and the ads
on television become louder, more shocking and more vulgar, and so do
the shows between the ads."

Coping with not just two or three but the hundreds of ideas that
shoot by our eyes and ears every day may be far more damaging to our

ability to concentrate and make judgements than we think. When I worked in the lab as a full-time scientist, I'd wake up in the morning thinking about a single research idea. Then I'd go to work and think and read and talk *only* about that idea. I'd often go back to the lab at night to keep working on it. And this could go on for days and weeks. It was the most exhilarating, creative time of my life. But today I find that my time is fragmented by more and more demands: faxes, voicemail, email and letters; the insistent pull of television, newspapers and radio; not to mention my family and friends.

"Part of this vicious spiral is that we get distracted . . . . And when we get distracted, we can't concentrate as well," Shenk maintains. "And we don't communicate as well, and so forth. As information speeds up, it gets to a point where it's literally impossible to think sceptically. The way we think sceptically is that first we examine a claim, and then we accept it tentatively. And then, a millisecond later, we ask ourselves whether or not that is a true or false thing. Well, information is moving so quickly that we don't get to that second part. We tend to accept things much less critically, and I think that is a terrible thing — not just for consumers, but for society. I think it's going to have all sorts of repercussions, some of which we don't even understand."

Dying for Attention

> *The most significant events happening on the face of the Earth today do not make the evening news. The evening news is predictably about plane crashes or murders or political shenanigans. It's not about how 30,000 children died today because they lacked clean drinking water. That happens every day, so it's not news.*
>
> — ELIZABETH MAY

Physiologists have long known about an interesting property of neurons. When a nerve cell is stimulated, it gives a measurable electrical response. But if the neuron is repeatedly stimulated, the intensity of the response gradually diminishes until it no longer registers. This is called habituation. You notice it if a loud bell goes off. You are startled at first, but very soon, if the ringing continues, it recedes in your awareness until you hardly notice it. In many ways, we have an analogous response to stories about the environment. Many environmental problems, like global warming, are what can be called slow-motion catastrophes. We don't see blood and people dying right away, so after a while we become habituated to the warnings. Like everybody else, environmentalists need ever more spectacular stunts to garner attention.

As described earlier, Lois Gibbs runs one of the most effective grassroots citizens' groups in the world, the Center for Health, Environment and Justice, out of Maryland. They fight waste dumps and the spread of dangerous toxins like dioxin in the environment. She says, "Some media are our allies. But the other side of the coin is that the corporations, the polluters, can buy as much media as they want. So for us to get media, we have to do something — I call it performing for the media — to get their attention. But it's a Catch-22. If you behave outside the ordinary in order to be newsworthy, you can be dismissed as a bunch of crazy, hysterical radicals, whereas the governor, health commissioner and corporations can get serious coverage or just run ads. So it has to be very fine-line organizing that we do. We take people who are outside the norm and get them jumping around to get the media's attention, and then we have the other folks — the moms, the dads, the straight, conservative people — give the presentation."

Paul Watson is arguably the most radical of environmentalists. He was a founding member of Greenpeace, but he left twenty years ago to form the Sea Shepherd Society, an organization dedicated to the protection of marine mammals. Watson is probably best known for leading the fight against the slaughter of baby seals in Quebec and Newfoundland. More recently, he sank the entire Icelandic whaling fleet, an act for which he was put in prison. He's out now, and he claims he does these things not because he's angry, but because of the way our media culture has developed. Watson tells a story about calling the foreign editor of the *Los Angeles Times* because the paper had carried a piece on the whaling industry that suggested that since people weren't protesting anymore, they had become receptive to the idea that whaling would be resurrected. "I called him up and I said, 'That's absolutely ridiculous. People aren't supportive of whaling,' " Watson recalls. "And his answer to me was very telling about how the media view these things. He said, 'Well, you'll just have to go out and sink another whaling boat, won't ya?' In other words, that's the only way they're going to pay attention to us. They're not going to give you the time of day if you're trying to talk about something based on information and statistics or whatever. They're only interested in dramatics."

Watson has studied the media, and he claims that most reporters and their bosses are interested in only four things: sex, scandal, violence and celebrity. A story isn't a story unless it has at least one of these elements, and when you have all four, as the O.J. Simpson trial did, you have a super-story. When it comes to his type of activism, Watson says, "We just play the game. When I did a campaign against aerial shooting of wolves in British Columbia in 1984, we had a story that carried the headlines across Canada for two weeks. The reason is that we had all four elements: sex, scandal, violence and celebrity. We had the violence of them shooting the wolves. We had the violence of the threats against us. We had the scandal of a provincial environment minister [forced to resign after being accused of] taking a bribe from a big game–hunting organization in return for giving them what they wanted. And to round that all off, I simply recruited Bo Derek as our spokesperson.

"And I remember at the press conference a reporter saying to me, 'Oh, come on now. What's Bo Derek know about wolves? I mean, what do you have her here for? . . . She's an actress, for God's sake.' And I said, 'Yes. You make the rules. And we play the game. You must have just graduated from journalism school or something, because if I had the best wolf biologist in the world and called this press conference, there'd be an empty room. But because Bo Derek is our spokesperson, it'll be the headline of your newspaper tomorrow—and there's nothing you can do about it.' "

Watson says his theatrics work "like an acupuncture needle. We stimulate responses to try to get people thinking. My role as a conservationist is to say things that people don't want to hear and do things that people don't want to do. I'm here to rock the boat. To sink the ship, if necessary. But my main focus is to make people think, 'Why is he doing that?' Because there's a problem, that's why."

Manufacturing Consent

The Earth Summit itself was greenwash on a grand scale,
because it gave the false impression that important, positive

*change was occurring, and it failed to alert the world to the
root causes of environment and development problems.*
— JED GREER AND KENNY BRUNO, IN *GREENWASH:
THE REALITY BEHIND CORPORATE ENVIRONMENTALISM*

There's another obstacle to getting heard — it's called greenwashing.
It involves using the rhetoric of the environmental movement to cloak
ecologically damaging activities. Elizabeth May, head of the Sierra Club
of Canada, one of our country's most vital environmental watchdogs,
says that there is a deliberate effort to play on public exhaustion and distraction. "Nobody has energy for outrage when you're trying to raise
kids, hold down a job, pay your mortgage. Maintaining that extra thing
in your life — wondering when the government is going to do something to protect us from pesticides, radioactive fallout or whatever — is a
strain." She says that when politicians are pressured by the voting public,
a very clever government, industry and corporate approach is to say, "We
heard you. Uncle! We're so sorry! *Mea culpa!* Now we're all going to be
environmentalists together. There's no difference between Sierra Club
and Monsanto; we're all just so concerned about the environment. We
are parents, too, after all."

May got into environmental issues back in the late 1960s and early
1970s, when as a college student she tried to stop the spraying program to
kill the spruce budworm in Nova Scotia. The practice has been linked to
the destruction of salmon streams, as well as to Reye's syndrome, a deadly brain swelling in children. She says that when governments began to
create the environmental departments and international agreements that
were supposed to protect biodiversity, stabilize carbon emissions and regulate polluters, most of us breathed a sigh of relief and went back to worrying about our mortgages. The problem is that a lot of it was just talk.
Information without action.

"It's not that it was nonsense at some level. Sometimes, with political
will, these globally agreed-upon treaties could do something. But a very
good way to shut down public outrage is to say, 'We've dealt with it. We're
on top of it now.' The public goes back to sleep, which they wouldn't
do if the media stayed on the case. The media have to say, 'Right, the
government says they're committed to this; they've said they're going to
take action. Let's see what happens when the rubber hits the road.' "

Perhaps because, as Neil Postman contended, there is little indication of cause and effect in today's information, there seems to be very little demand in newsrooms for follow-up stories to see whether people really act on what they espouse. And information without an invitation to action is ultimately disempowering.

Besides all the inertia, babble and entertainment working against informing the public about vital issues like environmental degradation, despite the empty promises and inaction, there is also a conscious and commercially supported effort to create *disinformation*. Today, as more people are realizing, there's a whole industry devoted to selling messages, whether the message is to buy some consumer item or to vote for a certain politician. The industry is called public relations, and its employees now considerably outnumber actual journalists.

So if a product has negative environmental repercussions, a company doesn't have to change the product so much as it has to change the information getting out about it. Josh Karliner, head of the Transnational Resource and Action Center (TRAC) in San Francisco, is also author of *The Corporate Planet*. He contends that PR firms went into high gear when environmentalists started to threaten the way their clients did business. "The environmental movement put a tremendous amount of pressure on [those in] the corporate world. It cost them billions of dollars. It forced them to change some of what they've produced and how they produce it. And there's a potential that the environmental movement will foster even more thorough-going change. So the public-relations counsellors of these big, global, public-relations corporations have counselled their corporate clients to sit down at the table and make friends with the environmentalists, and promote a joint approach to solving some of these problems."

It all goes back to the 1980s again, when public pressure for change was at its height. "These corporations," Karliner explains, "at the behest of their public-relations firms, began to pour hundreds of thousands of dollars into the environmental movement. This new approach was summed up in the slogan Cooperation, Not Confrontation, and it seemed to augur a new age of harmony, with responsible corporations working out compromises with sophisticated and economically realistic environmentalists. But when the information highway and the global economy superseded the environment in people's minds, there were big changes .... Suddenly, all these same corporations that [had been] . . . talking about sustainable development were lobbying the halls of Congress to completely gut the environmental regulations of the country."

We have to remember that corporations, especially global ones, are in the business of making it as easy and as profitable as possible to produce their goods. They cannot have the environment and the public good as a central concern because those matters can conflict with their primary goal: realizing bigger profits. Corporations must do battle with anyone and anything that threatens their bottom line, and that includes environmental groups. And the only weapon environmental groups have is to reveal the PR spin ecologically destructive industries are using.

Lies, Damn Lies and the Public-Relations Industry

No one talks about the fact that there's an international, $35-billion-a-year propaganda-for-hire industry. And what I learned in writing the book is that even our most exaggerated parody of this industry couldn't anticipate how cynical and pervasive it really is.

— JOHN STAUBER

John Stauber is a lifelong environmental activist and the director of the Center for Media and Democracy in Madison, Wisconsin. He and his colleague Sheldon Rampton co-authored a wonderfully arresting and blackly funny book called *Toxic Sludge Is Good for You*. They have a follow-up out now called *Weapons of Mass Deception: The Uses of Propaganda on Bush's War in Iraq*. Stauber says that after years of trying to figure out why the environmental message was so hard to get out and so often got twisted, "it dawned on me that every issue I cared about or worked on — whether it was an environment or a community issue, a peace and social-justice issue — every time, on the other side of that issue, was a vested interest, usually somebody making a profit off the status quo."

He decided to do some serious research into how these interests were protecting themselves from the various civic movements designed to legislate, regulate or otherwise change the way they were doing business. He says that using the Freedom of Information Act, he found "a motherlode of insider public-relations documents that revealed how the propaganda-for-hire industry works." This industry is worth about $35 billion globally, and it employs 150,000 paid practitioners in the United States alone. Its mandate is well summed up in a brochure produced by the PR giant Burson-Marsteller. "The role of communications is to manage perceptions which motivate behaviours and create business results," the company explains. "We are totally focused on this idea as our mission. Burson-Marsteller helps clients manage issues by influencing, in the right combination, public attitudes, public perceptions, public behaviour, and public policy."

Josh Karliner, who runs a monthly magazine called *PR Watch*, says, "[Burson-Marsteller is] a subsidiary of Young and Rubicam, which is one of the largest advertising firms in the world. They have offices in dozens of countries around the planet. They serve as spin-doctors, public-relations counsellors, to large transnational corporations. And they even have a department that provides emergency response. They advertise themselves as the ones who will help corporations with image problems, especially if they've got consumer activists, labour activists or environmental activists breathing down their necks." A list of Burson-Marsteller clients is extremely revealing. "They worked with Union Carbide during the Bhopal gas disaster," says Karliner. "They worked with Exxon during the *Exxon Valdez* disaster. In the 1970s, they worked for the dictatorship in Argentina. They worked for Ceauşescu in Romania. They worked for the Mexican government in the NAFTA negotiations. And they worked for the United States oil and chemical industry in their efforts to roll back the Clean Air Act in Congress." In Canada, they worked for Hydro-Québec, promoting the James Bay Two Project, and they formed the British Columbia Forest Alliance, a group that campaigns against restrictions on logging and actively works to discredit environmentalists.

Other public-relations companies like Montgovern, Duchin and Bisco even maintain dossiers on social activists, people who might be, as John Stauber says, "working to stop an incinerator from being built in their community, or it might be people concerned about the health effects of factory farming or dioxin-contaminated products."

A few years ago, Ron Duchin, head of Montgovern, Duchin and Bisco, addressed the National Cattlemen's Association (the same organization that sued Howard Lyman and Oprah Winfrey) on how to divide and conquer environmental and social grass-roots activists. He said, "Activists essentially fall into categories — the radicals, the opportunists, the realists and the idealists." Duchin described the "radicals" as people who want to change the system. They "have underlying socio-political motives, and see multinational corporations as inherently evil. These people do not trust the federal, state and local governments to protect them and to safeguard the environment. They believe rather that individuals and local groups should have direct power over industry . . . . Their principal aims right now are social justice and political empowerment." These are the people Duchin wants to stop, because their objectives clearly threaten those of his corporate clients. In Duchin's prescription, "The way to stop these people is to isolate them from the other activists."

The "idealists" also get involved in civic, justice and environmental issues, but they are very different. Duchin says they "want a perfect world, and find it easy to brand any product or practice which can be shown to mar that perfection as evil. Because of their intrinsic altruism, however, because they have nothing perceptible to gain by holding their position, they're easily believed by both the media and the public, and sometimes even politicians." But Duchin believes these people are vulnerable. "If idealists can be shown that their position and their opposition to an industry or its products causes harm to others and cannot be ethically justified," he explains, "they're forced to change their position." He goes on to say, "They need to be educated." And generally this education process "requires great sensitivity and understanding on the part of the educator."

John Stauber, who quotes extensively from Duchin's speech in his book, gives an example of how this education might take place. "A typical corporate rap might be to sit down with a church group or other concerned group of people and say, 'Look, you know no one is more concerned than the people at XYZ corporation about chemicals in the environment. After all, we have kids too. And we live in this community, and we're just like you. You've got to understand — if these radicals have their way, we'll be put out of business. And if we're put out of business, we can't provide jobs to the community. And if we can't provide jobs to the community, then there's going to be a depression in this town. Everybody's going to suffer. Nobody wants that. So let's sit down and work for a common solution.'" Sounds reasonable. Except that there will still be pollution. Or an incinerator, albeit a smaller one. Or something else that the group didn't want.

The third category of people comprises those Duchin calls the "realists." These are people who "can live with trade-offs, are willing to work within the system and aren't interested in radical change." Duchin adds, "The realists should always receive the highest priority in any strategy dealing with the issue ....The credibility of the radicals will [then] be lost." This is because Duchin's realists will strike a deal, some kind of compromise favourable to industry, that will end up making the so-called "radicals," however rational and just their demands, look extreme.

The final group is made up of the so-called opportunists. Duchin says they become involved with an issue for fame or glory or personal aggrandizement, and they, of course, are the easiest of all to deal with. "Radicals" like Lois Gibbs can't be bought with money or thrown off track by disinformation. They can only be discredited. And, to a degree far beyond even my own understanding, as both a journalist and an environmentalist, they often are.

Looking for the Terrorists

Environmentalists fall into two categories: socialists and enviro-religious fanatics.
— RUSH LIMBAUGH, IN *THE WAY THINGS OUGHT TO BE*

When we started the radio series on which this book is based, we wanted to look at all sides of the environment issue. We thought it would be a good idea to get in touch with radical environmentalists, groups like Paul Watson's Sea Shepherd Society or the loose confederation of "tree huggers" known as EarthFirst! We'd heard in the media about tree spiking and other acts of "environmental terrorism," and so we pictured these groups in camouflage outfits, monkey-wrenching power lines and spiking trees. What we found was rather different.

Josh Karliner, of TRAC, says we are all victims of a strategy to "marginalize some of the most progressive elements of the environmental movement, to divide and conquer the environmental movement, and to make the progressive groups look extreme." There's a name for this strategy: brownwashing. Paul Watson, who has been called an environmental terrorist, knows all about it. He says, "We do . . . target property. We sink whaling ships. We confiscate drift nets. Our targets are all material that is being used to illegally destroy life, and I have no problem with that at all. Martin Luther King once said, 'You cannot commit an act of violence against a non-sentient object.' And that's where our values come into conflict. Society at large puts more value on private property than it does upon life." He tells a story to illustrate this point. "A ranger in Zimbabwe a few years ago was severely criticized by human-rights groups for shooting a poacher who was about to kill a black rhino, the reasoning being 'How dare you take the life of a human being, just to protect an animal?' And his answer, I thought, really summed it up and exposed our hypocrisy. He said, 'If I were a police officer in Harare, and a man ran out of Barclay's Bank with a bag of money and I shot him in the head and killed him on the spot, you'd call me a hero. You'd pin a medal on me, and say, 'Good boy!' " As Watson points out, we will kill to protect property. We will kill even to protect paper money. And we are honoured for doing it. But protecting the natural heritage of the nation for the benefit of future generations is not considered as valuable as a bag of money.

This is not to say that people in the environmental movement condone killing of any kind. Quite the contrary. "One of the things that I'm most proud of in the conservation movement is the fact that no conservationist, of any non-government organization that I ever heard of, has been responsible for killing or injuring another human being," says Watson. "It's an unblemished record of non-violence. But the media call us violent and terrorists and extremists. And when environmentalists get killed, which is often, people say, 'Well, they shouldn't have been in the way.' "

Watson points out that Chico Mendez was murdered for trying to protect the rain forest. Dian Fossey was murdered for her efforts to save mountain gorillas. Francisco Perrera was murdered by agents of the French government when they blew up the *Rainbow Warrior,* which was protesting atomic-bomb tests. Yet few people, especially in the mass media, condemned those acts as terrorism. Organizations like EarthFirst!

are classic, direct-action, non-violent protest groups that organize sit-ins or sing songs while blocking logging roads. They do not resist the police, even when being beaten or pepper-sprayed. Their members are regular people, most of them women, from all walks of life: teachers, firefighters, librarians, you name it. In short, they are the progressives that the PR people worry about.

"I don't know of any environmental terrorist organizations," Watson asserts. "Never heard of such a thing. To me, Exxon or Union Carbide are environmental terrorists. But you know, the language gets polluted through media probably just as much as the air and water do. Anyway, what is a radical conservationist? It's a contradiction in terms. A conservationist is a conservative. So I am a conservative. And probably an arch-conservative because I'm a more aggressive conservationist than most. I represent the majority, because I represent all of those untold numbers of children that have yet to be born. And I am not about to give in to the so-called majority rule of the minority who happen to be alive today."

Watson invented something that most environmentalists are reluctant to claim as their own: tree-spiking. But he says this act, too, is the victim of disinformation. "We first did it in 1982 on the south base of Grouse Mountain, facing the city of Vancouver, British Columbia. We spiked 2,000 trees and informed the sawmills. Those trees are still there because the sawmills wouldn't buy them. It costs too much money to remove the spikes." Watson claims the tactic is safe because there's no point in spiking unless you warn the loggers — and besides, chainsaws have guards to protect loggers from hitting bits of metal (which are often embedded in trees in any case). But the most convincing reason for claiming it's safe is that, to date, no one's been hurt by it.

"The one case that's trotted out is George Alexander, at the Louisiana-Pacific plant in northern California, who was almost killed by a fragment of shrapnel in a tree," says Watson. "What they don't tell you is that it wasn't a tree spike, it was a piece of metal, which is not that uncommon — a nail, gunshot, that kind of thing. And it was Louisiana-Pacific's negligence that resulted in the injury. They refused to put up a proper shield to protect their workers. George Alexander himself agreed with me when I talked to him about it."

Watson also says he doesn't advocate tree-spiking to hurt or terrify loggers. Tree-spiking means downtime in the sawmills, which can cost the industry big money, anywhere from $20,000 to $50,000 an hour. "When you cut into their profits, they notice. But, of course, they get Burson-Marsteller and the PR companies to put forth a volume of information about how this is terrorism. I'll tell you what terrorism is. Terrorism is destroying forests. Terrorism is destroying the carrying capacity of thousands upon thousands of species. Terrorism is depriving our children of their heritage. That, to me, is what terrorism is."

The Monkey-Wrench Gang

> *We use all the tools in the toolbox, short of violence.*
> — EARTHFIRST! CREDO

Most people do not condone destruction of property. Many feel uncomfortable even with direct action, and believe that civil protest should be organized within the system. But the fact remains that the word "eco-terrorism" has been used freely to describe small citizens' groups organizing petitions and engaging in boycotts; peaceful, legal, public demonstrations; and sit-ins. Today, there are serious efforts in post-9/11 legislation, like the U.S. "Patriot Act," to lump these kinds of organizations in with actual political terrorist groups such as Al-Qaeda. And even well before 9/11, of course, many participants in such activities were beaten, gassed and pepper-sprayed for their attempts to protect the environment or to fight for civil rights. Members of the media don't often tell those stories.

When the airwaves, the networks, the newspapers and the vast advertising industry are all controlled by vested interests that must protect their ways of doing business, citizen protest becomes difficult and easily slandered. Chad Hanson is a national director of the biggest American environmental group, the Sierra Club, a very conservative organization with a large number of Republican members. He says, "The message that's been promoted by industry in the media — that somehow there are all these radical and violent environmental organizations out there — is absolutely ridiculous. Completely untrue. Most so-called radical groups are just promoting positions that the public already supports. This includes EarthFirst! I have never seen any EarthFirst! activist engage in anything that could possibly be construed as violent behaviour. Ever.

This is an absolutely peaceful, non-violent community. The only violence that goes on out there is violence perpetrated against citizens, of all types, by industry and often government . . . . And I would like to see the media cover this more."

A perfect example of Hanson's claim is the Headwaters Campaign in northern California. For years, EarthFirst! and other citizens' groups have been trying to save the last hectares of ancient redwoods and old-growth forest in the region from a corporate raider that began liquidating the forest about fifteen years ago. EarthFirst! lobbied, circulated petitions and wrote letters. Finally, a group of mostly young women went to the offices of the local senator and the company itself. Karen Pickett, a fifteen-year veteran of EarthFirst!, was with them. She says that this was a classic demonstration, with signs and banners. Once in the offices, the women sat in a small circle and then "locked themselves down," as they call it, with a length of steel pipe and a bracelet so they would be very difficult to move and so the demonstration could be prolonged in order to attract the media attention they needed so badly.

When the sheriff's men came in, the women continued singing and chanting. The police warned they were going to use pepper spray. There have been eighty recorded deaths from pepper spray; it can be dangerous for asthmatics, and it can cause permanent eye damage. When the women didn't unlock themselves from the pipe, the police didn't just spray them from aerosol cans. "They actually took Q-tips full of the stuff and swabbed it directly into the protesters' eyes," recounts Pickett. "They took a video of themselves doing it, while the young women cried and screamed and begged them not to do it any more." The women, who were still locked together, were helpless even to put up a hand. Although the local media picked up the story, no one else in North America heard much about this particular display of violence. And the police still maintain that their actions were necessary to "defend themselves" against the dangerous demonstrators.

Who Do You Trust?

The real key is information, getting the right information so you can make decisions today about yourself, your family and

your community. Should you or should you not drink that
water? Should you or should you not allow your children to
play in that yard? You need to know these things.

— LOIS GIBBS

Organizations that are trying to get vital information out to the citizenry
have to use information technologies, but these technologies often have
other priorities. Increasingly, ownership of print and electronic media is
concentrated in the hands of a few tycoons, who determine the perspective and output of their property. Environmental and human-rights
groups must also run the gauntlet of the public's boredom, distraction
and media overload. On top of that, media outlets themselves may not
want to carry information that could antagonize advertisers, program
sponsors or owners.

I know from firsthand experience how rough the private sector can
be when anything critical is broadcast or printed. A few years ago, *The*
Nature of Things was to carry a special look at Canada's forestry
practices. Before the program had even been completed or pre-screened,
the forest industry took out newspaper ads condemning it and calling for
public support. At the same time, the CBC found itself under enormous
pressure to have the program cut or watered down. The Canadian Imperial Bank of Commerce announced that it would not advertise on *The*
Nature of Things because of the episode. It was crude hardball, but to the
CBC's credit, the program was not modified and was broadcast to wide
public acclaim. Since then, however, similar efforts have greatly affected
even the CBC's desire to tackle really difficult issues on a regular basis.

Ultimately, the entire question of information boils down to: Who
should you believe? If programs that discuss an issue like global warming
are sponsored by gas or car companies, we might want to think twice. If a
scientist makes a declaration about the safety or hazard of genetically
modified food, a new drug, the burning of fossil fuels or clearcutting
forests, we should find out the primary source of support for that expert.
A whole strategy has grown up around this idea of demanding that
all experts provide "full disclosure" of their sources of funding. Often
an organization that seems to be grass-roots has, in fact, been set up
and funded by industry, so it's just as important to see who funds such

organizations and who's on their board of directors. On the other hand, a group that has little or nothing to gain monetarily by its stance, and that raises concerns about children, quality of life or health issues, just might be what it says it is.

And there are a lot of groups like that. Many activists will tell you that there is nothing that frightens polluters more than a little old lady with a sign at a demo. Lois Gibbs says, "Corporations are scared to death of us because they can't deal with us ... we will not compromise. We will not give up on our children. We will not give up on our homes. We will not give up on our schools and our communities for somebody else's profit. So they just walk away from the table, saying, 'You're unreasonable, hysterical, radical people.' But we really aren't."

She says most communities, even the most disadvantaged or impoverished ones, can win many of these fights. "This isn't a legal battle, because it's perfectly legal to poison people and poison the environment. It's not a scientific battle, because you can have lots of scientists arguing about whether grey is grey or whether it's really black and white or something in between. These are *political* fights." And they're fought, like most political battles, at the level of information. To illustrate this point, Gibbs recounts the story of a member of her centre, a PTA mom in Texas, who introduced a resolution at her local elementary school to support efforts to eliminate the release of the dangerous and ubiquitous carcinogen dioxin into the environment. At the very next PTA meeting, people from the oil and chemical industries showed up to state their side of the issue. Nevertheless, the resolution still passed. The woman then took it on to a state convention of the PTA, where she again planned to introduce the resolution. "The petrochemical industry went crazy," Gibbs recalls. "They called all of their plants, all of their workers, all of their managers, to tell them to get their own wives off to the PTA to vote against the resolution."

At the convention, the industry bought space to set up an exhibit, which was rather unusual at a PTA convention. On the floor, a battle ensued when industry supporters argued that the resolution would wreck the Texas economy. But the resolution passed, and the members of the Texas PTA voted to support programs to eliminate dioxin discharges. The expensive, frenzied effort of the chemical industry to control a non-binding PTA resolution gives some idea of the extent to which powerful interests will go to attempt to control information in our society. That's why information is still so important, even if so much of it is banal, silly or superficial. Since this book was first published, a new prime-time PBS series, Bill Moyers's *Now,* has sprung into existence to discuss in detail these very issues. And unlike most mass-media programs, it does provide intelligent, complex, political information.

The Centre of the Universe

The message you get from the consumer world of television all the time is: "You are the most important thing on Earth. Nothing is more important than you." If we don't get past that idea . . . then we're unlikely to ever deal clearly with either environmental or social-justice issues.

— BILL MCKIBBEN, AUTHOR OF
THE END OF NATURE

If we are to use information in a way that empowers us to make informed decisions, we have to understand the current nature of information and the technology that has been developed to bring it to us. Information has always been the key to our survival. But today, as never before in our evolutionary history, the information reaching our brains is filtered, fabricated, shaped and controlled by powerful human forces with agendas that differ from the common good. Information dissemination is not something we can leave to the marketplace, to commercial interests or even to Hollywood. On some level, we all have to make sure that genuine survival information continues to be freely available, alongside all the diversion and entertainment.

Paul Watson asks, "Why do I sink ships? Because nobody will listen to me if I don't. And I'm trying to convey a message. That message is, 'We're in some very serious trouble.' The possibility of losing biodiversity, and therefore the ability of the environment to sustain us, is very, very real. In 200 years, the human species could be extinct, unless we take action now and stop worrying about whether we're going to get a new car or whether some reality show is going to come up again in the fall. We're so engrossed in trivialities, we never get down to thinking about what is real in this world."

What is real in this world is the human impact on it. We're moving too fast and demanding too much. Our extraordinary technologies could do a lot to help us and the world's biodiversity survive, but we're not very good at predicting and controlling their effects. They take us down paths that distance us from understanding the kind of planet we need to live on, and they have nothing to do with the productivity rates of the natural services we need to survive. "We had no control about where we came from, but we certainly do have control about where we're going," says Paul Watson. "The world that is being shaped now is the . . . world of the future. Will there be elephants here 2,000 years from now? We have to make that decision. Will there be redwood trees? It's up to us. We spend most of our time entertaining ourselves and trying to avoid reality. But the reality is, we've become gods. We are making these god-like decisions as to whether the planet will survive, whether our own species will survive. And that is an awesome responsibility. It's one we should be facing up to, instead of involving ourselves in circuses and trying to entertain ourselves."

CHAPTER 5
UNNATURAL SELECTIONS

We've taken human genes and put them into salmon. We've taken the fluorescent genes from fireflies and put them into tobacco plants. It is very important to understand that we are crossing species boundaries at will. There is no time in history, that I'm aware of, where flounders mated with tomatoes, where humans mated with mice, where salmon mated with chickens. This is a completely new arena.

— ANDY KIMBRELL, INTERNATIONAL CENTER
FOR TECHNOLOGY ASSESSMENT

Genes could be called the pre-packaged destiny of every living being. Every detail of the hereditary makeup of each of the Earth's organisms has been honed by natural selection so that it can survive in its environment. Whether it's the camouflage of a zebra's stripes, the sensuous floral structures of an orchid, the social behaviour of ants or the hummingbird's extensive tongue, all of life's characteristics have been refined through evolution to achieve an exquisite balance with the organism's surroundings. As biological beings, we, too, are products of evolution; every aspect of our physical makeup, as well as much of our behaviour, is the result of hundreds of thousands of years of selection and adaptation, and enables us to survive in the places and circumstances in which we find ourselves.

When, about 10,000 to 12,000 years ago, some people discovered the efficacy of domesticating plants and animals, they began a revolution that transformed us from nomadic hunter-gatherers to farmers. Very quickly, those ancient farmers recognized one of the most basic principles in genetics — like begets like. By applying that principle, people learned to breed animals and plants for desirable traits, such as bigger kernels on corn, thicker wool on sheep or greater speed in horses. Human tinkering with genetic destiny has been going on all over the world for millennia, and it has been making gradual but sweeping changes.

However, in recent years a transformation of this process has occurred. It began with the recognition that genes could be pinpointed as a molecule within cells. That molecule was deoxyribonucleic acid, or DNA. The basic structure of DNA was proposed by Francis Crick and James Watson in 1953. Their elegant model, which explains how DNA can encode information, replicate and mutate, is a double helix in which pairs of four kinds of structures called "bases" are arranged in a linear sequence. The four bases can be thought of as letters in a four-symbol alphabet. The words in the biological language are spelled out three letters at a time, thereby giving us sixty-four possible arrangements of the four letters. (In other words, four possible letters at each of the three positions, or $4 \times 4 \times 4 = 64$ different combinations.) Hundreds of such triplets strung together make up a sentence, or a gene. A simplified form of life like a virus may have as few as ten genes, while a fruit fly has 10,000. When this book was first published, geneticists were sure that human beings had 100,000. The total human gene number turned out to be only a third of that, hardly more than many insects. As we said in chapter 1, this surprise underlines how much we still have to learn about the science of genetics.

A component of basic biology that has never been challenged or disproven is that the scientific definition of a species is that individuals belonging to it can only have fertile offspring with other members of the species. They cannot interbreed with other species. Two creatures that seem basically alike — say, a red-headed and a downy woodpecker — cannot produce viable offspring, even if they should mate. The reason is that each one evolved under different circumstances, and therefore has its own private genetic code, like no other species'. It means that each species has its own habits, mating dances, nests, diet and ecological niche. It also means it has viruses, bacteria, parasites and other symbiotic guests specific to its own genetic makeup. Other species may share them, too, but viruses or other infective agents that have evolved specifically to live in, say, a chicken or a chimpanzee, do not readily infect another species.

Over time, viruses and their hosts evolve a relationship; the host can tolerate the virus without being quickly overwhelmed, while the virus tempers its effects so that the host isn't killed off too quickly. The species barrier, which has kept different species from mixing their genes, has also restricted the range of hosts for disease bacteria and parasites. We don't know exactly why this barrier exists, or all the details about how it works, but it is the very clear foundation on which life's complexity rests.

In the past thirty years, genetic tinkering has leaped from breeding into an entirely new realm. We have not only acquired an understanding of how genes are structured and function but we have also gained tools to manipulate DNA. DNA can be chemically extracted from virtually any organism, specific sequences can be isolated and the sequential arrangement of letters in those fragments determined. Those DNA pieces can be replicated millions of times, attached to DNA from any other organism, then inserted into yet another life form.

For billions of years, life has followed the principle that each living organism exchanged DNA only with others of its kind. But today, scientists can deliberately bypass species barriers and introduce foreign DNA into an organism at will. This is called horizontal gene transfer. Mae Wan Ho's book *Genetic Engineering: Dream or Nightmare?* voiced deep concern within the scientific community over this practice. She says, "Genetic engineering, or biotechnology, is based on recombining genes from very widely different sources. And also transferring genes between organisms that are not just different species, but are in different kingdoms . . . . So that species that had no possibility of interbreeding, and very, very, very low probability of exchanging genes in nature, are now unrestricted due to these new laboratory operations." That is the basis of biotechnology. We now live in a world where new life forms created by human beings are replicating among us.

Over the past two decades, the fantastic benefits promised by the proponents of genetic engineering are always *on the verge* of coming true. Scientists propose putting pharmaceuticals, such as a vaccine against childhood dysentery, into common foods like bananas, thereby "saving millions of lives." Fish or cows that grow bigger and faster on less food are another research area. A lab just outside Montreal, now closed, spliced spider genes into goats in an effort to create new, super-strong fibres that are expressed by the genes in the milk. Food could be designed to contain more nutrients, like vitamin A, and fewer "bad things," like cholesterol. Mice have been genetically engineered to grow human ears, and pigs are being bred to grow human organs for transplants, though for now those transplants are considered too dangerous to use. Scientists are even isolating DNA from species now extinct, such as mastodons, Neanderthal people and the killer flu virus of 1918. These are just a few of the coming genetic-engineering miracles, and they've been chosen from an enormous list. And because of the incredible promises, stock in biotech companies was, throughout the early and mid-1990s, the darling of speculators, who invested billions of dollars in this industry. Today, however, biotech, especially agricultural biotech, has taken a beating. Monsanto has lost two aggressive and initially successful CEOs over unexpected resistance to its products and has reorganized twice; these days its stock is at a serious low. Pharmaceuticals are doing far better, but following the IT market crash of 2000, they're certainly no longer a way of cloning money. Because these products have proven to be more problematic than their early hype, and because investors have learned from the information technology crash to think twice about speculation on projected earnings in hot new technologies, biotech is not the exciting investment it used to be.

The spectacular stories have also been accompanied by disturbing developments, such as the report that scientists have created headless tadpoles as a kind of prelude to generating other headless animals for use as organ and tissue banks. Israel and Iraq both tried to develop weapons designed to home in on specific genetic targets. Today, even though the packaging doesn't inform us of this, at least 70 percent of our modern, processed foods contain parts of modified organisms, new genetic inventions. These organisms are, in fact, having "unexpected effects."

Crops genetically engineered to expel the natural insect toxin of *Bacillis thuringensis,* or Bt, are leaving large amounts of this material in the soil, with unpredictable effects on soil organisms. The same plants have been implicated in the decline of monarch butterflies, lacewings and other desirable insects across North America. The target insects are also rapidly developing resistance to this strategy. Entire species of important, useful crops like Canadian canola and Mexican corn have become contaminated with other genetically engineered organisms via both pollen drift and spilled seed, to the extent that there are fears that "pure" varieties of these crops will soon disappear, with possibly disastrous effects on food safety and genetic diversity.

The use of "Roundup Ready" varieties of canola, corn, soy and potatoes has brought other problems. These crops are genetically engineered to resist being bathed in Monsanto's patented herbicide, Roundup. From Mexico and Canada to India and New Zealand, farmers attempting to grow normal varieties have had entire crops contaminated with GE pollen and seed, resulting in many legal battles. Also, whole new strains of wild plants related to these crops are acquiring the resistant genes and creating new races of "superweeds" that will require applications of far more powerful, expensive herbicides. Because of this history, the Canadian Wheat Board is doing everything it can, including going to the courts, to halt the introduction of GE wheat. In Alberta, the county of Smoky Valley is not only opposing the introduction of such wheat into the province, it's demanding that all trial locations be made known to the public, something the government has so far refused. The county intends to spearhead a drive to make the entire province GE-free. And to date, no one has done serious, double-blind medical research on the effects the consumption of modified organisms, bacteria, viruses and foreign cells may have on human health, except Dr. Arpad Pusztai, whose distressing saga of honest science confounded by political and economic manoeuvring is told later.

These scandals and worries are only some of the reasons why the countries of the European Union have made their de facto ban on biotech products more official. The United States, Canada and Australia, the prime producers of GE technology, launched a formal complaint in June 2003 with the World Trade Organization against Europe's stringent controls on agricultural biotech, but long-awaited European rules released a month later have outraged GE promoters even more by mandating that GE food carry labels to let the consumer know that modified organisms are present. In addition to a clear and stringent labelling system for consumers, the EU has also included measures, albeit weak ones, that attempt to track contamination in pure agricultural or wild plant varieties so that effects on the environment can be monitored. In the United States and Canada, companies regard labelling laws as detrimental to the sale of their products. They insist consumers do not need to know that engineered organisms are present in their food.

As a geneticist, I have to say that the European Union's use of the precautionary principle seems mild and only prudent, is being implemented in many other parts of the world, like Japan, China, western Australia and New Zealand, and is being urged by petitions and popular sentiment in Brazil and even some U.S. states such as Oregon. Canada recently announced a "voluntary approach," which means food producers only have to label the presence of genetically engineered organisms in their food if they feel like it. While the technological dexterity of biotechnologists is truly remarkable, most practitioners in this industry are carrying out commercially driven manipulations on complex organisms whose makeup and functions we barely understand. The reasons why the release of these human-created organisms into the environment and into our own bodies is so problematic is the basis for the rest of this chapter.

A Personal Declaration

> *The point is not that* science *is bad — the charge that is too often made by the green movement and by journalists in the popular media — but that there can be* bad *science that ill serves humanity.*
>
> — MAE WAN HO, GENETICIST

There are those among my colleagues who suggest that I am a heretic because I was once deeply involved in genetic research but have since raised questions about much of what is being pursued today. The practice of genetic research was the great joy of my life for twenty-five years, a period when I ate, slept and dreamed about experiments and ideas in genetics. At one time, my lab was the largest genetics-research group in Canada, and I am proud of the work that we did. It continues to be a great source of satisfaction that a genetics book I co-wrote with Tony Griffiths has become the most widely used introductory genetics text in the world. Genetics has been my passion, the source of my friends and colleagues, and the most challenging and rewarding intellectual activity of my career.

It is precisely because genetics has been my life's passion that I also care about its future. As a science that goes to the very heart of all life's forms and functions, genetics has enormous implications; it is full of promise to benefit and improve human lives, but equally heavy with potential to destroy and cause untold suffering. For those who care for the long-term flourishing of genetics, it is as vital to raise questions and anticipate problems as it is to proclaim the potential benefits.

Val Geddings is also a geneticist. He specialized in the area of evolution, and he has worked as a genetics adviser for the United States Department of Agriculture, the World Bank and the Office of Technology Assessment of the U.S. Congress. He once represented the United States at the biosafety negotiations of the United Nations' Biodiversity Convention, the body that was formed to hammer out international regulations on genetically engineered organisms. Today, Geddings represents private industry as a vice-president of the Biotechnology Industry Organization, an umbrella group for more than 700 biotech corporations worldwide. He says public fears about the safety of biotech are misplaced, because DNA, the material we're splicing from one organism to another, is the same chemical substance throughout the living kingdom.

Geddings says, "What's a gene? A gene is a stretch of DNA. What we know from genetic analyses of chimpanzees and humans is that 98.6 percent of the nucleotide-base sequences in chimpanzees are identical to those you find in humans. Why is this? That's a function of our descent from a common ancestor. It illustrates the degree to which different species are related to one another. And that's the beauty of this part of genetics — that we're all related. There isn't any violation of the natural order, taking a gene from one place and putting it someplace else. It might have come from there in the first place, and it's just going home! There's a vast amount of genetic similarity, in a very high degree, between any organism and many, many, many other organisms. It doesn't make any biological sense to talk about moving a gene from one species to another as a violation of natural order.... Everything's related to everything else. We're all united by descent from a common ancestor."

That sounds convincing, even inspiring and, of course, it is true that all living things are related by a shared evolutionary history. But human genes won't make a chimpanzee, even though most of them are identical to the chimp's. What Geddings's statement ignores is any recognition of *context*. Natural selection doesn't act on each and every gene separately. Selection is based on the phenotype, the ultimate expression of all the genes acting in concert to create an organism that functions in its surroundings. Individual genes are like the separate instruments in an orchestra. Each plays its part according to the score, and taken together, they can make beautiful music. Music is music, you may say, but what would happen if you stuck Elvis Presley into the middle of a Beethoven sonata? Or substituted a twelve-year-old playing marching-band trumpet for Louis Armstrong playing jazz? That's what genetic engineering does. Technically speaking, a trumpet *is* a trumpet. A gene *is* a gene. But taking a gene from one species and sticking it into another is placing it into an entirely different genetic context. What biotechnologists do represents a fundamentally different way of making new genetic combinations. And it *is* unnatural.

The Most Elegant Science

> *Information coding, decoding and storage, that was all beautifully and simply stated by Watson and Crick. And that paradigm of what a gene does, how it stores information and is replicated and so on, is pretty much solved. But what's gone on now is the illegitimate application of that legitimate notion of genetics, to this illegitimate area, of reducing complex behaviours to single genes. It's just a mistake of the first order.... It's great science. It's absolutely great science. But it's flawed. It's the wrong assumption.*
>
> — RICHARD STROHMAN, EMERITUS PROFESSOR OF MICROBIOLOGY

The word "engineering" conjures up images of roads and bridges and buildings, all designed and constructed to precise specifications. But as a geneticist, I can assure you that *genetic* engineering is based on trial and error, rather than on precision. For instance, if I want to insert a gene from a fruit fly into a daffodil, I can't pluck out just that gene and set it down exactly where I want it to go in its new home. The technique just doesn't work like that. Some geneticists even use a kind of molecular shotgun to blast the genes into cells. And they never know exactly where they'll end up.

Christine von Weizsaecker, a molecular biologist who is also the president of a German environmental group called Ecoropa, thinks that one reason biotechnology is dangerous is its imprecision. She uses a metaphor to describe the crudeness of gene splicing: "These genes . . . are thrown like sand on a piece of paper. Let's say you have a favourite poem — and the genes of plants and animals and humans work together in the same consistent and forged way as a good poem. Now, if you say, 'I have a favourite poem, but one of the words that I really like is lacking from that poem.' So you cut out the new word you want from somewhere else, and say, 'I want to insert it into that poem.' What genetic engineering allows you to do is to make fifty copies of the word, let's say 'love' or 'wind' or 'flower.' And you throw them on the page of your favourite poem. Now, what are the chances that you really have a better poem? It may even be that that piece of word obliterates one of the words that is essential to the poem. Or that they come into completely different context. Like, you wanted the word 'love' included in the poem, and all of a sudden it reads, 'There is no love,' because it's contextualized. So there is no control over the genetic context in the gene, or over the environmental context in which it finds itself."

Brian Goodwin, a theoretical biologist teaching at Schumacher College in Devon, England, says, "The assumption is that a characteristic can be transferred from one species to another simply by moving the gene. The problem is that genes are defined by context. For example, a gene that in a mouse produces a hormone regulating growth will have one effect in the mouse, but the same gene producing the hormone in a human being will have a very different effect. So as we move genes from one species to another, we will keep getting unpredictable effects that simply could not have been anticipated."

That sounds surprisingly pessimistic, given all we hear about this wonderful new science. After all, genetics *does* allow us to predict the pattern of inheritance of genes, describe how they're expressed and transfer characters from one genetic strain to another — but all *within* a species. Once we cross species barriers, we're in brand-new territory. We have absolutely no idea what might happen. Goodwin explains, "The problem is, this is experimenting as you go along. Nobody has any confidence anymore that they can exactly predict the consequences. They just *hope* that they're only going to get small effects, the effects they want, and that they're not going to propagate really serious changes in the other constituents of that organism, that could be damaging to human health or dangerous for ecological systems."

Goodwin believes that when it comes to biotechnology, there's a time-lag in our thinking. Most of science has moved away from a focus on isolated elements to the analysis of complex patterns. Biology, and especially commercial gene technology, is an exception. As the ultimate expression of scientific reductionism, biotechnologists view a gene as an isolated unit that will express itself in exactly the same way, no matter where we choose to put it. But one of the great lessons of twentieth-century biology is that at every level — from the human genome to sub-atomic particles — there are interactions that confound reductionist expectations. For example, cotton plants have been genetically engineered to produce a protein from Bt, which acts as a natural insecticide. The gene was supposed to confer immunity from that great scourge of cotton plants, the boll weevil. However, in many cases, farmers who have bought and planted the engineered seeds found that not only did they still have to spray against pests, but the plants also showed deformities like twisting, few leaves and impossibly tall heights. Other plants had deformed and unusable cotton bolls. The engineered cotton plants yielded shocking monsters. The gene obviously functioned in its new surroundings in unpredictable ways.

The industry has defended itself by blaming lowered yields (so common that the term "yield drag" has been coined to describe a phenomenon that dogs GE varieties) on "fluctuations in climate, different soil types," or unexpected changes in weather. But that's precisely the whole point of the criticism. The genetic diversity we used to enjoy was painstakingly developed by farmers over thousands of years to respond to every possible condition. There were crop varieties that did best on hillsides, others that produced in wet valleys; there were some that could cope with drought, and others with cold. You planted a bit of each and hedged your bets. The modern, industrial method is to treat every field, hollow and hill as the same tabula rasa on which engineered seeds and chemical combinations will produce identical yields. But this method utterly ignores, one might say even heroically challenges, physical reality. No matter what we do, some summers are wetter or drier or hotter than others; no matter how many studies we make, new diseases and pests keep appearing, fostered by complicated relationships in each separate ecosystem. Trying to ignore these "seasonal, local or climatic" fluctuations is ridiculous; they are a given in farming. The terrible surprises that have greeted industrial varieties of crops like rice and cotton in countries of the Third World are only a hint of what awaits us if we continue to place limitations on the varieties available to farmers and put our eggs in the extremely limited (and expensive) GE basket.

Bali serves as a perfect example of how biotech methods can fail. It is probably the most completely rice-dependent nation in Asia. At the beginning of the so-called Green Revolution in the mid-1960s, a new variety of rice, IR-8, was introduced by the proud, high-tech agronomists of the World Bank–funded International Rice Research Institute (IRRI). IR-8 matured in less time and was supposed to produce an amazing 6,500 kilograms of grain per hectare. The Balinese government aggressively marketed this grain to their farmers, setting up credit so the farmers could afford its heavy inputs of fertilizers and pesticides. By 1977, 70 percent of the south-central terraces of the country were planted with the new wonder crop. But IR-8 turned out to be susceptible to the brown planthopper, which destroyed 2 million tonnes in Indonesia in 1977 alone. So the IRRI scientists came up with IR-36, which was supposed to be so much better that it was legally mandated. Balinese farmers were forbidden to plant anything else, and countless useful varieties of native rice were lost forever. The government also went into serious debt with the World Bank in order to improve irrigation for this crop.

But by 1979, IR-36 was falling prey to the viral disease tungro. PB-50, a new variety quickly parachuted in, turned out to be vulnerable to rice blast. By the mid-1980s, farmers were locked into a struggle to stay ahead of the latest rice pest with new, ever more expensive varieties. So despite paper gains in yields, they were poorer, and also found themselves unable to raise eels, fish and ducks in their paddies because of all the poisonous pesticides. Testicular cancer, a disease closely linked to pesticide exposure, also began to rise among the paddy workers. The farmers petitioned for the right to return to their old methods, which they were eventually allowed to do. (See our book *Good News for a Change*, pages 163–69.)

A similar scenario has been unfolding in India with a variety of cotton Monsanto genetically engineered to resist insects. Initial distrust by local farmers, which resulted in vandalized experimental fields, gave way in some of the poorer states to grudging acceptance. But yield drag, crop failures and continuing pest problems, as well as escalating expenses, has led to widespread outrage, controversy and even riots across the country. As one journalist notes, "Cotton farming in India is far different from that in the U.S. Monoculture cotton is not grown on an industrial scale or season after season . . . . Indian farmers alternate crops, and this is what wreaks havoc . . . . The large question is: Is the country inadvertently inviting a greater disaster on the environmental front for a dubious technology that has not been consistent in its dependability even in its country of origin?"

If all this is an agricultural improvement, it's hard to see who, besides the GE producer and patent holder, benefits. In fact, there is now increasing evidence that pests are developing ways not only to survive the built-in pesticide engineered into the cotton, but to actually thrive on it. Researchers in Venezuela found that the larva of the diamondback moth, "an increasingly troublesome pest in the southern U.S. and the tropics, grew at a 56 percent higher rate than on normal plants. Bt transgenic crops could have unanticipated nutritionally favourable effects, increasing the fitness of resistant populations."

The hybrids of the Green Revolution and the new genetically engineered varieties continue to reap money and glamour among professionals in both microbiology and agriculture. Compared to traditional seeds, these new hybrids and GE varieties are limited in number and are constructed so that they can't produce without expensive industrial inputs such as large fields and rigidly timed applications of chemical fertilizers, herbicides and pesticides. GE varieties even require complicated *refugia,* buffer zones that are supposed to protect non-GE crops but which studies have shown farmers rarely understand or implement. Soon they will require even more chemicals, designed to release the desired gene expression in "Terminator" varieties. Increasing numbers of analysts believe that the reason why complications and failures of these varieties are downplayed — and their advantages emphasized — has to do less with their performance and more with how money circulates in most research institutions.

Today, there is little of the "pure" research that brought me into science. With both corporate and government funders hungry for results they can brag about in yearly reports to their shareholders and funding agencies, researchers have to come up with "product" as relentlessly as if they worked for a private business. Too often they do work for a private business.

Dr. Richard Strohman is a renowned molecular biologist who is now an emeritus professor and past chair of the prestigious department of molecular biology at the University of California at Berkeley. In his retirement, he has begun to lend his reputation and expertise to the analysis of the hazards and pitfalls of the practice of genetic engineering. He says flatly, "Academic and corporate researchers have become indistinguishable. Every country in the post-industrial world now requires scientists to show that their work is going to be useful in the day-to-day world. This insistence that everything be greeted with immediate usefulness is counterproductive beyond measure."

This approach is incredibly short-sighted and is sure to mean that important insights are ignored because they don't seem useful at the time. When I was a graduate student, we had to read difficult but beautifully documented papers on "jumping genes" by Barbara McClintock. We all believed the phenomena she was studying were a quirk of the special properties of the corn seed. Only decades later was it shown that in fact jumping genes can be found in most organisms, and they are now a standard tool of genetic engineers. If the current criteria for practical results were used decades ago, McClintock's research on jumping genes would never have been funded.

Strohman goes even further: "The problem is that in order for the application to be profitable, which it has to be in this culture, you've got to have your eye on magic bullets. You've got to be able to boil down some very complex behaviour — let's say a human disease or some ability for corn to produce more protein per ear — to a single entity that you can identify, isolate, and then put in a bottle or into a feed bag or whatever. That's how you make money. And when it works, it's wonderful. But the trouble is that it often doesn't work. And when you put a biological entity out into the environment, or into a human being, and you're not completely certain — and you can never be certain in this business, in my opinion — your ability to do damage is very, very high."

Jumping Genes

Fears that genes for antibiotic resistance could jump from genetically modified foods to bacteria in the gut may be fueled by new research from the Netherlands. The results show that

DNA lingers in the intestine, and confirm that genetically
modified (GM) bacteria can transfer their antibiotic-
resistance genes to bacteria in the gut.

— DEBORA MACKENZIE, "CAN WE REALLY STOMACH
GM FOODS?" *THE NEW SCIENTIST*, JANUARY 30, 1999

Genetic engineering, as it is being practised for the marketplace, may be
built on faulty assumptions that are deeply disturbing. Equally worrisome are the actual dangers inherent in gene-splicing techniques. The
heart of biotechnology is the ability to work with pure DNA in a test
tube. DNA is the primary building block of life, and it doesn't matter, in
a test tube, where it comes from. So scientists can snip a bit of DNA from
any living source and then insert it into another organism. Geneticists do
this under controlled conditions to try to achieve a predictable result: an
engineered organism that has new and, they hope, useful properties. But
even DNA isolated in laboratories doesn't always travel alone. It can
include what Mae Wan Ho calls genetic parasites, elements of DNA such
as viruses, plasmids (rings of DNA that are separate from a cell's chromosome) and transposable elements (sequences of DNA that can change
position in the genome). When we snip a bit of DNA out of an organism,
we can also include the parasites — about which we know very little.

Mae Wan Ho says, "Natural genetic parasites, like viruses and other
transposable elements, proteins and plasmids, are naturally specific to certain species. They have species barriers, genetic barriers. A virus from a
pig will not generally attack humans, and so on. What genetic engineering does is to destroy these species barriers. So when you join these viruses and transposable elements from widely different sources, you create
similarities to all these different species. You're levelling the gene-transfer
barriers that naturally exist. I really am worried about this, because since
1993 an increasing number of publications [have reported] that horizontal gene transfer is responsible for new and bold pathogens arising."

Horizontal gene transfer, the tool on which biotechnology is built,
can also happen in nature, but very rarely. It's a good thing it's rare,
because we now know it can be dangerous. In December 1997 and
again in January 2004, officials in the Far East panicked over a deadly
infection spread by a virus found in birds. They exterminated millions
of chickens because the chicken virus had horizontally transferred to

humans. Many new flus originate in Asia because in Asian countries like China, pigs, ducks and chickens are raised in close proximity. As a result, viruses specific to different species can infect the same animal simultaneously. When that happens, the genes of the different viruses can be shuffled to create a brand-new disease, which can infect the people living in the barnyard as well. Such pathogens can be extremely dangerous, since the host to the new virus has not built up any defences to temper the attack. The flu pandemic of 1918, which killed more than 22 million people worldwide, was such a disease. AIDS is now known to be a virus that originated in chimpanzees and jumped to humans who ate the chimps or accidentally exchanged blood with them. Mad cow disease is almost certainly the result of horizontal transfer of an infectious protein long known to kill sheep. The SARS epidemic of 2003 that came out of China has been linked to the civet cat; it, too, is an example of horizontal gene transfer.

Seen in this context, genetic engineering, whose entire intent is to increase horizontal gene transfer, creates more and more opportunities for these kinds of unwanted species jumps. Mae Wan Ho says, "What makes me so worried about the whole thing is that, within the past thirty years, we have at least thirty new diseases arising, including ebola, AIDS, hepatitis C, Lyme disease, SARS, hantavirus and so on. And we didn't know where they came from. But within the past few years, they have now been attributed to horizontal gene transfer; that is, they come from other species and have jumped to us. Now there may be no connection between the commercialization of genetic engineering and this escalation in new diseases that have jumped species, but I think we need a proper public enquiry to look into it." Her fears are not without foundation. Just a few months after Ho and I spoke, scientists reported in a prestigious journal, *The Proceedings of the National Academy of Sciences U.S.A.*, that a genetic parasite belonging to yeast, a favourite subject for genetic engineers, had suddenly jumped into many unrelated species of higher plants. Obviously, if parasites of yeast are making jumps into entirely new phyla, many other kinds of genetic parasites can as well.

In the early 1970s, geneticists, including me, were enthralled by their newfound abilities to create novel gene combinations. One scientist proposed removing a gene from a cancer-causing virus and inserting it into E. coli, the common gut bacterium found in our intestines. He wanted to study the virus gene's expression away from the other genes in its genome. At some point, a colleague wondered whether such a bacterium might become a cancer-causing organism that posed a hazard to human beings. Suddenly, scientists were confronted with the very real possibility that they might create a nightmarish organism. It led them to what I believe was their finest hour: all geneticists voluntarily suspended such potentially dangerous experiments until there had been a wide discussion of hazards and ways to reduce risks. Eminent scientists, politicians and members of the public met under a blaze of media interest. Eventually, it was decided that potential escapes of hazardous organisms could be minimized by categorizing experiments into different levels of hazard and prescribing containment facilities that became more stringent at each level, rather like nuclear-containment facilities. As well, viruses and bacteria were deliberately "crippled" with defective genes to reduce their ability to survive outside lab conditions.

In those days, Mae Wan Ho was the biosafety officer in her recombinant DNA lab at the Open University. She says, "We never thought it was safe just to dump everything down the drain, or even in the bin. We had to incinerate everything. We had to autoclave all the waste. We had to do everything in a flow hood . . . . And my colleagues who are still working in human genetics are still carrying out these procedures very carefully in the recombinant DNA lab. Meanwhile, private biotech companies are releasing these things into the environment in field trials and out of their labs."

So what happened to that moratorium? As new tools and techniques were developed and refined, the potential for exploration of new areas and commercial opportunities escalated. Scientists began to put pressure on governments to relax the safety measures because they were cumbersome and expensive. The media began to focus on the incredible benefits and economic opportunities to be realized from biotechnology. At around the same time, in the late 1970s, genetic engineering went from scientific research to product output. The biotech commercialization drive began in earnest. Once private industry saw a potential for making money, it began to fund huge labs and large-scale projects. And when it did, the moratorium withered away.

Meanwhile, public fears and scientists' worries subsided or were ignored, and most genetic engineers convinced themselves that the hazards were conjectural or imaginary. At one public meeting I attended in

Cambridge, Massachusetts, the eminent Harvard molecular biologist Marc Ptashne testified that recombinant DNA technology posed less of a hazard than owning a pet dog. It was an astounding statement given the totally new status of the technology and our complete ignorance of the ramifications of gene manipulation at that time. Unfortunately, Mae Wan Ho and her colleagues have some bad news. She says, "Evidence has come to light within the past few years that these so-called crippled strains can actually sometimes survive quite well in the environment. Or they can go dormant for a while and then reappear, having acquired the new genes from their friends out there in the environment to enable them to live. The ideology of competition, the 'survival of the fittest' that you have in reductionist science is nowhere more inaccurate than in the microbial world. Members of the microbial world are very good at sharing their genes."

Seduced by the Lab

Scientists devote their lives to making conditions predictable, stable and harmonized. They just don't have a clue of the conditions in other parts of the world. They never think of what a biotech lab might be like in the hands of a desperate, totalitarian government. In a way, science trains people to be naive. And in a way, it's touching and loveable. But it's not very good for taking precautions and anticipating risks.

— CHRISTINE VON WEIZSAECKER

Back in the late 1980s and early 1990s, Ricarda Steinbrecher was a young geneticist working in a prestigious King's College lab in London on the genes that control hemophilia. Just as I did more than thirty years ago, Steinbrecher found her research mesmerizing and a joy to perform. But like me, she couldn't ignore niggling questions and concerns about how this science was being applied. She became particularly worried about how genetically modified organisms were being released into open fields with little or no investigation into their impact on the health of the environment and the diverse web of life they would find there. Eventually, Steinbrecher left the lab and offered her considerable knowledge and expertise to various non-governmental and environmental organizations that are trying to find out more about the possible effects of genetic engineering. Today, she helps them analyze and evaluate data, as well as educate the public on the issues. She misses the joy of lab research and says, "Genetics is a fascinating field and it's beautiful. I'd love to learn more about it and so would a lot of others. But that doesn't mean we should produce something we do not know much about yet and put it into the open environment. That's just the opposite of what all logic and science says."

So why have scientists embarked so enthusiastically on the path of corporate genetic engineering? The allure of money cannot be the full explanation. In part, it's because of scientific attitudes and enthusiasm. And it's because of human curiosity. There is a wonderful sense of community in a lab — a collective of talented people sharing an interest and feeling the thrill and power of probing nature's secrets. It's seductive and addictive, but it's also sometimes so much fun that you can actually forget the implications of what you're doing. When I was a geneticist studying heredity in fruit flies, we would refer to the insects as "bags of chromosomes." We bred them only to follow genes and chromosomes, and we seldom looked at them as whole, living organisms. To study gene expression, we bred them so that legs grew out of their heads and wings out of their eyes. I have filmed in labs where human organs were being grown on mice, and where rats were sewn together so they could exchange blood through criss-crossed arteries. Ricarda Steinbrecher says, "Where does a scientist draw the boundary? When should you stop being fascinated and experimental, and when should morals come into play? This is why constant interaction and checking with society is so crucial, so vital. Because, as a scientist, it's very easy to be so into one's own experiments that there's no boundary left. The only boundary is the boundary of imagination."

The kind of genetic engineering going on today desperately needs more societal scrutiny. In their obsession with specific goals and fascination with their technical dexterity, scientists seldom reflect seriously on the social or environmental ramifications of their work. Christine von Weizsaecker is a molecular biologist with laboratory experience. She says the scientists who play with ebola vectors and antibiotic-resistance genes are dumping their failures in the food supply because they don't see things the way the general public might. "I think it has to do with one-track minds. People think of risks in their own contexts. [Scientists] work in labs with Petri dishes; with a homogeneous environment; with exactly the same temperature, standardized nutrient broth and mono-cellular organisms, not several types of them, but just one type. And in this environment, all the normal accidents, changes, unpredictable things that the future may bring just don't come to mind."

Brian Goodwin concurs. "The education of biotechnologists is one of the narrowest in the world," he says. Biological education has been getting more and more specialized over the years, but with the introduction of recombinant DNA technology, the situation has worsened. Industrialized countries have put a lot of money into training scientists in this area because it's seen as the technology of the twenty-first century. "The scientists are extremely creative and very competent within an extraordinarily narrow range," explains Goodwin. "But often they don't know about organismic physiology. They don't know about whole organisms. And they certainly don't know about the ecosystems that they are potentially threatening."

So the problem for society is that even though these are "good" scientists, they have become more like technicians. Their training is so specialized that they really are, as Goodwin says, "simply unable to see the potential consequences of the sorts of products they're producing. And they are not required, within the conventions of science, to take responsibility for their actions." Goodwin goes on to argue that it's absolutely essential that laws, liability and insurance are devised to make scientists in these fields more directly responsible for what they create. That is the only way that the training will be forced to come in contact with the real world it affects.

Genetic Pollution

Scientists from the government-funded National Institute of Agricultural Botany have discovered the first genetically modified (GM) superweeds in Britain, following the spread of pollen from a genetically modified trial crop to wild turnip plants. The hybrids were produced after plants in a field of

*wild turnip crossed with a nearby test-site of genetically
engineered oilseed rape. Some of the "Frankenstein" plants,
which had inherited their GM parents' herbicide-resistant
genes, were able to breed.*
> — MARIE WOOLF, *THE INDEPENDENT*, APRIL 18, 1999

Scientists like Mae Wan Ho, Brian Goodwin, Ricarda Steinbrecher and
Christine von Weizsaecker are worried about unexpected risks, such as
the inadvertent introduction of a virus from one species into another,
that could have devastating consequences. But there's a longer-term
potential with deliberate experiments that is even more disturbing.
Genetically engineered organisms are not just kept in a lab or in growth
chambers; they have been, on a very wide scale, intentionally released
into the environment, mostly as agricultural crops.

And yet we have ample evidence about the hazards of exotic organisms in a new environment. Rabbits in Australia, zebra mussels in the
Great Lakes, cats on tropical islands, purple loosestrife spreading across
North American waterways — all these species were benign or useful in
the places in the world where they belonged, but out of the surroundings
in which they evolved, where they existed along with other organisms to
keep them in check, they have become pests that are devastating local
native wildlife and costing billions of dollars a year in lost crops, eradica-
tion programs and restoration. If that's what wild creatures that are displaced from their natural habitats can do, what can be expected from life
forms with genes spanning species barriers, life forms that are therefore
exotic anywhere on Earth?

Many of the so-called "first generation" of genetically engineered
products were designed to help farmers live weed-free. "Roundup
Ready" soybean, canola, corn and potatoes were genetically engineered
to resist Monsanto's patented weedkiller, Roundup. This was a strategy
by Monsanto and other chemical companies to keep farmers dependent
on their products. The expensive patented seed and the herbicide were
sold as a package, and these kinds of package deals are growing.

Roundup destroys all green, leafy plants. Before the genetic engineers developed Roundup Ready seed, the pesticide had to be applied
early in the season, before the emergence of the main crop, so it wouldn't
be killed. Now farmers can drench their food crops with herbicide all

season long, and the food plants won't die — though the crabgrass, milkweed and lamb's quarters that surround the crop will. Many so-called weeds are crucial food for other species, as milkweed is for monarch butterflies.

We're destabilizing nature's balance to accommodate a sophisticated but, one can argue, largely unnecessary technology. There are simpler mechanical ways to deal with weeds, including no-till farming, mulching and companion cropping. But none of these Earth-friendly methods can be patented to make big money. They're not splashy, high-tech, instant fixes. They evolve over years and generations because methods have to be developed at a local level by individuals who care about and observe the crops, soils and wild plants of each locality.

Industrial farmers may think of weeds as competitors for sunlight, space and nutrients, but one person's weed can be another person's, or species', food. Vandana Shiva is a physicist from India. She heads two agencies concerned with agricultural crops and food production and also works with the Third World Network on environmental and justice issues. She has written many books and articles on the politics of food and on biotechnology. She points out that many people get their nutrition from so-called weed species: "In India, at least 80 to 90 percent of the nutrition comes from what the agricultural industry terms 'weeds.' . . . [Industrial agriculturalists] have this attitude that the weeds are stealing from them, so they spray herbicides like Roundup on a field that has sometimes 200 species that the women of the area would normally use in various ways as food, medicinal plants or animal fodder. At the moment, about 40,000 children in India are going blind for lack of vitamin A, only because industrial farming has destroyed so many wild field plants, the sources of vitamin A that were available to the poorest person in the rural areas. With biotechnology, they are increasing this lunacy."

Herbicides and pesticides are powerful metabolic inhibitors — that's why they are used. And most have also been shown to be carcinogenic. Roundup is supposed to be a relatively mild compound, but it has been shown to be associated with illness, poisonings and serious allergic reactions, and is highly toxic to fish. In fact, it's rated the third most poisonous agrochemical in California. A government study released in June 2003 in Denmark shows that Roundup, the glyphosate used lavishly on the popular crops genetically engineered to resist it, is ending up in the Danes' drinking water. Even when "sprayed by the rules," there is a concentration of .54 micrograms per litre. "This is very surprising, because we had previously believed that bacteria in the soil broke down the compound before it reached the ground water," researchers stated, although their surprise came from reassurances provided by "the producers' (Monsanto's) own research." The contaminated water has been found not just in the soil, but in wells in several districts around the country, including Copenhagen. Part of the problem is the vast increase in the chemical's use because of GE crops; it has doubled in the past five years alone. However, when Danish Environment Minister Hans Christian Schmidt brought out restrictions on glyphosate use in response the following autumn, the producers of the chemical, including Cheminova, Syngenta and Monsanto, all condemned the move as "unacceptable," claiming sample levels weren't deep enough to affect drinking water. The minister stood his ground, also using the word "unacceptable" to describe a weedkiller in groundwater "in such concentrations . . . Danes should be able to put the coffee on in the morning without worrying about pesticides." Crops engineered to be resistant to this herbicide are repeatedly dosed with it throughout their growing period; conventional plants would die if they were sprayed with such chemicals. So we have been eating, in our soy and canola, season-long residues of this herbicide, and to this day no tests whatsoever have been performed to find out if that's safe.

We have to ask ourselves if we are getting benefits from these new technologies that are in any way commensurate with their risks. Vandana Shiva says, "There's really a struggle going on between the view that imagines the whole world is a marketplace, and the thousands of cultures that think that we are part of an Earth family. It could be called the web of life. It could be called a democracy of all beings. But once you're able to recognize that you're in that democracy of life, and you accept the fact that all beings have to live, then you shape a whole different set of alternatives and technologies. If you assume you can push every troublesome boll worm to extinction, every troublesome beetle into extinction, then you shape the technologies that you think will make them extinct.

"But, of course, they're more hardy than we are. They're more resilient than we are. And as is being said more and more, it's the small beasts and the small organisms, especially the microorganisms, that are going to prove to be much more superior in the silly war we have declared on them. The best we can do is accommodate. The best we can do is give each other space. And through that giving of space, we create the sustainable systems that also look after our needs."

Feeding the World

Biotechnology is one of tomorrow's tools in our hands today. Slowing its acceptance is a luxury our hungry world cannot afford.

— MONSANTO PUBLIC-RELATIONS STATEMENT, REPORTED BY THE AFRICA NEWS SERVICE

[This] monopoly of agricultural-development funds by a narrow set of technologies is dangerous and irresponsible . . . . It is only too obvious to concerned scientists, farmers and citizens . . . that we are about to repeat the mistakes of the insecticide era, even before it is behind us.

— HANS HERREN, DIRECTOR OF THE INTERNATIONAL CENTER OF INSECT PHYSIOLOGY AND ECOLOGY, NAIROBI, KENYA

Throughout human history, we have used new technologies to increase food yields, always seeking more food for less input and less effort, whether it was a horse to replace human muscle power or a plough as an alternative to sticks. After the Second World War, a revolutionary new kind of farming was developed, one based on the wide use of petrochemical fertilizers, herbicides and pesticides. It also depended on growing large fields of single crops — or monocultures — and the use of heavy machinery. At first, the results seemed miraculous: yields per hectare were doubled and tripled. By the 1960s, it was hailed as the answer to world hunger and it was called the Green Revolution. But by the 1980s, those miracles had begun to evaporate. The increased yields, particularly in the Third World, were illusory.

Vandana Shiva has spent many years studying the legacy of the Green Revolution, and she says, "The issue of more food through the Green Revolution is a construction. It's an illusion. If you reduce food to just wheat and rice — and more and more land has been converted to just growing wheat and rice — and you define rice and wheat as food, then, of course, you've grown more food, because you have stopped sowing the millets and the legumes that fed the people, the oilseeds and the straw that fed the cattle and the soil."

Shiva continues: "You actually have lower biomass production in the monoculture fields of the Green Revolution than [on] a typical polyculture farm of indigenous agriculture. When you get rid of your straw and it stops being edible to cattle, sooner or later you have to use other lands to grow that fodder. So if your land used to be providing enough fodder for your animals, as well as food for the family, you now have to use two or three times the amount of land to grow the same amount of food, plus fodder, for that same unit of a farming family, which has to include its animals. Over time, it means that people stop producing fodder for their cattle. Over time, it leads to the cycle that we've seen take place in England, where farm production of fodder shifted to off-farm purchase of cattle feed. The last round of that was rendered meat, ground up from infected carcasses of sheep and cows, being fed back to the cattle. And we get mad cow disease. That's how the food-security system of industrial agriculture works."

Up until recently, because of the explosion of human populations genetic engineering was promoted as the only means of averting mass starvation. Dick Goddown was the vice-president in charge of food and agriculture for the Biotechnology Industry Organization (BIO) when we interviewed him for our radio series back in 1997. He stated the promise of biotechnology: "I think that it's the best hope we have, as denizens of this planet, of being able to feed the people who are going to be on it. I don't see anything else on the horizon, and I do see a real, very definite promise in biotechnology. I see very definite evidence of the increased yields that come from genetically engineering plants. And producing more, particularly more of a staple crop, is going to be a subject that we're all going to have to deal with as the population of the Earth increases, perhaps doubles, in the next forty years, and goes from the more than 6 billion people [who] are on the planet right now to somewhere between 8 and 10 billion. Therefore, the potential for agriculture is just phenomenal. Biotechnology will be agriculture and agriculture will be biotechnology in the very short future."

This type of rhetoric stops many well-meaning First World environmentalists in their tracks. How can we stand in the way of feeding starving and desperate people, even if the only solution being offered has so much potential for environmental and health problems? This is Vandana Shiva's response: "Oh, my God, the charity argument! That's on every level a deception. First of all, the kinds of things they're producing don't feed the Third World. Tomatoes don't feed the hungry. Not even soybeans. Soybean goes to feed the pigs and the cattle of the North. It's not the staple of Third World people who are going hungry. Most of the research is not about increasing production. All the investments in agriculture are about increasing chemical sales and increasing monopoly control."

If we look at the major products of biotechnology that characterize the entire industry, we find agricultural crops like corn, soybeans, cotton and potatoes that have been engineered to resist expensive, patented herbicides, or that have encoded the Bt insecticide so that the whole plant is poisonous to insects. These are crops for northern nations, and they require heavy machinery, huge fields and copious chemical fertilizers, all of which are beyond the means of farmers subsisting on small landholdings. Third World staples that primarily benefit the poor are not primary targets of the biotech industry. "There's another important reason why this would not solve the hunger problem," Shiva explains. "All this is taking place in the private domain, by corporations that are not in the business of charity. They are in the business of selling. The food they will produce will be even more costly. It will not reach the entitlements of the poor."

She continues: "The fiction of population is amazing. Because every time we are told that we need to shift to new forms of producing food, nobody informs the public that the methods being proposed will use more resources to produce the same amount of food. And therefore, we'll actually create a deeper crisis, in terms of poorer populations. If you take the Green Revolution in industrial agriculture, 300 units of input — that is, energy from fertilizers, machines, pesticides and so forth — were used to produce 100 units of food. For organic, biodiversity-intensive agriculture, internal-input agriculture, you need only *five* units of inputs for the same 100 units of food."

Many find it hard to believe that traditional systems — recycling waste back into small, multi-species fields, for example — could be more efficient than high-tech modern methods. But all energy, whether in food, muscle power or fuel, represents stored sunlight. And much more stored sunlight is used up by burning petrochemicals for transportation, creating fertilizer and running heavy machinery than in companion cropping, composting and mulching. So when the total amount of energy used to grow and move a crop is weighed against the amount of energy actually in the food, modern agriculture is lavishly inefficient compared with traditional small-scale farming. Perhaps the lesson is that to get at the roots of hunger, we have to examine sociology and politics, not buy into another quick technological fix.

The Parable of the Golden Snail

This rice could save a million kids a year [from going blind].
                                                    — *TIME*, JULY 31, 2000

How many cases of blindness could be averted right now if the industry were to divert its river of advertising dollars to a few ... ridiculously obvious, unglamorous, low-tech programs?
— MICHAEL POLLAN, "THE GREAT YELLOW HYPE," *NEW YORK TIMES MAGAZINE*, MARCH 4, 2001

In a 1999 issue of *The Nation*, journalist Peter Rosset told the story of how a group of Filipino farmers reacted when asked about their feelings concerning genetically engineered rice seeds. Like many paddy growers, these farmers have long been supplementing their diets with creatures that can live in the flooded fields — eels, ducks and, in this case, snails. Imelda Marcos had the bright idea to bring in a gold-coloured South American snail that was said to be more productive, so the farmers would be better nourished. No one liked their taste, though, and the project was abandoned. Unfortunately, a few escaped and ended up driving the local variety almost to extinction. This not only wiped out the farmers' former food supply, but forced them to use toxic pesticides to keep the new snails from eating their young rice plants. "So when you ask what we think of the new GE rice seeds, we say, that's easy," their leader says. "We think they are another Golden Snail."

In 2000, the International Rice Research Institute in the Philippines proclaimed yet another of its many technological rice-breeding break-throughs. The institute called it "Golden Rice." The rice is engineered to contain a gene to express beta-carotene, making it actually yellow. The idea was that it would cure the serious vitamin A deficiencies that plague much of the Far East, causing eye problems and blindness, by providing a new source of the needed nutrient in the Asian diet. This very early announcement (the variety was far from perfected) seemed to fascinate the media and rapidly gained the international centre stage. ABC and CNN extolled the rice's new properties, presidents and prime ministers called for its immediate perfection and distribution, and *Time* put it on its July 31, 2000, cover, accompanied by a story that clothed its origina-tor, scientist Ingo Potrykus, in the mantle of a Jonas Salk. Sometime later, when it was noted that Salk hadn't patented his polio vaccine with an eye towards profit, the major funder of the wonder seed, the Rockefeller Foundation, announced that Golden Rice would not, as all previous genetically engineered varieties had been, be accompanied by a hefty royalty fee. It would be "free" to all.

By 2004, however, Golden Rice was still not on the market, and its reputation had sagged badly. Dozens of scientists and NGOs had been able to attack it effectively for many reasons. Even its major funder, the Rockefeller Foundation, publicly admitted that promoters' claims about Golden Rice "saving hundreds of thousands of children a day" were "seriously overstated." However, Ingo Potrykus, its inventor, is still push-ing for Golden Rice to be planted all over the Third World. He even wants its opponents to stand trial before "an international court . . . to justify the pain and suffering they are inflicting on so many people."

Potrykus's indignation must be weighed against his product's rather noticeable design flaws. Besides possibly insurmountable problems with the variety itself — whether the beta-carotene it expresses would be in a form that could be taken up by the human body, how much rice a per-son has to eat to create an effect (some estimated as much as seven kilograms a day) and finally whether there could be dangers of too much vitamin A (poisonous in larger quantities) getting into the food supply — there is another surreal quality about the exciting announcement. This is the fact that vitamin A is readily obtainable in its most digestible form in cheap, simple foods like almost any sort of leafy greens and even unhulled regular rice. The main problem is one of education, which the new yellow rice, more expensive among other things, would also need for acceptance; pure white rice has long been the only acceptable staple in Asia.

Michael Pollan, a *New York Times* regular and author of *The Botany of Desire*, asks dryly, "Why not simply mount a campaign to persuade people to eat brown rice? . . . or grow green vegetables? . . . What about handing out vitamin A supplements to children so severely malnourished their bodies can't metabolize beta-carotene? As it happens, [such programs] are already being tried, and according to the aid groups behind them, all they need to work are political will and money." He asks, "So to what [question], then, is Golden Rice the solution? The answer seems plain: to the public-relations problem of an industry that has so far offered consumers precious few reasons to buy what it's selling — and more than a few to avoid it."

The Best Way Not to Find Is Not to Look

We may never know. After Aventis pulled its application for approval, the government stopped testing to see if StarLink was a health concern.

— BILL HORD, "CONTROVERSIAL BIOTECH CORN NEARS EXTINCTION," *KNIGHT-RIDDER TRIBUNE,* OCTOBER 21, 2002

Most people remember the recall of StarLink corn, a GE variety bred with a possible human allergen that was never supposed to enter the human food supply. But its insect-killing protein, Cry9c, was discovered in everything from Taco Bell tacos to corn syrup. This caused a massive recall of food products that destroyed the crop year for a lot of farmers, and many customers abroad, like Japan, ceased buying U.S. corn. The incident ended up costing the developers, StarLink Logistics and Advanta USA, $110 million in class-action lawsuits.

In the late 1990s, Canada was the world's number one supplier of rapeseed (also called canola), a miracle hybrid that rapidly became the market's most popular cooking oil and was beloved by health enthusiasts for its low cholesterol and light flavour. Today, as discussed farther on, the market for Canadian canola has nearly collapsed, because we've lost

our European and Asian buyers. It can only be sold in Canada and the
United States because pollen outcrossing, with genetically engineered
varieties containing a gene that makes the plant resistant to Monsanto's
best-selling herbicide Roundup, has contaminated even organic canola in
Canada. Worse than that, a theoretical worry voiced by independent
experts that canola's many wild relatives might pick up the same gene
through cross-pollination has come true. Herbicide-resistant GM canola
is popping up everywhere across the Canadian West and is now a major
weed problem in the country.

Monsanto looks forward to having GM wheat on the market within
three years, despite protests from Canada's largest agricultural unions;
accusations of conflict of interest within Agriculture and Agri-Food
Canada; worries about the spread of a specific fungal disease, fusarium,
via the herbicide Roundup; and warnings from major importers such as
Japan and South Korea that they won't buy it. Just as it did with canola,
Monsanto is belittling these worries, saying the U.S. market will accept
the product and citing its own studies that show the spread of wheat
pollen in fields "is minimal, and management of volunteers will not be a
significant issue."

The idea of environmental or food contamination with the genes
used to induce pest resistance pales, however, next to the possibility of
pharmaceutical contamination. In 1999, when we interviewed geneticist
Val Geddings, a vice-president of the industry organization BIO, he
enthused about the possibility of putting vaccines into bananas, because
they would somehow be easier to transport and administer than pills and
injections. Like any parent, I wondered whether making a powerful medicine resemble a delightful dessert would be such a good idea. It seemed
to me that we should be drawing a clear line between when we're eating
and when we're taking a drug. This has remained a deep concern for people analyzing the effects of Golden Rice. But what the geneticists really
liked was putting other organisms to work for us, like machines. They
realized we could conceivably grow human organs for transplant in pigs,
insulin in cows, medicinal enzymes like trypsin in plants. The hardwork-
ing cells of these other organisms could manufacture our drugs for us.
The only problem would be keeping these truly "organic" drugs out of
the normal food supply. And so far the industry isn't doing a very good
job of that.

In November 2002, the *Washington Post* reported that ProdiGene Inc., a Texas biotechnology company, mishandled dangerous gene-altered "pharma-corn" not once, but twice. First, the USDA and the FDA quarantined 500,000 bushels of soybeans in a Nebraska warehouse, because they contained a detectable amount of leaves and stalks from gene-altered corn that had been planted the year before and that had volunteered among the soy. The government agencies, while congratulating themselves on keeping this material out of the food supply, refused to identify what industrial or pharmaceutical protein was in the corn. Then it was learned that 63 hectares of Iowa corn had been incinerated for similar reasons in September 2002. Finally, it was revealed the corn had been engineered to produce an experimental type of insulin. Anyone familiar with the ravages of diabetes can imagine what such a drug — and an experimental one at that — might do if eaten as food, and it was in an open-pollinated crop. Although most of these potentially dangerous proteins would be destroyed in the human digestive tract, the *Washington Post* article pointed out that "a few can survive long enough to ... cause health problems."

The industry claims such situations are just minor glitches but, in any complex human activity, human error is inevitable, especially as a technology becomes more widespread and accepted and handlers become more habituated and blasé. There is concern about loss of control in a less-regulated country, like Russia or Mexico, where the industry is also trying to test such crops. Jane Rissler, senior scientist with the Union of Concerned Scientists, says, "If a company cannot be relied upon to perform such a simple task, keeping pharma-corn out of soybeans, how can it be trusted in the far more complicated process of keeping drugs out of cornflakes?" John Cady, president of the National Food Processors Association, voiced business concerns when he called the error "alarming" and said it "very nearly placed the integrity of the food supply in jeopardy." But although the FDA said it was confident that no biotech corn made it into the food supply, its track record in the larger field of widespread and disastrous genetic pollution isn't reassuring.

In the fall of 2002, a researcher in a state university lab in East Lansing, Michigan, lost two vials and application notes for a new, genetically engineered bacterial disease more deadly to pigs than anything yet known. "[It] could devastate the pork industry if replicated and released," the *Wall Street Journal* warned. The material was apparently stolen, and police treated the event as a potential terrorist threat. So far nothing seems to have happened connected with this new disease, but the fact that researchers in minor labs all over the world are actively developing lethal pathogens and, being human, are leaving them lying around to be lost or stolen, points to a need for stricter controls over this industry. The authorities admitted that the only reason they paid attention to the loss was because of the fear of terrorism gripping the United States at the time. "If this had happened thirteen or fourteen months ago, we wouldn't have thought twice about it," said Ron DeHaven, an official for the U.S. Department of Agriculture. So whose job is it to think about such things?

Assessing the Risks

It's easy to be carried away by a wonderful idea, especially one that can make a lot of money.
— BETH BURROWS, PRESIDENT OF THE EDMONDS INSTITUTE

The revolution in gene technology is happening quickly. Even though I'm a geneticist, I was unaware of the number of gene-engineered products that had already reached the market when I started working on our radio series and this book. If you eat packaged breakfast cereal, ice cream, baby food, canola oil or non-organic soy in any form, you're probably ingesting genetically modified organisms (GMOs) every day. That means the GMOs are being replicated around us and are being absorbed by our bodies as we eat them. They can't be avoided easily, because in Canada and the United States, there are no labelling requirements to indicate what we're ingesting or what creatures they came from in the first place.

In early 2004, Canada was finally joining almost every other country in the world (except the United States) and getting ready to label genetically engineered foods. The latest polls have shown overwhelmingly that Canadians — 91 percent — want GM foods labelled as such. And an almost equally high number of Canadians — 88 percent — want these labels to be mandatory. "Voluntary systems just don't work and consumers don't trust them," stated Peggy Kirkeby, vice-president of issues and policy at the Consumers' Association of Canada, which released the latest figures. Under the current suggested voluntary labelling guidelines, even if they choose to label their products, food producers will be able to claim they are GE-free under fairly low standards. While Europeans won't tolerate GE contamination of more than 0.9 percent in a food, Canadian levels are set at 5 percent. Moreover, the new, voluntary law will only come into force "if an international standard on verification is adopted" — in other words, only if there's any way to make sure companies aren't lying.

Meanwhile, in China, a country that's regarded as being highly autocratic and immune to citizens' needs, full GM labelling came into force in July 2003. Although stressing the label contains no value judgement as to its safety or healthiness, Wu Jianfan, director of the Beijing Genetically Modified Agricultural Organisms Office, said, "[This] small label shows respect to consumers." The Chinese are particularly concerned about GMOs, almost ubiquitous in non-organic soy products, because soy is such a large part of their diet. In North America, though, common engineered products include tomatoes with a fish gene, soy and corn with cauliflower viruses and petunia genes, enzymes made by genetically engineered bacteria and yeasts in our breads, prepared foods and soaps, and increased risks of pesticide ingestion in all Roundup Ready products.

When people realized that genetically modified foods weren't being labelled in Europe, a furore ensued until labelling laws were implemented. Now, huge supermarket chains and food giants like the European branches of Nestlé, Unilever and Burger King have banned genetically engineered products. But this hasn't happened in North America. If genetically modified foods do have deleterious health effects, the rest of the world can watch North America for signs of what these might be.

Christine von Weizsaecker warns, "We will be eating new things. We always have tried new things, but at a completely different pace. . . . If over a few years, you flood whole continents with new substances, you are going to pay a hell of a high price for your learning. If everything is flooded, who will then be able to prove what the cause is? And you know how much suffering came from the time-lag between the first signs of endocrine disrupters, or thalidomide, and the real, final scientific proof." In other words, the risks are huge as we perform a massive global experiment, using ourselves as guinea pigs.

Genetic Pollution II

I only know that I am afraid.
— OLGA MOLDONADO, ZACATECA INDIAN FARMER

In 1970, 15 percent of the entire American corn crop was destroyed by a terrible blight. The National Academy of Sciences blamed this plague entirely on genetic uniformity; the plants from Florida to Nebraska, in one acre out of four, were "as alike as identical twins." Losses would have been absolutely catastrophic if scientists hadn't been able to rush to the birthplace of the crop, the highlands of southern Mexico, to get new genetic material from traditional corn varieties and develop seed that could resist the new disease. This is something science does often — head off to Ethiopia to research the ancestors of wheat and millet, to Peru and Central America for corn, potatoes and tomatoes. The ancient "land races" of corn, as they are called, in the state of Oaxaca alone include more than sixty different varieties — corn that is not only yellow and white, but pink, blue, black and purple.

Like other early forms of food crop species, these land races constitute "the world's insurance policy," as Mauricio Bellon, director of the economics programs at the International Maize and Wheat Improvement Center, the world's most famous public research facility on these grains, puts it. "These genes are the basis of our food supply.... We need this diversity to cope with the unpredictable ... as climate changes, new plant diseases and pests evolve, diseases we thought we had controlled come back." Today, because of increasing instances of genetically engineered varieties that we were warned about by our contacts in 1999, it looks as if this door to our future has now been closed. The land races of Oaxaca, the genes that provide the basis for all the fancy breeding and science of new corn hybrids, have become contaminated, probably forever, with new living organisms fashioned for commercial purposes in an industry biotech lab. And that industry doesn't even have to face liability for the damages it has caused.

Olga Moldonado, quoted above in a long article by Mark Schapiro in *The Nation* and whose story was also discussed on Bill Moyers's *Now* television series, was one of the first smallholding farmers in this remote region of Mexico to make this discovery. In 2002, she noticed that some of the corn in the tiny plot of about 200 plants that provides sustenance for her family *did not have the hardiness to which she was accustomed.* Yields were off, not dramatically, but enough that she took her plants to be tested in a small lab in the nearby town of La Trinidad. There, University of California microbiologist Ignacio Chapela discovered that fifteen of the twenty-two corn samples he ultimately tested from similar plots in the area had traces of transgenes. That is, they contained foreign genetic matter, living engineered organisms that are now impossible to extricate from the plants' tissues, roots, leaves and seeds.

Since transgenic corn had been banned in Mexico in order to protect the land races, this was big news, so big that early reports in *Nature* by Chapela and his colleague David Quist were roundly attacked by the biotech establishment as being both mistaken and impossible. In an unprecedented move, for which it has been charged by NGOs with showing industry bias, *Nature* condemned the original study it had printed and even refused to publish Mexican government findings that reconfirmed the contamination. Transgenic contamination from seeds intended for human or animal feed but planted by mistake is what probably happened in Oaxaca and is what Zambia feared enough to refuse GMO food for its starving population. This was claimed by the industry to be impossible, or at least highly unlikely.

Just as our interviewees warned in 1999, the U.S. Environmental Protection Agency (EPA) and the Canadian Food Inspection Agency rely on the biotech companies themselves to submit studies on the potential of their own products to do harm, and there are still no laws requiring these government agencies to do follow-up inspections or independent monitoring. When the pollution of ancient Mexican land race corn was irrefutably proven to be true, the biotech industry switched from indignant denial and began hinting that because farmers like Olga are now growing their product, they owe the companies royalty fees. Dr. Mike Phillips, executive director of the food and agriculture department of the biotech industry umbrella organization BIO, also did something else our sources had predicted in the early years of this technology. He blamed the situation on the same government that had tried to control it, saying, "It really is incumbent upon the Mexican government to step up the process and get their regulatory system in place so that [they] can begin accepting these products and give farmers the opportunity to choose."

The choice will be pretty easy, since nobody seems to be protecting the non-GE varieties. The EPA web site itself admits there are still no long-term monitoring rules for these crops. Between November 1998 and June 2002, even in their place of origin, the economically and technologically rich United States, twenty "Conditional or Experimental Use" permits were granted for trial plots for the most common GE corn, the one that expresses the insect toxin Bt in all its cells. Not one of these sites was inspected until August 2002. A surprise visit from EPA officials to only two random plots being grown by Mycogen of Dow Chemical and the Pioneer division of DuPont in Hawaii discovered that both were in violation. The two companies were cited "for defying requirements intended to protect surrounding fields from the drift of genetically altered pollen." The next time a massive blight hits one of our food crops, it's difficult to imagine where we'll be going for help.

Don't Worry, Be Happy

> *Could we learn from past experience? Please? [Could we learn] that you don't trust the company that's making the product to tell you about the problems it has? How many times do we have to beat our heads against this brick wall before society in general learns that this is not a good way to do business? Yes, it may cost a few dollars right now, but down the line, it could save your life.*
>
> — ELAINE INGHAM, SOIL PATHOLOGIST

All kinds of genetically modified organisms are on the market or waiting in the wings. How do we know they're safe? We don't, and won't for years, even as exposure is increasing. How are they licensed? By government agencies that often make their decisions without public input or inspection, and often after heavy lobbying by biotech industries. Who does the testing? Mostly the companies that are developing the products. Testing by the company with the most at stake in marketing the product is the way it has been done for years; only when a government suspects a problem, or is invited to investigate by the company, does it conduct its own tests. And since government downsizing started in the 1980s, fewer and fewer labs and scientists do this work outside private industry.

Elaine Ingham is a soil pathologist at Oregon State University who also founded Soil Foodweb Inc., a soil research agency. She thinks there are serious problems with the testing regimes in most countries. "The testing systems are inadequate. It's a conflict of interest. We're expecting the people who are going to gain profit from these organisms to be honest about the dangers and the difficulties of those products. If they do want to make a huge, enormous profit, it's an immense conflict of interest to expect that they're going to be honest about it. We're all human beings, and we know we don't function that way . . . . We need independent testing agencies to carry out all of this kind of testing. You know, you'd think we would learn from the chemical industry."

Back in the 1940s and 1950s, we were told that all the new petrochemical products — herbicides and pesticides like lindane and DDT — were safe. "All these chemical companies said, 'Oh, we've tested it. Don't worry, there are no negative effects,'" says Ingham. Of course, now, fifty years later, countless people, especially children, have suffered cancer and deformities, and we've spent billions on toxic-waste clean-ups. Today, the companies that assure us that biotechnology is safe are often exactly the same ones that gave us DDT, PCBs, Agent Orange and all of the other toxins that created health problems as they leaked out of landfills and into water courses—Dow, Monsanto and others.

What about the government institutions that are already set up? In 1998, a scandal shook the United Kingdom when a researcher went public with the dramatic results he got from early tests of feeding genetically engineered potatoes to rats. He found their growth was affected, some developed tumours and some showed significant shrinkage of the brain after only ten days of feeding on genetically modified potatoes. A protein or common virus used in the gene-splicing technique in the engineered potatoes was implicated. Dr. Arpad Pusztai announced his findings to the public because he felt it was his duty as a scientist working for the tax-supported Rowett Institute. He was immediately forced to "retire" by the government and his institute.

Like many whistle blowers, Pusztai's subsequent story is sad. Public pressure forced the Rowett Institute to make a show of rehiring him, but he was soon encouraged to leave. His research was attacked, despite the impeccability of his data. He suffered two heart attacks and stayed quietly at home for four years, breaking his silence finally to reveal that in May 1999 there had been a break-in at his home where he and his wife, also a GMO scientist, kept their research. "The only things taken were some bottles of malt whisky, a bit of foreign currency, and the bags containing all their research data." Another break-in at the Rowett Institute a few months later only affected Pusztai's old lab. He also gave the press new information concerning two calls from 10 Downing Street that went through to his boss, Philip James, after Pusztai's first, disturbing announcement. This story, which seems to come straight out of a John Le Carré thriller, has been confirmed by Robert Orskov, OBE, and Stanley Ewen, another respected colleague. All three scientists claim that "Monsanto called Clinton, Clinton rang Blair, and Blair rang James [Pusztai's boss]."

These accusations and counter-accusations are only minor examples of the extent to which the biotech industry has found allies in the political arena. And, despite initially funding it, ever since Pusztai's data proved embarrassing to the industry, the Blair government has continued to try to push GE food and crops onto a British public that has remained singularly unenthusiastic about eating them. What is less sad is the fact that Putszai's announcement of tissue damage has reverberated; the British public refused to take any more chances with GMOs, especially after what happened with mad cow disease. In that case, people had also been assured by government scientists that beef fed with rendered animal products was safe. Britain now has stringent labelling laws and laws controlling experimental fields using GE technology.

As for what kind of damage these new foods really can wreak, we just don't know, since countries still depend on the manufacturers for their safety data. However, what little independent research that has been undertaken since 1999 has supported the need for greater vigilance. For example, in July 2002, the results of the world's first known trial of GM foods on human volunteers were published. British researchers demonstrated that genetically modified DNA material, in this case herbicide-resistant genes, can survive digestion to make a home in the human gut. The research, commissioned by the British Food Standards Agency, showed that "a relatively large proportion of GM DNA survived the passage through the small bowel," not in normal subjects, but in all the volunteers whose lower bowels had been removed in colostomies, and that "bacteria had taken up herbicide-resistant genes from the GM food at a very low level."

Although these trial results were based on a small and unusual sampling, Professor Michael Antonia, a senior lecturer in molecular genetics at King's College Medical School, reviewed the work and said it was significant because "everyone used to deny that this was possible . . . . It suggests that you could get antibiotic marker genes spreading around the stomach, which would compromise antibiotic use." This is one of the main contentions that has prompted the EU to label these foods and limit crop production. Antibiotic-resistant "marker" genes are widely used in GM technology to allow gene splicers to identify the presence of the added DNA in the target organism. Although the industry claims it can phase out these markers, it hasn't done so to date. Because of the continuing controversy and lack of knowledge, increasingly influential institutions, including the British Medical Association, have demanded a moratorium on commercial planting of GE crops on the grounds that "insufficient care is being taken to protect public health . . . there has not yet been a robust and thorough search into the potentially harmful effects of GM foodstuffs on human health."

A Transcalculational World

> *If you introduce a foreign gene into any genome, then your first assumption is that something unexpected is going to happen. You'll be extremely lucky if you have an effect that's restricted to the particular character you're after — say . . . a protein that produces a certain effect on the organism, such as producing an insecticide and therefore protecting it from insect attack. You'll be extraordinarily lucky . . . because genomes are integrated. They're organized. You have to understand them as wholes, not just as collections of parts.*
> — BRIAN GOODWIN, THEORETICAL BIOLOGIST

Biotechnology hasn't yet lived up to its PR claims. Medical gene therapy is admitted to be an almost complete failure so far. Even the poster child of genetic engineering, bacterially produced human insulin for diabetes, has been causing comas and other life-threatening problems in around 10 percent of diabetics. Far more worrisome is the fact that reports of these side effects appear to have been suppressed by the industry. Genetically engineered products that are present in everything from packaged foods to detergents have recently been linked to serious allergies. And virtually *all* genetically engineered agricultural products have been accompanied by problems and failures. Non-pest insects like lacewings, ladybugs and monarchs are being killed by the Bt-engineered crops. Pollen carried away from crop fields by bees and wind is transferring genes to weedy relatives of genetically engineered canola, such as wild turnip. The controversy surrounding these and other products has become so widespread and serious that it has adversely affected stock values and set off a chain of lawsuits against Monsanto and other manufacturers.

Dr. Richard Strohman thinks there's a fundamental problem that underlies the entire application of genetic technology, namely, that genes don't exist in a vacuum. "Crudely put, it's garbage in, garbage out. It's garbage because the assumptions are wrong," he says. What Strohman means is that genetic engineers are not only trying to apply hereditary laws derived from studies *within* species to the transfer of genes *between* species, they also assume the laws that apply to conditions determined by single genes apply to most hereditary diseases or traits. Strohman discusses specifically the work on human diseases, but his ideas apply across the board to all areas of biotechnology.

Monogenic diseases (diseases caused by one gene only), like Huntington's chorea or cystic fibrosis, are very rare. But because genetic fixes may work with monogenic diseases, it has been assumed that they could be widely applied to diseases in general. Strohman says, "Why doesn't this single-gene analysis work for complex things like most common cancers or most heart diseases or cardiovascular problems — the big killers and the diseases that take the quality of life away from people in a premature way? Because these are all what we call multifactorial disorders. They have many causes. They utilize many genes, and they're not caused by any one gene . . . . It's impossible to follow the effect of one gene in this matrix of interaction. We say such things are *transcalculational* — that is to say, hopelessly complex."

When first interviewed, most of the scientists quoted in this chapter expressed a variety of concerns that at the time were still speculative and unsubstantiated by data. In fact, what has actually happened since those interviews has been far more dramatic. Every single conjectural possibility has come true. The commercial products of genetic engineering, released with increasing rapidity, have turned out to be even more virulent and dangerous than the scientists had imagined. There is now widespread evidence of exactly the type of instability and environmental pollution that critics warned might occur if genetically engineered products were released. The warnings that were raised are being verified in the real world, showing once again that the one-gene, one-trait, reductionist assumptions of commercial biotechnology are dangerously out of date. It is time to call for a moratorium on genetic releases until more workable paradigms and institutions are in place to assess them.

We need to cast off the simplified models of living organisms proposed by reductionist science, and to realize that they are only caricatures of the real world. Strohman says, "Molecular biologists and cell biologists are revealing to us a level of complexity in life that we never dreamt was there. We're seeing connections and interconnections and complexity that is mind-boggling. It's stupendous. It's transcalculational. It means that the whole science is going to have to change. And this reductionist biology of trying to locate complex things in a single anything . . . is not going to work. We're going to have to accept the challenge of complexity. We're going to need to train scientists differently. We're going to need to mix biologists with engineers and mathematicians with ecologists. They're going to have to learn about the world in which organisms live."

Billions of dollars have now been invested in biotechnology corporations, and their products are flooding markets around the world. The challenge now is to bring informed critical voices to the discussion about the future of this crucially important area. It is precisely because the technological dexterity has become so powerful that unhurried, serious debate must counter the headlong rush to profit from these new tools far too prematurely.

CHAPTER 6

YOUR MONEY OR
YOUR LIFE FORMS

*Genes are the green gold of the twenty-first century, every bit as
important to power in the next era as oil and rare metals and
minerals were to the Industrial Revolution.*

— JEREMY RIFKIN, HEAD OF THE
FOUNDATION ON ECONOMIC TRENDS

When a private or public biotechnology company isolates a gene and
inserts it into another organism, it can then claim ownership of the resulting "new" gene and start making money. That means all of life — every
bird and dog and bee, every bit of tissue inside our bodies — is up for
sale. Genetically modified mice, sheep, fish, insects, plants and bacteria
have already been patented, and companies are claiming royalties on
them whenever they're used. Even people are fair game — the government of Iceland recently sold the DNA sequences of 70,000 Icelanders
to a private company.

Jeremy Rifkin, head of the Foundation on Economic Trends in
Washington, D.C., and author of many books analyzing the impact of
science and technology on our culture, says, "All over the world, global
companies are scouting for rare genes in microorganisms, plants, animals and the human population that may be of some value commercially
in agriculture, in energy, in bioremediation, in material development, in
pharmacy, in medicine. There's a great struggle going on around the
world as to who will control these genetic resources, this rich reservoir of
genetic information that took millions of years to evolve."

The search for DNA resources doesn't end with plants and other animals. Now we're prospecting for gold inside ourselves. As we mentioned in the first chapter of this book, what should have been big news when the Human Genome Project announced it had succeeded in encoding the entire blueprint of our species was our phenomenal misunderstanding about how many genes we would be working with. Geneticists were sure there would be 100,000, but found fewer than a third that many. It turns out this shocking news, which should have highlighted our fundamental ignorance about gene activity and expression, wasn't really the primary concern of this enormous undertaking. It has become clear that its more urgent purpose was the application of our new insights into the marketplace. As Rifkin puts it in his book *The Biotechnology Century,* "Already corporations are seeking patents on every one of these human genes as they're isolated. Within just a few years, a handful of companies will literally control the genetic blueprint of our species. They will have claimed [every one of these] genes as their private property in the marketplace, meaning that any human being that needs access to those genes for medical, screening, or therapeutic purposes will have to pay for that right. The genes of the human species will be controlled as intellectual property, as patents, by a few global companies."

Business corporations can now control these genes through the medium of patenting. But how and when was the patenting of life allowed? It's due almost entirely to a judicial decision made in the United States almost twenty-five years ago. Rifkin was in the courtroom and describes the event: "In 1980 the United States Supreme Court, by a ruling of 5 to 4, said that General Electric could claim a microorganism that had been created in the laboratory as a patented invention. That set the precedent for companies all over the world to begin to control the entire genetic commons in the twenty-first century. Only seven years later, in 1987, the U.S. Patent and Trademark Office issued a chilling policy statement in which they said, 'Any organism on this planet, any gene, any cell-line, any organ, any tissue, is potentially patentable as a human invention.' Here, in one regulatory stroke, the U.S. patent office reduced the entire living kingdom, the genetic commons, to the possibility of private property owned and controlled by a handful of multinational corporations."

Rifkin and his organization appealed this narrowly won decision. He says, "Interestingly enough, the appeals court said this microorganism looked more like a chemical than it did a honeybee or a plant or a flower. In other words, they looked at the new construct, and it didn't look like an organism. Had General Electric brought a chimpanzee, a bird, or a pig, it would have never gotten the patent. It shows you the anthropocentric nature of this decision. This is a very thin reed upon which to base the enclosure of the genetic commons of this planet."

The idea of patents was first conceived by Thomas Jefferson. The purpose was to stimulate innovation by rewarding inventors with exclusive rights to their creations for a limited time. When that period was up, *the invention would revert to the public domain, where it would continue to benefit society.* These days, there are also strong commercial reasons for a company to use patents — to recover its research investments, make a profit and, of course, secure the new market for itself.

Genetic engineering involves a manipulation of DNA in a naturally occurring organism, and that change means something new has been created. That's how living creatures are rationalized as inventions. The control that patents confer on their holders is the exclusive legal ownership over what they may know, discover or invent. This concept of owning ideas, as well as things, is also called "intellectual property." And today, the structure of many trade agreements, like NAFTA, the FTAA and so forth, obligates signatories to follow this U.S.-based patenting regime and protect "intellectual property" like any other business investment.

Dick Goddown, a former vice-president of the Biotechnology Industry Organization (BIO) already mentioned, is an enthusiastic supporter of such patents. In our 1998 interview, he said, "Patents are the lifeblood of biotechnology. Why? Because they are, in a very real sense, property. They demonstrate what is yours and not someone else's. And therefore they allow you to attract the venture capital, and to attract the stockholders, so that you have the investment, so that you can complete the work and get [a medical] product to the patients, for example . . . . A patent, I dare say, is one of the best social compacts that we've ever arrived at. In exchange for the exclusive right to use and produce your invention for a period of years — and it has been seventeen in the United States and now it has changed to twenty — in exchange for that exclusive right, the inventor has to spread his invention on the public record for all others with an interest to be able to share his intellectual perspicacity, his advance, his spark, that magic that he's brought to light. And it has been a compact that has benefitted mankind more than anything else that I can think of."

In terms of sheer research output, Craig Venter can be called one of the most productive biotechnologists in the world. He developed new technologies that accelerated the rate of decoding the genes for the Human Genome Project. He runs his own genetic company, the Institute for Genomic Research, in Maryland. His name appeared on the first application to patent human DNA. Venter says, "Patents are very important in terms of developing new drugs. The genes for insulin were patented by Genentech and Eli Lilly, and that's why human insulin is now available around the world as a drug to treat diabetes. So millions of people have human insulin to use . . . . The only reason that it's available commercially . . . is because a pharmaceutical company got a patent on the process to produce it and make it. If it wasn't for intellectual property protection, there would be no human insulin now available as a drug. I mean, it's as simple as that. If people don't like gene patenting, they can send their drugs back to the pharmacy."

Most people in the developed world can afford to pay for expensive drugs like insulin. Adult-onset diabetes, which is tied to obesity, a lack of exercise and diets high in sugar, is a common disease in the industrialized world. Biotechnology is also heavily involved in developing *in vitro* fertilization and other means of making babies available to the infertile. Infertility, too, tends to be a problem that only the richer people of the world can pay to fix. And, of course, there's baby cloning, whose only moral defence is supposedly to help the infertile in a world already seriously overpopulated with uncared-for children. But the diseases that destroy existing young lives by the millions, like schistosomiasis, malaria and dysentery, are largely ignored by science. Their victims are poor, so there's little money to be made in finding a cure.

Phil Bereano teaches engineering at the University of Washington and helped found the Council for Responsible Genetics. Over the past twenty years, he's become an expert on the workings of this new industry. We asked why biotech companies concentrate on First World problems like adult-onset diabetes and infertility. Bereano says the answer is obvious when you think about how these companies operate. "Who in the corporations determines the design criteria for what form of genetic engineering they're going to have? It's the money-boys .... It's the guys who are dealing with the bottom line. The projects that are likely to make them the most money are the projects that are going to be developed, not the projects that correspond to some authentic ordering of human needs."

Not only do private companies decide which areas will be most profitable, they also take advantage of all the work done with public funds in universities and government agencies. With the advent of patenting, the companies can demand the profits without investing in all the research. Bereano says, "This technology of modern genetic engineering was largely created with public funds, with taxpayer funds .... And I think that because the purported benefits are for the public and, of course, the risks are going to be borne by the public, the public ought to be much more involved in making these decisions. Why do we leave these kinds of decisions up to private entities? Why do we say that they ought to be able to do whatever they want to do? We have to remind ourselves that many important biomedical developments were not patented, like the Salk vaccine. The March of Dimes and Jonas Salk were both against patenting. They felt that it was completely inappropriate for something that was developed with public funds and that was in such public demand."

Eli Lilly and Genentech did not do all, or even most, of the work that led to the development of genetically engineered human insulin. Many years of government, university and other publicly funded studies painstakingly built up the knowledge base that eventually made gene manipulation possible. Yet private companies claim *all* the profits. Now that we know that there are problems with genetically engineered human insulin, the companies should be willing to investigate the reported side effects and also accept liability for any damage that might be linked to their products. But that, as we shall see, is another story.

Patenting the Air

During colonization, we were told that our indigenous knowledge is unscientific and superstitious. Today the same

Western powers that rejected our systems are patenting and privatizing our knowledge.
— PRESS RELEASE OF THE RESEARCH FOUNDATION FOR SCIENCE, TECHNOLOGY AND ECOLOGY, NEW DELHI

For thousands of years, the people of India revered a beautiful evergreen, the neem tree. They made extracts from it to use as an insect repellent, they isolated oils for soaps and skin cream, they cooked its leaves to make medicines. They even used its twigs as toothbrushes. The tree belonged to everybody, and was part of their culture. Not anymore. Biotechnology companies have genetically modified the neem and its components and are collecting royalties on patented preparations derived from it. What was once free and available to everyone has become private property you have to pay to use.

There are now patents pending on almost anything you can imagine, including an ancient South American staple grain, quinoa; peasant-developed Indian chickpeas; and kava, a plant sacred to and much used by the New Zealand Maori. Patent applications have also been made for various Indian recipes, like a popular fried dumpling, the pakora. And a patent was applied for to give the "inventors" of Dolly the sheep, cloning rights on all mammals, which includes us, although the company went belly-up before it could collect. Attempts have even been made by the U.S. government and private companies to patent cell-lines of indigenous peoples or of individual, unknowing patients.

Andy Kimbrell, a lawyer who runs the International Center for Technology Assessment in Washington, D.C., tries to force U.S. regulatory bodies to face up to the implications of new technologies and regulate accordingly. Kimbrell is suing the EPA over biotechnology issues. He says, "It's quite a remarkable circumstance; what we are experiencing is an ethical and legal free-fall, because there is no longer any understanding of what is patentable or why. We're seeing the corporate enclosure of literally the entire living kingdom, including the entire animal kingdom, with no real restraints."

Things that have been owned communally for many generations are called a "commons" — like common land. In the eighteenth century, the English government caused enormous suffering when it confiscated lands that had been used communally by villagers for at least a thousand years and redistributed them among the aristocracy so they could farm sheep as a cash crop. This was called "the enclosure of the commons." Now we are witnessing the enclosure of the very stuff of life — the enclosure of the gene commons by private business corporations. All the corporations have to do is alter a single gene in a living organism; they can then say it's a new organism, patent it and claim it as theirs.

A key example of how closing the gene commons threatens the sharing of scientific information was reported by the *Los Angeles Times* early in 1999. There is now a strain of *Staphylococcus*, the most common infectious agent in hospitals, that is resistant to every antibiotic. Despite the danger this strain poses to public health, government and university scientists were unable to prise information about the bacterium out of the private companies that were trying to decode its genome. The companies claim they've spent millions, and with a patentable product on the horizon they are not about to give up their information because of public needs.

The delay in making these data public "has slowed research by four or five years," according to Dr. Olaf Schneewind, a leading *Staphylococcus* specialist at UCLA. He says he understands the industry's position, but it doesn't have his sympathy. The resistance genes from this bacterium have the capacity to spread rapidly, and the disease has already killed a man in New York. Government and academic scientists argue forcefully that basic genetic information should be publicly available. Francis Collins, head of the government-funded Human Genome Project, says, "I believe science moves forward in unpredictable ways, and with something as basic as the instruction book of organisms, the more people who have a chance to look at it, the better the likelihood that a key insight will occur. The question we ought to be asking is, 'What is good for the public?' I'm not sure that all this gold rush is going to serve the public's needs very well in the long run."

Because of the Staph crisis and the industry's refusal to help, the U.S. government has had to commit more money to repeat what has already been done and decode the bacterium's DNA. No one thought that if it came down to a public health crisis, the industry would refuse to help. Since the information was available in the private sector, government agencies simply assumed they could get at it if they really needed to. They were wrong. We can only imagine what would happen if private corporations were working on the genome of the West Nile, SARS or any bird viruses in the event of more epidemics.

Force Feeding

"For better or for worse, we were right" is the cryptic summary of U.S. Agriculture Secretary Ann Veneman of the past decade's attempt to force GMOs into the marketplace without consumer labeling or adequate testing.
— TOM HAYDEN, "GLOBALIZATION AND GMOS," *THE NATION,*
JUNE 23, 2003

What does it mean when basic commodities in life, bacteria and other one-celled organisms, viruses and other disease vectors, right up to food crops like corn and canola seed, domestic animals like lab rats, sheep and pigs, and even wild animal and plant species like trees, fish and insects, whose movements and reproduction cannot easily be held in check by humans, become patented, privately owned business investments? On the one hand, it means you can't get at something you need because it's privately owned, as in the Staph crisis. On the other, it means you'll be forced to accept something you don't want, because a private owner has a vested interest in making it your only choice. One scenario for what happens when you genetically engineer a food staple necessary to a continent's survival played out in southern Africa late in 2002. It made for good press; the March 2003 edition of the British magazine *The Ecologist* covered the situation the most bluntly in a cover article entitled "Eat Shit or Die: America Gives Africa a Choice."

When you have a product you've put a lot of research and development funds into, as the biotech industry is always reminding us, you have to find ways to market it and get your money back, even if only on paper. Genetically engineered food crops such as soy and corn have crashed on the world market — most countries won't buy them. Since 1999, Europe and much of Asia's de facto bans on mostly Canadian, U.S. and Australian GE corn, canola and soybeans have resulted in serious market failures for these crops. The United States has lost an estimated $800 million on corn sales alone because of the rest of the world's dislike for genetically modified varieties, just as Canada has lost its pre-eminence in the canola market because it can no longer ensure Europeans its crop is GE-free. This makes these commodities more than available to the U.S. government as "surplus" for their food-aid programs. On paper, the crop is sold to the U.S. government, even if in reality there is no desire for it on the market and it is intended to be given away.

Since 1999, the American foreign-aid arm USAID has been dumping genetically engineered grains that are unmarketable in Europe and Asia on people suffering famine in Ecuador, Zambia, Bosnia, Mozambique or India. This food grain contains live, genetically modified organisms that can invade an area's existing crops and ecosystems and which has unknown effects on human health. So at least four of those countries, India, Ecuador, Bosnia and Zambia, have said "no thanks," preferring to take their chances with hunger rather than eat GMOs. For this attempt at internal ecosystem protection, they have been vilified by the United States as being "immoral" and letting their people starve.

This controversy has heated up as African nations have become more aware of the contamination problems that might face them when these patented genes arrive in their countries. In late 2001 and early 2002, five southern African countries, including Mozambique, Swaziland and Lesotho, reluctantly accepted GM grain aid, the only food they could get on short notice, but asked that it be milled before distribution so seeds wouldn't end up being planted and contaminate local crops. Even countries the grain travelled through asked that the container cars be specially wrapped to prevent spillage of what they regarded as a toxic substance. The reason the African countries were forced to take the GM grain had nothing to do with a shortage of non-genetically engineered grain. A story in *The Guardian* points out that "the U.S. is unique among major donors in that it gives its aid in kind, rather than in cash. The others pay the world food programme, which then buys supplies as locally as possible. This is cheaper and better for local economies. USAID, by contrast, insists on sending, where possible, only its own grain."

During an ensuing 2002 crisis in Zambia, Malawi and Zimbabwe, in which only 1 to 2 million tonnes of grain were needed, more than 2.3 million GM-free tonnes were readily available from neighbouring African countries such as Kenya and Uganda, with an additional whopping 65 million ready and waiting to be sent from India. But USAID doesn't give starving countries money to buy their nearby neighbours' surpluses, which would, of course, also greatly help those neighbours. USAID isn't really in the business of providing aid to the poor. As its web site proudly announces, *"The principal beneficiary of the U.S.'s foreign assistance programs has always been the U.S. Close to 80 percent of USAID's contracts and grants go directly to U.S. firms. Foreign assistance programs have helped create major markets for agricultural goods, created new markets for U.S. industrial exports, and meant hundreds of thousands of jobs for Americans."* Despite this rigid policy of no money for non-GM food, there was cash available for other things, notably 50 million U.S. dollars for Zambia to buy GM grain and 100 million in the general USAID budget to bring biotechnology to any developing country that wants it. One of this aid organization's stated objectives is to "integrate GM into local food systems."

With these kinds of figures, the force-feeding charge is hard to refute, and Zambia's refusal to go for all that food and money is even more astonishing. But, in fact, Zambia did get out of its 2002 drought-related famine without the GM grain, and by early 2003 was having trouble selling its own grain in a local market awash with imports. In March 2003, India also rejected a corn-soy blend for eight of its famine-beleaguered states, for fear this "free food" would contain genetically modified ingredients. Indians were particularly concerned that the emergency food might contain traces of the banned StarLink corn and announced they were prepared "to manage without it."

Global Pressures

> *The fight over GMO food is a major part of the battle against U.S. efforts to dictate policy on all aspects of international trade and development. Does the United States have the power to impose trade terms favorable to itself on thirty-four Latin American countries with 800 million people, producing more than $11 trillion in goods and services? That is what will be debated in Sacramento and fought out [at WTO meetings] in Cancún and Miami. Get ready: the empire is being renegotiated.*
>
> — TOM HAYDEN, "GLOBALIZATION AND GMOS," *THE NATION*,
> JUNE 23, 2003

The notion that you can patent life went global in the mid-1990s. The WTO passed sweeping new rules under the General Agreement on Tariffs and Trade (GATT) that changed lives around the world. Under these rules, countries cannot use national laws to prevent anyone from patenting living organisms within their territories. Chee Yoke Ling, a lawyer and activist who operates out of Malaysia, says, "The whole fight in the eighties and into the early nineties under the GATT trade regime was to have an international agreement on intellectual property rights. Until then, countries had a national choice on whether they wanted to sign an international agreement or not. Whether a patent or a new invention would be protected in my country would depend upon a balance of national policy, public interest and corporate interest, with national laws being passed accordingly. For instance, in many countries, like India, pharmaceutical products were exempted from patent protection because [these countries] felt that for their national public health interest, drugs should be cheaply available. With such regulations in place, thousands of people who died of HIV-AIDS in Africa because they couldn't acquire generic drugs might still be alive. Food products might also be exempted. Because of food security, we would not want to have patent laws that protected one company's seeds over, say, locally developed varieties. But now all that has changed. Under the intellectual property rights agreement at the World Trade Organization at the international level, which all the countries have signed onto . . . we are obliged now to change our national laws and honour all these patents."

Interestingly enough, most of the world's biodiversity — that is, the raw material of genetic resources that biotechnology requires — is in the southern hemisphere. But most of the biotech companies are in the industrialized nations of the North. This is why many in the Third World began calling the patenting of life "biopiracy."

Vandana Shiva, quoted in chapter 5, says biopiracy is nothing more than a modern version of the old colonial regimes. "Patents are a wonderful mechanism of collecting rentier incomes from things that nature does freely, that people do freely. They force people to make payments for what has been theirs, and their right: the public commons. In the 1920s, the British placed a tax on salt, salt-making, as a way of financing their armies. And Gandhi went out and said, 'No, this has been given freely to us by nature. We make it with our intelligence; we need it for our survival. We will not pay you the taxes you want, because you have not created the salt, and we must have it to live.' In a way, life patents are thousands of times worse than the salt taxes of the British regime of those days. It's like patenting air! Because it's the biodiversity in life forms that makes all life possible. It's as vital to life as air itself. It's a condition of our life process. And when our life process and elements of our life process start to get patented, and someone can make money every time we try to use them, it is really the ultimate of organized greed. It has not only generated . . . an ethical crisis, but is really threatening to generate a crisis of survival."

In India, people may be more aware of this than anywhere else. Over the past ten years, Indians have been staging huge demonstrations, have burned test plots of genetically engineered crops and destroyed the offices and storage facilities of the major corporations involved, mostly Monsanto. And out in the countryside, the farmers of India are fighting back in a unique and pre-emptive way. Vandana Shiva runs an organization called the National Program for Conservation of Native Seed Varieties, which fights life patents. She says, "In every village where we work, every farmer we work with takes a pledge: 'This is a legacy I have received from my ancestors, and I make a promise to myself that I will ensure that this legacy passes onto future generations. I will protect it; I will defend it; and I will not allow it to be privatized.' Through this, we are setting up these regions, which we call patent-free zones, where the people basically recognize that these are laws that are so immoral and unjust that they don't have to be obeyed."

This isn't just a Third World problem. It concerns everybody. And it creates ethical as well as economic and safety issues. Because of biotechnology, we have to ask questions we have never before had to face. Who owns trees or seeds or dogs or the secretions of your pancreas? Ferdinand Magellan and Christopher Columbus claimed ownership of the New World simply because they had sailed up to it. Is finding a medicinal plant in a Third World country and claiming ownership of it in order to charge royalties any different? That's exactly what biotechnology companies do.

There is a startling lack of clarity in the patenting regime. As the lawyer and activist Andy Kimbrell says, "They've patented sheep and fish that have been genetically altered, [but] they've patented a number of animals that have not been genetically altered. Rabbits that are used for HIV testing, nematodes and various insects that have been taken from various parts of the world and brought to the United States have all been patented. The only interventive steps have been that they've been taken from one country, where they were indigenous, to another country, where they are not. That's supposed to be the inventive step. So you have what I like to call biocolonization; transport is now the only interventive step you need for the patenting of life." As a result, the U.S. government and a large number of private biotech companies are embroiled in international disputes. India, Ecuador, and Thailand have all been fighting expensive legal battles trying to hold on to their indigenous plants and breeds.

Holding on to What You've Got

> *What enterprise is more noble and more profitable than the reclamation from barbarism of fertile regions and large populations? To give peace to warring tribes . . . to draw the richness from the soil, to plant the earliest seeds of commerce and learning — what more beautiful ideal . . . can inspire human effort? The act is virtuous, the exercise invigorating and the result often extremely profitable.*
> — WINSTON CHURCHILL, *THE RIVER WAR*

> *Obviously, the biotechnology industry is not trying to feed the hungry [or better their lot]. That's just their advertising theme. They are trying to feed themselves.*
> — *BIOTECHNOLOGY = HUNGER*

The most balanced, reputable and globally recognized organization on earth that deals with food production and therefore food security is a branch of the United Nations called the Food and Agriculture Organization (FAO). It hosts a yearly electronic conference on Biotechnology in Food and Agriculture, and in 2001 the meeting concerned the impact of intellectual property rights (IPRs), or what we would term patents, on food and agriculture. There has been a lot of controversy about how modern, chemically based agriculture, biotech and the patenting of animals, food crops and traditional medicines have affected the poor.

The argument from the industries and governments producing these technologies is that they will bring the Third World out of their "undeveloped" stages in order to live better, more like people in the First World. But the FAO conference found that "the majority of participants considered the impacts of IPRs to be primarily negative for the developing world." The participants included university, government and research specialists, independent consultants, and patent lawyers from eighteen different countries, and they made some very interesting points.

One of the first is that the kind of ownership a patent gives a typical biotech company often doesn't have the "aim . . . of commercialization, but rather to stop others from using the technology. It's a negative right. It permits the owner to exclude others from practising the invention." This has led to a very large number of what might be called "preemptive patents" that are having a chilling effect on the advance of knowledge. When a scientist or researcher working for one of the thousands of private biotech companies around the world discovers a new gene or gene process, a way of manipulating a gene or a protein sequence, or just about anything really, regardless of whether he or she can imagine how it can be used, it gets patented. This is a "just in case" measure, a hedge against the slight possibility this tiny finding turns out to be an important component of some future lucrative technology that hasn't been developed yet. The ownership of this material then means, of course, that other researchers can't continue their investigations involving that gene without either paying royalties, or more commonly, at all, since the "owner" will jealously try to guard the property until he or she figures out what to do with it.

Besides this threat to research, there is the question of gross inequality. When it comes to patenting life forms like plants and animals, the Third World, rich in unknown species, is at the wrong end of the biopiracy gun, and is increasingly losing control over its patrimony to private industry. India even had to take a Texas biotech company to court to prevent it from patenting basmati, the traditional rice developed by Indian peasants over thousands of years. The country was only able to save some of its rights. The same thing happened over the neem tree mentioned earlier. Only after expensive legal battles were the Indian farmers who developed the many uses of this plant able to retain at least some of their rights over it. Mexicans were recently outraged when a company patented the traits of a popping bean long used by indigenous groups; they lost their battle in patent court.

In global governance organizations like the United Nations and the World Bank, much discussion is raging over how to ensure that developing countries don't continue to lose this very one-sided legal arrangement, but so far very little has been actually done. Regulating institutions such as the Biosafety Protocol and the Convention on Biodiversity have failed to pass any enforceable guidelines leading to benefits for Third World countries. They also haven't been able to restrain Western capital from gold-hunting speculations on various forms of life. This rush by developed countries for the "green gold" of the poor South, as patentable genes have been called, is reminiscent of the excesses of the Edwardian era when Winston Churchill, quoted above, rubbed his hands in glee over the untapped wealth of the new colony of Uganda.

One FAO conference participant pointed out that the effect of patenting on the agricultural sector breaks "the traditional access and benefit sharing system previously implicit in agricultural research, i.e., that developing countries provide free access to their genetic resources and receive the benefits of the research in developed countries for free (as during the Green Revolution)." This has been replaced with "an asymmetrical system where access to genetic resources was still free, but the benefits of research were not — all access and no benefits." Some participants even charged that patents were "really used by developed countries to dominate and continue their exploitation of developing countries" in a new form of colonization. Many mentioned Genetic Use Restriction Technologies (GURTs), discussed later in this chapter, "that essentially constitute a private regulatory system that bypasses [the entire patenting regime] as well as government authority." Inequalities of "ownership" of these "new resources" are glaringly revealed in the 1999 United Nations Development Program (UNDP) Human Development Report, which shows that developed countries hold 97 percent of all patents. Over 80 percent of the 3 percent of patents that appear to be held in developing countries are actually in the hands of residents in the North, while 95 percent of the patents we're talking about, agro-biotech-related gene patents that control access to food crops and therefore survival, are held by only five top biotech companies. Obviously, the opportunities for the few to control the many are extremely strong in this area. In the light of all this, the FAO conference participants demanded either a completely new patenting system, or serious strategies to minimize its currently overwhelming negative impacts.

This is hardly a minority assessment of the situation by non-industry government and expert observers. Even Ismail Serageldin, a chief economist at the World Bank, highlighted in a July 16, 1999, article in *Science,* the fact that there are serious fears about the way intellectual property laws work that "will lead to the monopolization of knowledge, restricted access to germplasm, controls over the research process, selectivity in research focus, and increasing marginalization of the majority of the world's population." In an unusual, heartfelt editorial vein, he added, "These concerns cannot and must not be ignored." To date, however, nothing has been done about these problems with the present patenting regime, even as they affect research in the First World.

Rights and Responsibilities

> *GE producers . . . should be financially liable for potential losses that farmers suffer as a result of genetic contamination. Seems reasonable, doesn't it? But in the U.S. and Canada, it's the producers who are suing the farmers!*
> — GREENPEACE PRESS RELEASE, JULY 2, 2003

It's clear that holding patents offers substantial benefits to corporations: enormous profits and an almost limitless field of material out of which to create products. So what are their responsibilities? After all, genetically engineered crops and animals could have negative effects, both on the people who consume them and on the environment in general. In the preceding chapter, we looked at only a few of the potential hazards. There are also indications of some negative effects of genetically engineered foods and drugs on human health. We've already described the feeding experiments of Arpad Pusztai. A recent study in Sweden links exposure to glyphosate, the herbicide tied to genetically modified crops, to an increased risk for non-Hodgkin's lymphoma. In varieties engineered to be protected against insects, there is now evidence that the insecticidal Bt protein persists in both soil and mammalian digestive tracts and could have a cumulative poisonous effect. Several groups in Germany are calling for the investigation of the mysterious deaths of a dozen cows fed on GM grain for five years. The grain producer Syngenta has even awarded the farmer partial compensation for his loss. And we haven't even mentioned biological warfare. As long as corporate biotech labs around the world are so uncontrolled and unpoliced, there is nothing to prevent some of them from creating new forms of anthrax or ebola for the terrorist purposes we hear so much about.

If any of these frightening possibilities becomes widespread — and some are already a reality — who is liable for even the accidental damages these new products could cause? We all assume the inventors, those companies that hold the patents and are demanding all the rewards of their research, will be held responsible. In fact, under patent law they are not liable at all. Vandana Shiva has made a study of the situation. "Patent law is a very interesting form of property," she says. "With all other forms of property, you have ownership rights; but along with them, you have responsibilities." If you build a building that falls down or a car engine that blows up, you can be sued by both the buyers and government agencies. This is not true in the case of patent owners. "Patent law is a kind of ownership law in which there is no liability," says Shiva. "There's no responsibility. There are only rights to exclude others from the use of whatever is your patent."

As things stand, the responsibility for any damage that biologically engineered products may cause falls upon the human victims or their governments. In fact, the current wording of the international Biosafety Protocol, which was negotiated under the auspices of the United Nations' Biodiversity Convention, specifies that the importing country, not the patent holder or even the exporter, is liable for any problems that may arise. And if those problems affect nature or wild creatures — well, no one is responsible at all.

Some of the reasons for this situation are those familiar concerns: free trade and global competitiveness. Countries that have become signatories to GATT and the WTO are forced to reduce trade barriers and open themselves to new products in order to remain involved in international trade. One example of what can happen involves the hormones used widely in Canada and the United States to make beef cattle bulk up faster. A European panel on veterinary drugs says at least one of these hormones is carcinogenic. The European Union wants to keep the hormones out of its citizens' diets. The case was taken before the WTO, which found against the EU. In June of 1999 the EU said it would not accept the hormones, so it is subject to sanctions. Ever since, the EU has been losing about $128 million in trade concessions each and every year. If this cost proves too much, and the EU eventually buckles and allows these products in and then discovers that, in fact, they really are dangerous, the manufacturer won't be liable. The regulatory agencies that tried to keep the products out in the first place will be the only bodies that can be held responsible for having allowed them in! It's bizarre. The biotech industry is totally off the hook.

Vandana Shiva points out another reason for the double-think surrounding the issue: "Governments are helping the companies get away with definitions that treat a genetically engineered crop the same as the patent for regulatory purposes. When it comes to getting a patent, you of course don't say a genetically engineered soybean is like conventional soy, but when it comes to safety regulations it's very commonly accepted that it is just the same. And that's the basis on which the U.S. FDA and the Canadian Food Inspection Agency give their clearances for safety. A genetically engineered potato with chicken genes in it will still be treated as a conventional potato, and therefore you will not look at the implications of the chicken gene in it. In fact, you're not allowed to look."

Our entire regulatory system, in terms of genetically modified foods, is based on something called "substantial equivalency." It's not based on any scientific investigation whatsoever, on nutrition or cell differences or on the possible effects of introducing a gene from cauliflower, a virus or bacterium into a food plant. The government doesn't look at anything that complex; if the new potato *looks* like a normal potato, that's good enough to classify it as "substantially equivalent" and exempt it from any tests or labelling. In fact, it's a term that seems to have been invented to decide whether or not to regulate hybrids, which are sterile seeds bred in a traditional way between different varieties of the same plant. The FDA simply decided, unilaterally, to use the same method when biotechnology became an issue, and many other regulatory bodies in countries like Canada have followed suit.

How common is it for international regulatory bodies to echo one another to such an extent? All that has to do with the delicate question of the degree to which Western governments and the biotech industry, especially in the biotech-producing countries, have become entangled.

It's perhaps enough to say that in the United States alone, people have moved from very high-level appointments in government regulatory agencies like the Food and Drug Administration, the EPA and the Department of Agriculture to very high positions in biotech giants like DuPont and Monsanto. Opponents of the technology call this the "revolving door"; people whose careers were made in government regulatory agencies are recruited by Monsanto and vice versa. For example, Mickey Kantor left his job as President Clinton's secretary of commerce and U.S. trade representative and became a member of the board of directors of Monsanto. Marcia Hale, Clinton's former assistant for inter-governmental relations, then took the corresponding job, director of international government affairs, at Monsanto. Linda Fisher, who used to be assistant administrator of the EPA's department in charge of regulating pesticides and toxic substances, became vice-president of government and public affairs at Monsanto.

So ubiquitous is the biotechnology industry's influence on government that there have been controversies and at least one scandal in other countries as well. In Canada in 1999, a Health Canada official claimed that Monsanto offered the department $1 to $2 million on the condition that the company be given approval to market its controversial milk-production hormone, rBGH, without being required to submit further studies. CBC Radio then reported that Agriculture Canada was permitting Monsanto to conduct its own research at government facilities. When asked by Conservative Party Senator Mira Spivak to explain his ministry's relationship with a corporation it is supposed to be regulating, and to disclose how much money the ministry has accepted from the biotech giant, the minister of agriculture refused to divulge the information — on the grounds of commercial privacy.

Yet again, in November 2003, the CBC reported that Agriculture Canada, the government department ultimately in charge of deciding whether the most controversial GM product of all, genetically engineered wheat, is safe for the environment, has itself invested nearly $2.5 million in the product and stands to gain even more money — 5 percent of the profits — if the variety is licensed by its own branch, the Canadian Food Inspection Agency. The Canadian Health Coalition, the National

Farmers Union, the Canadian Wheat Board and many others protested such a blatant conflict of interest outside the minister's office a week later. These groups are also enraged by government response to their fears about the link between fusarium blight and increased use of glyphosate or Roundup, the herbicide that would be used in very large amounts in conjunction with the new wheat. Agriculture and Agri-Food Canada wheat breeder Stephen Fox, who also chairs one of the bodies involved in the registration process of the new Monsanto wheat, says that body won't even look at evidence showing the herbicide fosters this devastating blight: "We only consider the merits of the plant species itself."

Monsanto still indignantly insists that all its products are safe, even after Canada's ban and Europe's moratoriums. But it is instructive to remember that Monsanto Corporation was involved in the production and sales of DDT, Agent Orange and PCBs, which it also defended vigorously before they were banned. Monsanto even launched attacks against Rachel Carson and her seminal book *Silent Spring* in the 1960s, taking out full-page ads that mocked and denigrated Carson. The company continues to produce many substances banned in North America, like Alachlor and the extremely hazardous Butachlor, for export to countries whose laws are not as strict as ours. And Monsanto is a major producer of dioxins; the EPA has estimated that dioxins cause up to 3 percent of all cancers. Monsanto has been charged with manipulating data in studies it funded in the 1980s that exonerated dioxins of all danger. Despite a track record that includes four major convictions for various offences, such as neglect and widespread dissemination of misinformation, and including a $108 million liability finding in the case of the leukemia death of a Texas employee — despite ranking by the EPA as one of the largest corporate generators of toxic emissions into the U.S. environment and an increasingly shaky business situation — Monsanto continues to have influence in high places.

As more and more products of such chemical and biotech companies are introduced into the food chain without rigorous government testing, it will become very difficult to trace any negative effects back to their cause. Vandana Shiva says that "not only are we playing around with a very, very ecologically risky technology, we are playing with it in a context where we are prevented from knowing...what its risks might be. And we won't even know ten years down the line, because the early links in the chain of knowing are being wiped out. They're being erased."

Alice in Monsantoland

Protection of intellectual property is clearly not the only, or perhaps even the main reason for Monsanto's strategy in threatening to bring thousands of farmers into court. The purpose is to generate fear, and hence, compliance, in order to force the purchase of more GM seed.

— E. ANN CLARK, PLANT AGRICULTURIST

In 1998 Saskatchewan native Percy Schmeiser was coming to the end of a long career as a third-generation farmer in a difficult terrain, the Canadian Prairies, when he suddenly found himself being sued by Monsanto. The company claimed he was "illegally" in possession of canola carrying the gene conferring resistance to its herbicide, Roundup. Monsanto's team of roving inspectors, following a tip-off, insisted it had discovered the company's patented grain growing on Schmeiser's land. To an outsider, this lawsuit would seem to hinge on the idea that Schmeiser had unlawfully acquired the patented seed and had planted it without paying the required royalty to the "owner," Monsanto. After fighting the case through provincial, federal and federal appeals courts for five years, Schmeiser is now facing fines of $15,450 ($15 an acre), the value of his crop, $105,000, plus $25,000 punitive and exemplary damages. Court costs add another $153,000 to these hefty fines. Schmeiser also had an entire year's crop — and a lifetime of seed-breeding — confiscated by Monsanto. Pretty stiff penalties for the theft of some seeds. But what's most surprising of all is the fact that what he's been convicted of wasn't at all as simple as stealing patented material.

Percy Schmeiser is not a babe in the woods. He has been a politician, a Member of Parliament for the Reform Party and a local mayor and runs a successful farm-equipment dealership in Bruno, Saskatchewan. Many of his neighbours consider him a person more than able to look after his personal interests. Even though he's not entirely the simple, innocent farmer caught up in the intricacies of the modern world that the media have portrayed, his conviction still constitutes one of the most distressing legal precedents in Canadian history. Schmeiser was never convicted or even accused of stealing Monsanto's GE canola seeds. The grounds on which he was fined so heavily were simply that some plants containing Monsanto's patented genes were found in his fields — and he hadn't informed Monsanto. Whether they got there on purpose with the farmer's knowledge and assistance or not, the judge ruled to be immaterial. Schmeiser's position was that he didn't and, in fact, couldn't benefit in any way from the patented seed and therefore had no motivation to steal or plant it. He contends he didn't inform Monsanto because he thought he was the aggrieved party, since his crop had somehow been contaminated with the company's material. His grain had been contaminated with Roundup Ready canola, which he unwittingly sold as ordinary grain. Sales records prove he only sprayed Roundup, like many other farmers in the region, around his telephone poles. In fact, that's how he discovered his fields, seeded with the stock he himself had developed, were contaminated. Too many canola plants remained near the poles after the spraying.

In such a case, a Monsanto client is supposed to call the company and tell it to remove the volunteer plants polluting the fields. But if he does so, two rather nasty scenarios will unfold. The worst is that the company will decide to sue him for having their plants on his property in the first place. The lesser evil is to have Monsanto's people running through the ripening crop and tearing up plants to determine which have the Roundup Ready gene in them. Since there's no visible indication of the Roundup Ready condition, most of the crop would have to be destroyed, but there is no mechanism to recompense the farmer for the loss. That's why most farmers, even if they do notice, don't report pollen drift, which can occur over miles; spilled seed; leftover seed from other plantings, which can survive up to eight years in the soil; and the many other accidents that can contaminate crops.

Farmers would keep Monsanto very busy if they did report such things. Roundup Ready canola has become so ubiquitous that it seems to pop up everywhere, not just in pure variety canola fields, but in crops of corn, soy and wheat. In fact, it's now considered one of the most serious weed problems in western Canada. Monsanto's patented gene could be just about anywhere. That's why Monsanto, as reported in

"Blowin' in the Wind," a special report on CBC's *The National*, installed a hotline, where neighbours can turn one another in for a reward, and also hired roving "investigators." These investigators not only invade private property to test for the presence of the company's genes, but have been charged with overflying with helicopters and dumping Roundup, a systemic herbicide, on suspected farmers' crops to see if their plants will survive the assault. If canola plants are growing a few days later, Monsanto's investigators pay a visit to the "guilty" farmer. If he isn't guilty, an innocent farmer loses a big chunk of his year's crop. (See www.tv.cbc.ca/national/pgminfo/canola.)

Jeremy Rifkin prophesized in 1998 that agricultural biotech would fall by the wayside. Impossible numbers of legal cases would clog the courts, and he assumed this would cost the patenting companies millions in liability suits. Right now, although seventy-three farmers in the United States alone are countersuing Monsanto, a figure that doesn't include those who simply paid up, it has been Monsanto that got the ball rolling by suing them. Monsanto, at least, has found a way to make the victims of genetic contamination pay for the privilege, and for a very wealthy corporation, the costs of lawyers and legal wrangling are part of the cost of doing business. Individual farmers have to assume enormous costs by themselves. Plant scientist E. Ann Clark points out, "*Monsanto is the only biotech company that sues for patent infringement* [italics ours]. The others use other strategies, for example, embedding their patented traits in hybrid canola, which the farmer cannot save as seed anyway, to safeguard their patented genes." There is, in fact, only one way a western Canadian farmer who isn't currently growing GM canola can escape being harassed or sued by Monsanto when those genes inevitably show up on his property. He has to start growing that variety every year, no matter what Monsanto decides to charge for it.

It is a kind of madness that has driven Percy Schmeiser, now in his seventies and facing financial ruin, to continue his fight. The Canadian Supreme Court began hearing his case in January 2004. It was a shock that such an iniquitous case would be lost before the federal appeals court, but there is some hope in the fact that Canada's Supreme Court refused Harvard University's patent on a mouse genetically engineered for cancer research, the famous "onco mouse," on the grounds that a "higher life form" (i.e., a mammal) can't be patented. More to the point is a request from the Canadian Biotechnology Advisory Committee (CBAC), which has tried to institute some form of labelling in the country. In a statement, the committee asks that "the Patent Act be amended to permit farmers to save and plant seeds from patented plants." Even before Schmeiser's appeals decision, the CBAC affirmed that farmers like him, who end up with GE plants in their fields "through the adventitious spreading of patented seed or patented genetic material," should be considered innocent bystanders and not be liable for prosecution.

Sterilize the World

> *The aim of science is not to open the door to infinity, but to set a limit to infinite error.*
>
> — BERTOLT BRECHT

Biotech companies continually try to find ways to ensure their products don't escape their control, making certain they won't have to deal with both patent infringers and outside research. Ismail Serageldin, the FAO conference and all the involved anti-GE NGOs mentioned earlier are particularly concerned with one method the biotech industry is employing to get around these problems. It's pushing Genetic Use Restriction Technology, or GURT, the most infamous example having been aptly dubbed the "Terminator."

This technology has such devastating ramifications that attempts at commercialization were withdrawn in 1998, following a public outcry. However, research, much of it secret or untraceable, has continued, and recently more patents have been applied for from a variety of companies using a range of techniques. Basically, the seeds harvested from patented Terminator crops are genetically engineered to be sterile so they won't germinate the next season. They contain a kind of suicide gene that destroys their fertility. Some of them can be activated to produce crops that will become infertile with the application of a particular chemical during the growing season. Others are infertile to begin with and will only germinate the first time if the farmer applies another such chemical, called a "safener," which includes substances such as flurazole, napthinic anhydride, dicyclonon, oxabentrilil, fenclorim, cyometril, fluxofenim, furilazole or dietholate. Obviously, if GURTs are commercialized, many litres of these chemicals would be dumped onto fields and find their way into soils, the water table and food. As an Institute for Science in Society report dryly puts it, "There do not seem to be many publications reporting on safety tests on the safeners." The plants on which Terminator-type patents are pending include corn, potato, tomato, oilseed rape, most of the cabbage family, alfalfa, sunflower, cotton, celery, soybean, tobacco and sugarbeet — in short, most of our major food crops.

Dr. Mae Wan Ho, a geneticist teaching at the Open University in London, says such complicated chemical/genetic constructs, "all of which have to be precisely engineered and integrated into the plants exactly as intended . . . [are] beyond the capability of current technology. . . . Genetic engineers cannot control where they are integrated, either, thus multiplying the uncertainties and unpredictability of the GM crops produced." A major fear is that pollen or seed outcrossing would occur, making non-GM and organic varieties in the vicinity sterile. After thousands of years of attempting to increase our food supply and the flexible characteristics of each species of crop plants, it would be as suicidal as the Terminator gene to tinker with the fertility of nature. But the industry is so anxious for ownership of this material that it seems quite willing to do so. Perhaps from its point of view, it will only kill off competing products, i.e., other types of seed.

Dr. Martha Crouch, a biologist at Indiana University and author of the most accessible paper on the subject, says, "It is likely that Terminator will kill the seeds of neighbouring plants of the same species under certain conditions [wind or insects carrying the pollen]." The hapless owners of the newly sterile seeds will only find out the following year when they attempt to plant them and they don't germinate. "If many seeds die," she says, "it will make saving seed untenable for the adjacent farmer. Even if only a few seeds die, they will contain the toxin and any other proteins engineered into the GURT variety. These new 'components' may make the seed unusable for certain purposes."

Crouch points out that the effects of the toxin that kills the seed are serious. We don't know how toxin-laden seeds will affect the birds, insects, fungi and bacteria that eat them, to say nothing of stock animals and humans. An ornamental Terminator-type sunflower "could spread the gene to an oilseed variety, and then the toxin could end up in edible oil or in sunflower seed meal." Studies of allergenicity for such toxins have not been made, and as we've seen with other GE products, are very unlikely to be adequately tested for or guarded against. Mae Wan Ho also mentions two other possible dangers: "The Terminator lethal genes and gene products are known to be harmful to cells, including mammalian cells. Some of the most hazardous of these genes are designed to spread through pollen." That means simple human and animal health hazards are present and largely uncontrollable. Moreover, GURTs use the antibiotic tetracycline as a repressible promoter system to interact with the new recombinase gene. Use of this antibiotic declined enormously when it was discovered it had deleterious effects on human skin, teeth and bones, especially in children. But Terminator technologies would see it spread throughout the environment, as every seed will have to be soaked in the antibiotic to "set the cascade of toxin-gene activation in motion."

Up until now, for millennia, we've always eaten, stored and exchanged *live* seeds. Crouch asks, "Will the dead seeds be more or less easy to store? Perhaps they will respond differently to changes in humidity, or to infection with bacteria and fungi . . . . There also may be [significant] nutritional changes in seeds that are killed late in their development." Crouch points out further that even the industrial rationale for the GURT seed — to protect its owners' patent — may not work out, either, as treatment of each seed would have to result in 100 percent sterility in order to avoid producing pollen that would have a non-functional toxin gene. That extreme level of performance is impossible to imagine for any human technology, and with engineered *living* organisms there are bound to be surprises.

Crouch, like Ho, is a geneticist who is not dependent on industry funding. She fully expects that "the genetic components of Terminator will reshuffle during sexual reproduction" and produce any number of unintended changes, not just in the seed, but in other parts of the plant. One possibility is that "if plants were to carry silenced toxin genes . . . those genes might suddenly be activated again, causing seeds to die unpredictably in subsequent generations. By the time the phenomenon occurred, however, it might be difficult to ascribe the cause to Terminator." That means we could find ourselves in a future where very few of our food crops would grow and produce without the application of dangerous and expensive chemicals. Our choice of seed, once based on thousands of varieties carefully selected and bred by people all over the world, could be reduced to only those made by a few chemical companies. And if, for whatever reason, that seed fails, we wouldn't be able to go back to any pure, unadulterated or even wild varieties of our food plants, because most of those varieties would be dead or unreliable, infected by Terminator genes. And that's why, like many other concerned scientists, I, too, have signed the very strongly worded ISIS document demanding that "this highly hazardous and immoral development . . . must be stopped, and all Terminator crops that have been released . . . or are undergoing field trials must be recalled and destroyed."

Liability and Insurance

> *Genetically modified foods are among the riskiest of all possible insurance exposure that we have today. . . . And there is good reason. No company knows where the path of genetically modified foods is ultimately going to take us in terms of either human health or environmental contamination.*
>
> — ROBERT HARTWIG, CHIEF ECONOMIST OF THE INSURANCE
> INFORMATION INSTITUTE

We hear a lot about national security — efforts to protect the population of Western countries from biological agents and "weapons of mass destruction." So who or what is shielding us from the possible dangers of an almost unregulated biotech industry? Could we control one of these sterilization methods if it got out of hand and threatened to wipe out an entire species? If one of the new engineered food crops ends up with unexpected side effects, could we trace the source of the problem? Two points are very much at issue here: first, patent rights are not accompanied by social responsibilities; and second, the very nature of patenting involves industrial secrecy. As a result, obtaining detailed information about a product and its possible dangers would be as difficult as identifying and controlling the source of the problem.

We have been through similar situations already, and we must not forget what happened. The Union Carbide India Limited chemical leak at Bhopal, India, in 1984 was just such a disaster. Unlike the biotech companies, Union Carbide was insured, but the victims were never adequately compensated. India itself shouldered most of the burden of this catastrophe, for the simple reason that India was where the deadly leak took place. The factory was manufacturing a common pesticide, but the people trying to help the victims couldn't get full information about the chemicals involved because the poison was a patented, industrial secret. Chee Yoke Ling, a lawyer with the Third World Network, is all too familiar with this kind of story. "We know from a whole history of consumer rights and environmental rights that if you don't attach liability to the release of a product, when things go wrong, companies say, 'Oh, we're sorry. We didn't know it would do that.' Or they just keep quiet. They go on to the next product; they go on to the next country. Who has to clean up the mess? National governments are usually stuck with it."

The doctors working with victims during the Bhopal disaster needed all the information they could get. If they'd had it right away, they might have saved more lives and prevented more people from going blind. But a corporation doesn't have to reveal patent information, even to a national government. Chee Yoke Ling knows patent law, and she maintains that without strong regulatory controls companies left to police themselves will have the same response in a genetic emergency as Union Carbide India Limited had at Bhopal. "They will say, 'Oh, it's confidential business information. It's protected by a patent.' Crucial information that [is needed] . . . to make an assessment of whether a product is safe or not will not be there; most of the time, it will be blocked by the patent."

The biotechnology industry often portrays its critics as hysterical. These things haven't happened yet, so even talking about them is said to be overly pessimistic in the face of the bright biotechnological future. Unfortunately, history shows us that all kinds of innovations such as petrochemicals, CFCs, new drugs and nuclear power, heralded as perfectly wonderful new technologies, have turned out to be extremely dangerous. History informs us that caution is well warranted when it comes to buying into a powerful new technology. That's why the regulation of such materials was created in the first place, and we have to remember not to wait until suffering begins before we legislate. Yoke Ling says, "Strong consumer laws in countries like the United States or Canada. . . came because of disasters. . . . history shows us that liability is absolutely crucial to balance the rights of different sectors of society. Strong liability laws are also a very [powerful] incentive for companies to have high standards of behaviour, because otherwise they will have to pay the price."

How do commercial interests usually protect themselves from liability claims? Through insurance. In fact, in our society, the litmus test for safety *is* insurance. You can be insured for almost anything if you pay enough for the premium, but if the insurance industry isn't willing to bet its money on the safety of a product or a technology, it means the risks are simply too high or too uncertain for them to take the gamble. What does this mean for biotechnology? Suppose a modified virus escapes into the environment and begins replicating. It wouldn't be like other forms of pollution because it couldn't be cleaned up like an oil spill. It wouldn't even degrade over time like nuclear waste. And its interaction with other organisms in the environment would be completely unpredictable.

Brian Goodwin is a theoretical biologist at Schumacher College in Devon, England, and an independent scientist — that is, he doesn't depend on private industry for grants or position. His independence and biological expertise have made him a major voice in the United Kingdom in discussions about how to regulate this new biological technology. He reports on the alarming fact that "insurance companies will not sell insurance to biotech companies, because they know that the results of using genetically engineered organisms cannot be predicted. There is no satisfactory risk-assessment procedure. And the fact that insurance companies won't undertake this shows that this is a very, very unreliable technology."

Jeremy Rifkin concurs. "The fact is, and I think that the public doesn't know this, there is no insurance against catastrophic loss from introducing these organisms anywhere in the world. The insurance industry has spoken on this. They're not willing to place insurance premiums on the potential for loss in introducing these organisms. The reason is that there is no predictive ecology. There is no risk-assessment science by which to judge what might happen when you place one of these organisms in the environment. The insurance industry understands that, and that's why they're not willing to insure the products of commercial biotech."

In Canada, the National Farmers Union (NFU) is demanding that Ottawa make agro-biotech firms liable for damages caused by genetic pollution; otherwise farmers believe they will have to shoulder the liability themselves. The NFU's Saskatchewan coordinator, Stewart Wells, says that he and other organic farmers could lose their certification and their livelihoods as a result of contamination from pollen drift. He points out, "Tax money helps their research, but the profits are always privatized." Agronomist E. Ann Clark says that Canada is typical in that $700 million of taxpayer money is going annually to further biotech research, while almost none is earmarked for risk assessment. And many governments are so involved with the technology — Britain, Canada, and the United States — that it's debatable whether they are still capable of an objective search for possible problems.

Jeremy Rifkin is so sure horrifying scenarios are coming that he thinks genetic engineering in agriculture will be forced out of business because of its extreme danger. "I suspect that we'll end up rejecting genetic engineering in agriculture," he predicts. "We'll end up rejecting the release of massive numbers of organisms into the environment, because we're going to learn in the next few years that the environmental risks of extending genetic pollution across this planet far exceed any short-term benefit we might get from placing any of these organisms into the environment in the first place."

I am a geneticist who remains enthralled by the technological dexterity and insights inherent in splicing genes. But there are so many uncertainties in the process at this point in time that I can only hope we won't see a man-made plague on the order of other horizontal gene-transfer plagues of the past or suffer the loss of essential foods or animals that we can never replace.

What Are We Eating?

The hope is over time the market is so flooded [with GM products] that there's nothing you can do about it. You just sort of surrender.
— DON WESTFALL, VICE-PRESIDENT OF PROMAR INTERNATIONAL

Our health and well-being are directly related to the quality of our food. After all, when we take nourishment into our bodies, digest it, then incorporate it into our cells, food becomes us. Some of the main products of biotechnology are what the industry calls novel foods. Genes from a wide array of sources — viruses, bacteria, plants, mammals, reptiles, fish — are being put into our food crops. We've always assumed that regulatory bodies like the FDA in the United States and the Canadian Food Inspection Agency exist to ensure the safety of our food. Furthermore, as citizens of a democratic society, we assume that we have the right to choose what to put into our bodies. But in order to do that, we have to know what's in the products we may eat.

Andy Kimbrell has spent a decade pressuring the FDA to label genetically engineered food. He is angry about their response. "The FDA, in truly one of the major sell-outs to corporate interests that I've witnessed in my many years here in Washington, D.C., decided that the foods that have been genetically engineered should not be labelled, that they should not even have to go through what's generally regarded as safe certification, which is the basic requirement when you try to put a new element into the food chain." Kimbrell thinks it's very important, in terms of both health issues and ordinary democratic rights, that people know what's in the food they're eating. And as he says, "If the biotechnology industry thinks its food is safe or its food may be better than the food that is out there, well, it certainly has the money to advertise." But right now, in the western hemisphere, the fact is that there are dozens of novel engineered foods out there — engineered soy in our cereal, engineered canola in our mayonnaise, engineered potatoes in our fries — that are not labelled.

Andy Kimbrell told a story in 1998 that suggests that in Canada and the United States we shouldn't assume our interests are being taken care of. "In perhaps one of the most shocking examples, we discovered that the United States Department of Agriculture, without telling the public, allowed pigs into slaughterhouses and into the food chain that had been involved in experiments making them transgenic. These were animals that had foreign genes in every one of their cells, and that had been part of experiments by major researchers and corporations. . . . These were animals with human genes; these were animals that had a variety of crippled viruses in them. They did this without consulting Congress. They did this without making it public. And this is really a shocking example of how an agency is doing nothing to protect the public interest, and absolutely everything to protect private interests."

There have been two equally ominous cases since then. Six years after the first transgenic pig fiasco Andy Kimbrell described, several hundred transgenic pigs were again butchered and turned into sausages. These GMO-carrying pigs were created at the University of Florida and were sold in 2001 into the local food chain, even though, besides being transgenic, they had also been treated with barbiturates and chloroform. Most of the press seemed to think the incident was funny and ran joke headlines like "Wurst-Case Scenario." The number of people who consumed the transgenic pigs was few, and the incident was completely ignored by the FDA. But in February 2003, the University of Illinois was reported by the FDA itself to have marketed almost 400 pigs, all the progeny of transgenic animals. The FDA admitted the pigs hadn't been adequately tested for the presence of the introduced DNA but still claimed the incident was "isolated," and moreover, that "transgenic products [posed] no health risks." Andrew Pollack of the *New York Times* learned that some of the pigs had been engineered to express a gene for insulin-like growth factor, a well-known cancer promoter. ISIS pointed out that "cancers associated with high insulin-like growth factor include colorectal and breast. The transgenic pork poses a real and immediate danger of cancer promotion for consumers."

What does all this mean? As we feared when the issue first came up, our societies just don't have the sustained political will or the limitless wealth and expertise to constantly ride herd on such complicated and dangerous technologies. In fact, after the latest transgenic pig event in the United States, the FDA threw up its hands and dumped its regulatory responsibilities on academic researchers! In a very quiet announcement, as of May 13, 2003, all Land Grant universities, which do the majority of livestock research, are now responsible for overseeing GM plant and animal development. Melanie Loots, associate vice-chancellor for research at the University of Illinois, says, "Universities need to be aware that we [now] have a monitoring responsibility. How much of a burden that responsibility places on us remains to be seen." Her words were diplomatically measured, but a government that relinquishes its basic regulation duties to academic institutions with no experience, no authority to implement controls regarding the findings of its monitoring, and especially no resources to perform all this, is a nightmarish situation with few precedents in the modern world.

Many of these mounting regulatory problems reflect a clash between consumers' anxiety to keep tabs on the kinds and amounts of transgenic material they may be ingesting and the governments' unwillingness to help them do so. This could be easily addressed by the strict labelling that is law now all over Europe and practised in many other countries as well. Then, at least, allergenic reactions or ecosystem problems could be more easily traced, and GE technology could be either pinpointed or exonerated. But that solution hasn't appealed to North American governments. Canada has begun to institute a "voluntary" labelling guide, while the United States has nothing in the works. It's not that difficult or expensive to implement; much poorer countries, like China and Malaysia, are doing it. So what reasons do the FDA and the Canadian Food Inspection Agency give for this particular dereliction of public duty? In the late 1990s, the biotech industry and the U.S. and Canadian governments claimed it was impossible to determine whether a seed or a foodstuff contained any transgenic DNA. That might be true for sight, smell or taste, but it's always been easy to tell the difference by using a simple chemical test. Otherwise, Monsanto and other biotech companies wouldn't be able to tell who was using their patented products and be able to sue interlopers, or know if they had succeeded in invading one genome with another.

One reason why Europeans dislike GM crops is that they contain "antibiotic markers," genes with antibiotic resistance that are inserted into the genome of the new crop, along with the genes carrying whatever trait the patenter wishes to introduce. The biotech scientists merely drench the DNA-treated cells in antibiotics, and if any survive, then they know that the organism they are trying to alter carries both antibiotic-resistant markers and the desired trait that is linked to them. Europeans are concerned that the widespread insertion of antibiotic-resistant genes in our daily food will spread these resistance genes and increase the amounts and kinds of bacteria that are immune to antibiotics. But the fact remains that antibiotic marker-genes are present in nearly every biotech product, and that's how you can tell GE from non-GE food.

The biotech companies also said it was impossible to segregate non-GM from GM grains, because both are milled, hauled and stored in the same buildings, trucks and silos. However, when the Europeans slapped a de facto moratorium on GM food in 1999, even U.S. grain elevator operators quickly learned some lessons in segregation, in a desperate effort to retain their markets. Food processing giants such as ADM and A.E. Staley refused to deal with any soy or corn that wasn't GE-free. We already had our grains mixed up in North America, because here the biotech industry, with the help of our governments, started putting small amounts of GE material into our diets in the mid-1990s. When the industry tried to do the same thing in Europe in 1996, all hell broke loose.

That year European activists discovered that the normal soy shipments on their way to Europe had been intentionally mixed with genetically engineered soy, which was how the material had silently been introduced into the North American food chain. But Europeans didn't stoically accept their new diet. Greenpeace members across the continent chained themselves to mill doors, blocked shipments in the harbours with inflatable boats and dumped five tonnes of soybeans in front of the doors to Monsanto's Belgian office. Friends of the Earth joined in similar tactics and especially encouraged supermarket boycotts; in Germany, farmers and rural citizens pulled up so many transgenic plants in experimental fields that the plots had to be protected with armed guards. In England and Ireland, civil disobedience against GM resulted in several show trials that were extremely damaging, not only to the industry, but to the governments. Large sections of Italy, Austria and other countries pledged to keep GE products out of their food and fields. French farmers broke into a GE warehouse and ruined the crop by urinating on it. In less than two years, the food giants Nestlé and Unilever, as well as all the major supermarkets in the United Kingdom and many more on the continent, had pledged to refuse to market any genetically modified products or ingredients, and EU governments were forced to impose an unofficial moratorium on GE products.

So the current, progressive EU labelling law isn't the result of enlightened governments protecting their citizens. On the contrary, in Britain and Germany especially, governments did everything they could to convince people that GE foods were safe and that their mass inclusion in national diets was a non-issue. But the real sticking point is the continuing lack of research using animal subjects and real clinical trials that would demonstrate the safety of these new products, both to the consumer and to the environment at large. This lack of proof of safety, plus all the suspicious secrecy surrounding the introduction and presence of GE foods, has only stiffened the European public's distrust.

In Canada and the United States, a lack of the same kind of passion about food safety and quality, possibly because until 2003 North Americans hadn't had to deal with mad cow disease, has resulted in less of a public response. The entire issue of labelling hinges on the FDA ruling that transgenic food is "substantially equivalent" to regular food. The original decision was based on how hybrid crops were assessed and was simply extended to include GE products. Thus, governments in the United States and Canada have been able to maintain the remarkable fiction that even though a certain food contains living, genetic material, viruses, bacteria and other DNA from other plant and even animal species, that is, because a GE potato, soybean or ear of corn still *looks* like a normal potato, soybean or ear of corn, it doesn't have to be tested or labelled. On the other hand, when a food does contain that new gene or DNA from another organism, it's trumpeted as an important new invention and its proprietor has the right to collect a royalty every time the "product" is used and can sue anyone who tries to use it "unlawfully." In other words, the "new" food is exactly the same, except when it's different. It's an Alice in Wonderland scenario where our various governments' claims of protecting the food supply have become very surreal.

If You Don't Look, You Don't Find

> *The fusarium fungus can produce a range of toxins that are not destroyed in the cooking process, such as vomitoxin.... More lethal compounds include fumonisin, which can cause cancer and birth defects, and the very lethal chemical warfare agent fusariotoxin.*
>
> — JEREMY BIGWOOD, FUNGI EXPERT

As Michael Sligh of the Rural Advancement Foundation International (RAFI), the NGO headquartered in North Carolina, says, "it's very easy for the industry to claim, as it does regularly and loudly, that there has been no harm proven to be caused by agro-biotech crops. That's largely because in the main originating country for this technology, the United States, there has been no risk assessment done on this stuff at all. That makes it possible for them to report that there's no evidence of harm from their products; from a public point of view especially, no money or effort has gone into looking for any. Only very recently are they allocating a whole 1 to 3 percent of an enormous biotech promotion budget to go into risk assessment. But I haven't heard if even that is going into real double-blind, animal-use critical trials."

What scandals have hit the news — allergens in StarLink corn, the death of beneficial insects like lacewings and monarch butterflies because their food sources are being killed or contaminated by genetically engineered plant pesticides, the contamination of Mexican land race corn with GM traits, Arpad Pusztai's news of brain and organ shrinkage in rats fed GM potatoes — have all come from independent and even accidental studies, most of them underfunded and often suddenly curtailed after their first announcements. All of them have been vigorously attacked by the biotech industry and even by their own publishers and funders. Only the StarLink corn allergen and the contamination in 2003 of food crops with dangerous GE pharmaceutical crops have resulted in regulatory action by any government. In the case of the latter, as we described earlier, once the infected material was disposed of, the FDA turned around and dumped any future responsibilities on universities and academics. So whenever reports of harm do surface, it's always interesting to see if they can survive the combined pressure from government and industry sources that will be arrayed against them. The latest one to make the news, however, will be very hard to ignore.

Andy Coghlan, a reporter for one of the world's leading sources of scientific information, *The New Scientist,* excited media outlets in August 2003 with a story entitled "Weedkiller May Boost Toxic Fungi." In it he describes a five-year study of plant diseases headed by Myriam Fernandez of the Semiarid Prairie Agricultural Research Centre, run by Agriculture and Agri-Food Canada in Swift Current, Saskatchewan. Fernandez was quoted as saying that "in some fields where glyphosate (Monsanto's patented Roundup, used in combination with their genetically engineered crops) had been applied in spring just before planting, wheat appeared to be worse affected by fusarium head blight." This disease, which turns the grain heads pink, destroys 20 percent of wheat in Europe and cost Manitoba farmers, for example, over $100 million in 2002 alone. This blight doesn't just reduce yields; it produces a vomitoxin that makes infected crops "unsuitable for animal or human consumption." Fernandez's colleague, Keith Hanson, said that they had found "higher levels of blight within each tillage category when glyphosate had been used." In lab work, the researchers also discovered that two of the most damaging forms of the fungi, *Fusarium graminearum* and *Fusarium avenaceum*, "grow faster when glyphosate-based weedkillers are added to the nutrient medium."

This is extremely hot news for anyone using the many Roundup Ready crops on the market. It means that a dangerous fungal disease may be pushed out of control in the soils because of some kind of imbalance caused by the escalating application of this wide-spectrum herbicide. The news has already provided ammunition to the many Canadian groups fighting the introduction of GE wheat, which could become even more susceptible to blight if these findings are verified. With billions of dollars at stake in its product, Monsanto has rejected the newest studies, claiming the results were preliminary and that "there is a high level of confidence that the use of Roundup . . . does not have any negative impacts on soil microbes." However, over the past two decades, scientists from New Zealand to Africa have noticed a glyphosate-fusarium relationship in independent, small-scale experiments and have reported the results in almost fifty scientific papers in academic journals. Jeremy Bigwood, an American expert on fungi writing for the InterPress Service (IPS), reports: "They all describe an increase in fusarium or other microbes after the application of glyphosate."

Robert Kremer, a soil scientist at the University of Missouri, is researching the glyphosate-fusarium relationship in soybeans. His experiments with GE and regular soybeans revealed that Roundup "seems to stimulate fusarium in the plants' roots to such a degree that . . . *the elevation of fusarium levels [seems] to be glyphosate's secondary effect.*" Although the blight only showed up in soils and not on the soybeans in his experiments, Kremer fears an accumulation in the soil that could produce an epidemic wave flowing from one field to another. His fears are increased by the fact that fusarium isn't just an economic worry for farmers or a temporary inconvenience to consumers. When the mould passes into the food chain undetected, fusarium can kill. In 2002, at least four farmers in Iowa reported a mysterious 80 percent breeding failure in their sows, soon after beginning to feed them GE corn. Other possibilities, like disease and insemination problems, were ruled out, but when the corn was tested, it was revealed to have high levels of fusarium mould. As soon as the farmers switched back to conventional corn, the sows regained their fertility. Contamination with fusarium was considered responsible for thousands of human deaths in Russia during the 1940s, and as recently as 2001, fusarium "caused a series of deadly birth defects among tortilla-eating Mexican Americans in Brownsville, Texas, after the blight had contaminated corn" (see the previously mentioned IPS story).

Despite the important impact of this story, however, or perhaps because of it, only five days after the article first appeared in *The New Scientist,* a CanWest News Service reporter was frustrated by project leader Myriam Fernandez's refusal to discuss her group's findings any further. "We don't want to create more hassles," she said. "The data on the project should not have been discussed." Since the project is funded by Agriculture Canada, the reporter went to the ministry, where a spokeswoman said their concern is that Fernandez wasn't consulted before publication, and that they were drafting a letter urging *The New Scientist* to correct its story. It remains to be seen who was getting hassled and what kind of corrections will be demanded. René Van Acker, a university of Manitoba plant scientist who co-authored a recent report on Roundup Ready wheat for the Canadian Wheat Board, explains that, "If you asked a farmer near Winnipeg, what's your number one concern in wheat, they'd probably say 'fusarium.' That's why it's touchy."

Experts do say Roundup is less damaging to the environment and soils than many other systemic herbicides. They may emphasize the fact that Agriculture Canada and Syngenta, a biotech company headquartered in Switzerland, are both trying to develop wheat that's resistant to fusarium-head fungi. But that, as Michael Sligh of RAFI would say, misses the point. The fusarium increase in soils is most likely to be an indication of a fundamental and dangerous imbalance caused by the new practices of using herbicide so lavishly all season long. Letting such activity continue and simply putting a Band-Aid on one or two crops is hardly the way to deal with a potentially deadly food contaminant. Sligh says, "The new fusarium problem seems to be based on very good data, and I think it's just the tip of iceberg. Here's that harm they claim they can't find, finding them. You can't deal with it by developing ever more expensive, complicated seed-and-pesticide technologies. But they'll try, because ultimately, biotech is simply an extension of the old pesticide treadmills. Destroy one pest and create a disease; engineer out the disease and cause a soil problem; and so on. It simply fails to appreciate the bounds of nature and the holistic nature of agriculture." What he means is that soil organisms, plants, animals and environmental conditions all work together in incredibly complex ways that we don't understand to produce either food — or disease.

At this time, with so little knowledge of what we're eating and what materials are used in food production, if thousands of children were to end up being born with birth defects, would we be in a position to trace the causes so that we could exonerate genetically engineered food and the processes that accompany it? Without good research and strict labelling, certainly not. As Brian Goodwin puts it, "You would never allow a new drug to be produced without a clear label, without knowing what company produced it, without knowing exactly where it was produced and even under what conditions, what batch it came from and so on. Genetically modified foods ought to be put in the same category as drugs because of their potential harm. They're actually even more dangerous than drugs, because after all, we eat a lot more food during the course of our lifetime than we take drugs. Even if there are small effects, they can accumulate over years. And therefore people should have the right to say, 'I'm not going to eat genetically modified food because I have no confidence that this is going to be safe for the whole of my lifetime.' Now that's a kind of baseline, it seems to me, for consumers."

Bioweapons

We have reached a point now where there is an obvious need
for an international convention to control biological weapons.

— SPOKESMAN FOR PORTON DOWN,
BRITAIN'S BIOLOGICAL DEFENCE ESTABLISHMENT,
QUOTED IN THE *TIMES* (LONDON), NOVEMBER 1998

So far we've been talking about the unpredictable behaviour of organisms that were designed to be benign. What about the other side of the coin — organisms deliberately designed to do harm? If important crop plants are sterilized to ensure a high price for them, couldn't they also be sterilized to starve a political or ethnic enemy? If we study organisms to design cures, wouldn't we also acquire the ability to create new diseases? In 1967, a monkey virus escaped from a factory in Marburg, East Germany, and killed a number of workers. The virus, which was related to ebola, caused a horrible disease in which the victims drown in their own blood. A subsequent death a few years later in a Soviet-run research facility showed that the Soviets had seen the potential for using the Marburg virus as a bioweapon. They wanted to make it into a powder so it could be loaded into a missile aimed at the United States.

Undoubtedly, the Soviets weren't the only ones with such ideas. There's a lot of money to be made in developing biological weapons for a wide variety of clients. Unlike nuclear technology, biological weapons are relatively simple, cheap and extraordinarily portable and easy to use. It's a combination that could prove irresistible to terrorists or renegade countries. Bob Phelps of the Gene Ethics Network says, "We have computer hackers. Why not gene hackers? It would only take, for example, somebody to genetically manipulate an existing pathogenic organism and release it to have a new and perhaps untreatable strain of some disease."

The spectre of bioweapons use is so horrible that most of us don't want to think about it. But it was a major impetus and funding source for biotechnology in its early days. Biological warfare was discredited in the 1970s, and most countries in the world signed a convention banning it on the grounds that it was an insane kind of weaponry: whatever you infected the enemy with could turn around and get you. But Andy Kimbrell says, "With the advent of biotechnology during the Reagan administration . . . there was a whole renewal of interest in biological warfare. They realized that with genetic engineering, they could create designer weapons very specifically targeted at certain plants, animals and ethnic groups, through a greater understanding of the genetic makeup of those particular organisms. And they began a multi-billion-dollar program with over 120 university laboratories around the country, a program that was pretty much secret, to work and engineer the most dangerous pathogens known to man. We're talking about anthrax and botulism and venom. Every virus, bacterium and toxin you can think of that is quick-acting and deadly, they were working on it . . . . This use of biotechnology has been pretty much forgotten."

A recent scandal in Israel revealed that the Israelis were working on a disease weapon that would strike only ethnic Arabs. Most Israelis were horrified and demanded an investigation. But there is evidence that many countries, including South Africa during apartheid, have been looking into similar weapons. And, of course, like everything else in biotechnology, these weapons may not end up working the way their creators hope.

Many researchers have already noted the terrorist possibilities of supposedly benign GE agricultural technologies. Terminator varieties of important food crops are being engineered to be sterile, and can only be made fertile by patented chemicals. If this technology becomes widespread, certain trading partners could be denied the needed chemical, and large amounts of their food crops would fail. It's a powerful weapon to hold over another country. Or even more diabolically, why not make sure the Terminator gene is encoded in grain you sell or even drop on an enemy? Then the sterilization gene might spread throughout an enemy's entire staple food supply. Of course, like poison gas, such applications could backfire on the perpetrator. But those drawbacks have yet to stop the development and deployment of chemical or atomic weapons.

Canada has a bioweapons laboratory in Suffield, just outside Calgary, Alberta. Like the Americans, Canadians are told the laboratory is concerned only with defence against potential weapons that may end up in the hands of the nations' enemies. However, to test antidotes and other defences, the researchers first need the offensive bioweapons. So the distinction between offensive and defensive seems pretty illusory.

What to Do

We, the undersigned scientists, call upon our Governments to: Impose an immediate moratorium on further environmental releases of transgenic crops, food and animal-feed products for at least five years; Ban patents on living organisms, cell-lines and genes; Support a comprehensive, independent public enquiry into the future of agriculture and food security for all, taking account of the full range of scientific findings, as well as socioeconomic and ethical implications.

— WORLD SCIENTISTS' STATEMENT CALLING FOR A MORATORIUM
ON GM CROPS AND BAN ON PATENTS, APRIL 1999

The genetic engineering revolution has developed very quickly, but the European rejection of this technology shows there is still an opportunity to control it if we act now. Vandana Shiva predicts, "There's going to be a contest between coercive power, the manipulation of government systems, and the public awareness and public consent to all this. The consumers and citizens of the world are not nincompoops. The people of the Third World are not imbeciles. Our voices need to be heard too."

At the moment, because of the untested, unregulated nature of these new products, we can say for sure that only *one* group will benefit from their introduction: the companies that make them. These products have been put out on the market without public input and without labelling. What recourse is left to an individual who doesn't want to eat new, genetically modified organisms? Christine von Weizsaecker says, "If people want to choose, they now have to organize. They have to . . . defend a right that they always had, and that is to eat free-of-gene-tech food . . . . Labelling is crucial. But patenting is also crucial. Because if there is no patenting, the incentive to clog the market with thousands and thousands of these types of products would disappear." It was a big mistake not to challenge the ruling that allowed the patenting of life. But it's not too late to correct it.

The reason we have to consider such a sweeping change is that we humans really may have bitten off more than we can chew. In January 2004, *The New Scientist* reported on recent work with transgenic pigs that had human stem cells spliced into them when they were fetuses. These pigs were genetically engineered so that they can grow organs for transplant that won't be rejected by the human body. "What we found was completely unexpected," said Jeffrey Platt, director of the Mayo Clinic Transplantation Biology Program. "We found that the human and pig cells had totally fused in the animals' bodies." About 60 percent of the cells in the non-pig cells in the animals' bodies were hybrids, with chromosomal DNA that contained both pig and human genes. But the remaining 40 percent were *fully human.* Moreover, the extremely dangerous porcine endogenous retrovirus (PERV), present in almost all pigs, was also contained in the hybrid cells. "While PERVs in pig cells cannot infect human cells, those in hybrid cells can," the article notes. To say that "this suggests a serious potential problem for xenotransplantation" is to put it mildly. The moral implications of creating creatures that are significantly part-human, and of also engineering a new, highly efficient vector for disease, weren't discussed. But if you ask people outside the biotech research world what they think about such developments, they are very likely to be horrified and alarmed.

"We didn't get to vote on whether to take human genes and put them in animals, which they're doing through genetic engineering," says Andy Kimbrell. "It seems to me that we really can't call ourselves a democracy until we have instituted ways in which we can vote on the technologies that are introduced into our society. . . . Do we really want unlimited genetic engineering of humans, of animals, of plants, and the cloning and patenting of life forms? Do we really want our generation and the generations coming after us to view the entire animal kingdom as so many machines to be reprogrammed, cloned and patented? Is that the image of animals that we want for successive generations? . . . I think [that] if the answer to these questions is no, we have to have the courage to say no to these genetic technologies, the way we said no to nuclear weapons proliferation."

Literally thousands of new NGOs have been founded around the world to try to protect both the natural world and our most cherished social values from the ravages of genetic engineering and all the implications that surround it. Jeremy Rifkin's is the Pure Food Campaign. He says, "This is the first global movement in modern history where non-governmental organizations are coming together across national boundaries to say no to genetic foods for the twenty-first century. Health groups, environmental organizations, animal-rights organizations, organic and sustainable farmers, consumers, those involved in issues of social justice are all coming together in a new global network of resistance to the introduction of genetically engineered foods into the global marketplace. I think this is going to be one of the great social and cultural battles of the coming decade."

All around the world grass-roots, academic, human-rights and other groups are working on this issue. Canada has a number of such organizations, and some of them are listed at the end of this book. One of the most effective strategies worked for the British. Since government and industry weren't listening, citizens put pressure on food outlets and eventually got every major supermarket chain to divest itself of genetically engineered products. The lack of a market forced growers and the government to pay attention. Rifkin says, "We are asking every consumer on this planet to go to the grocery store, go to the supermarket, demand that the manager have a written policy that there are no genetically engineered foods in that store, and if there are those foods that they're clearly labelled. If that store manager cannot guarantee that they have labelled foods or that there are no genetic foods, then take your business elsewhere. I think the pressure has to be put on from the consumer end and from the farm end. If enough people do this in every country, this global boycott will triumph."

Once people's awareness is raised about food, we should be able to understand why we are hunting for genetic gold in our pets and our houseplants and our own hearts and tissues. We have to be able to define the kind of world we want to live in, and whether the commodification and marketing of life should be part of it. But we have to understand that our underlying economic, trade and ownership systems are a fundamental part of this issue. Vandana Shiva thinks we are trading in our own cells because of a very severe crisis of capital. We are, in fact, running out of real wealth in the world. As Shiva says, the current economic system, which drives the patenting of life, never recognizes the concept of having enough. "Capital," she claims, "like a cancer cell, needs to keep growing. If it stops growing, it must die. The need to keep growing is forcing capital to find new ways of accumulating [wealth] . . . . And it can stop only when this desperation destroys itself. My hope is that it'll destroy itself faster than it will destroy the world."

CHAPTER 7

FOLLOW THE MONEY

Economics is really politics in disguise. We need to unpack the whole thing and say, "Look, an economy is really nothing but a set of rules. Let's be up front about it. There's no actual science here."

— HAZEL HENDERSON,
FUTURIST

Over the past several years, I've noticed that when I discuss environmental issues, sooner or later someone always says that we can't *afford* a clean environment if we don't have a strong and growing economy. The economy — and the need to keep it strong and growing — has somehow become the most important aspect of modern life. Nothing else is allowed to rank higher. The economy is suffering; the economy is improving; the economy is stable or unstable — you'd think it was a patient on life support in an intensive-care unit from the way we anxiously await the next pronouncement on its health.

But what we call the economy is nothing more than people producing, consuming and exchanging things and services. Economists keep track of the money we exchange for goods and services, and then make judgements about how well we're doing. Hazel Henderson is a futurist and economic analyst who has been writing books about what's wrong with economics for the past thirty-five years. "An economy really is a fiction," she says. "You can't separate production, consumption and exchange and all of that from the rest of human life.

It was always a mistake of the economic textbooks to say that the economy is some kind of separate thing."

As a biologist, I've always felt that economics was flawed in another way: it pays no attention to the goods and services that are provided by nature — the air, water, soil and sunlight that produce everything we need for survival and all the riches we are able to accumulate. In a paper for the National Academy of Sciences U.S.A., the renowned biodiversity expert Norman Myers offered some conservative estimates of what nature does for us for free. These services — the economic benefits of which are enormous, though they are never included in our economic calculations — include our climate; the earth's biogeochemical cycles, such as the carbon cycle; all hydrologic functions, including sewage disposal; soil protection and production; crop pollination; pest control; and miscellaneous services, such as potential insights into medical and pollution problems that are provided by resistant organisms. Altogether such services provide annual benefits worth, if they could be duplicated by human-created services (which, of course, they can't), far more than the cash income of the entire Earth.

Robert Constanza is a leading scholar in an exciting field called ecological economics, which attempts to reconnect two disciplines whose profound interlinkages and interdependencies have been ignored. He led a group that refined Myers's efforts to put an actual dollar value on seventeen ecosystem services in sixteen different biomes. Astonishingly, they calculated that the services' annual worth was on average US$33 trillion. That compares with a total annual GDP for all nations on Earth of $18 trillion. And yet none of these services appears in the supply-and-demand handbook of mainstream economics.

Hazel Henderson points to another important part of our lives that isn't counted in the economic equation. She calls it the love economy, and it includes all the productive work that humans do that does not involve an exchange of money — things like raising families, doing community work, taking care of the elderly, being active in a club or charity. It may be impossible to put a price tag on these activities, but they are the very glue that holds societies together. In the Third World, with its traditional economies in which money does not play an overwhelming role, whole families are supported on home labour that is not part of any cash exchange: people raise vegetables and chickens; they barter and exchange with neighbours; they gather medicines and foods out in nature; they pool resources and labour. But none of this appears in the country's estimates of per capita annual income.

To illustrate her point, Henderson likes to imagine the economy as a three-layer cake with icing. She says, "The economists think only about what they call the private sector, which is the icing on the cake, and the public sector, the government and tax-funding, which is the next layer down. But they forget these two other layers at the bottom — the love economy and nature — which are holding the whole thing up."

Henderson isn't the only one who has noticed what economists don't include in their figures. Marilyn Waring, the youngest Cabinet member in New Zealand history, is known internationally for her book *As If Women Counted*. She recalls an experience she had when she was a representative from a rural conservative seat and served as chair of the parliamentary public expenditure committee of the New Zealand federal government. A corporation had applied for a mining licence in Mount Pironia, the largest forest park in her constituency. In its cost-benefit analysis, the corporation included all the benefits that would accrue to the area in terms of exports, infrastructure development, employment and taxes. But there were no costs listed. "This was the largest mountain in our constituency," Waring says. "It completely controlled the microclimate, giving us the rainfall that had made us the richest area of the country. It controlled the water table and our expectations of a temperate, agriculturally stable climate. It provided the habitat for all the pollinating species on which we were completely dependent for our year-round agricultural economy. And, of course, it provided lots of leisure and recreation."

The cost-benefit analysis never talked about the pollution that gold mining would bring to the watercourse. It never mentioned the seepage of arsenic and other toxins from the waste tailings. It didn't indicate what the toxins, the forest and water destruction, and the bulldozers and infrastructures would do to the habitat and the recreational desirability of the mountain. Waring found that there was "no way to express in the unidimensional formula of classical economics what the mine would *cost* our community."

Counting Money: The GDP

One of the main problems with the GDP is that it averages out income. If you average out income per capita, you could have a society with one or two billionaires and everybody else is homeless, and it would still look like it was great. This is how insane the GDP is.

— HAZEL HENDERSON

Economists must learn to subtract.

— SLOGAN FROM THE MEDIA FOUNDATION
AND *ADBUSTERS* MAGAZINE

Virtually all of the extremely important services that nature provides are completely ignored by conventional economics. The ozone layer, for example, shields all life from DNA-damaging ultraviolet radiation. But that crucial service, on which all agriculture, forest and marine industries depend, is not added to the balance sheet when economists figure out the profit margins for the industries that affect the atmosphere. The environment is also ignored when economists calculate the gross domestic product (GDP). This measure is accepted around the world as an indication of how well a country is doing. But is that what it really measures? The GDP totes up exchanges of money. Ralph Nader has pointed out that every time there's a car accident and people are hurt, the GDP goes up. An accident means ambulances, doctors and nurses, hospital beds, car repairs, even funeral services, all of it involving an actual dollar expense and therefore increasing the GDP. After the *Exxon Valdez* oil spill, to cite another example, the U.S. GDP rose by $2 billion; presumably, if we look only at the GDP, it's good for us to spill as much oil as possible, so we have more clean-up services. But surely the GDP is no indicator of an improvement in the quality of our lives.

We take the love economy for granted too. Marilyn Waring is particularly concerned with the fact that most economists do not see women's work as valuable. She says that in her rural riding, "male farmers know more than anybody the complete dependence of the family farming operation on women's work, both in the fields and in balancing the books. And, of course, anybody who knows farming knows that as soon as children are mobile and competent, they're engaged in the farming operation for at least some part of the day."

Waring gives an example of a wife on a sheep station who broke her leg just before shearing was to begin. New Zealand's Accident Compensation scheme didn't cover her because she wasn't considered "a part of the market." But for the farm to continue its operations, she had to be replaced by three "real, recognized workers": a shepherd; a rowsee, or clipper; and a housekeeper. The system ultimately recognized the shepherd and the rowsee and paid their wages, but it refused to pay any compensation to the housekeeper, since she was the woman's mother. The system recognized the love economy so far as to say that it was an "altruistic and community expectation" that the mother would help out. In some cases, then, the love economy *is* recognized and expected by economists — but not when they might have to allocate funding or subtract it from their totals. It's not only gender that's at issue here; it's cash. Economists don't recognize household labour because basically no cash is exchanged. There's even a famous economist's joke, which says, "When a man marries his housekeeper, the GDP goes down."

Hazel Henderson is involved with an investment fund called the Calvert Group, which is trying to come up with new criteria for measuring true economic and social well-being. She says, "We have twelve aspects for measuring the quality of life based on what voters all over the world, in democracies, say is important to them — education, health, recreation, environment, culture and some other things, like the poverty gap. People don't like it when the poverty gap gets too wide; it's very uncomfortable for us. And, of course, it creates a lot of crime."

In San Francisco, a group called Redefining Progress has developed a new economic indicator, the genuine progress indicator (GPI), which *subtracts* those activities that negatively affect the quality of our lives through social disruption or environmental degradation. These activities consist of resource depletion, including loss of wetlands, farmland and minerals such as oil; asymmetric distribution of income so that when the poor benefit more, it is added, and when the rich receive a disproportionate increase, it is subtracted; unemployment and underemployment; pollution; long-term environmental degradation; and disposable consumer goods.

At the same time, the GPI *adds* such elements as housework, non-market transactions and increased leisure time. Redefining Progress has found that while the per capita GDP in the United States rose steadily from an average of $7,865 in 1950 to $16,414 in 1992, the GPI was a far different story. The GPI rose from $5,663 in 1950 to a peak of $7,441 in 1969; then it *declined* to $4,426 by 1992. The GPI reveals that in striving to keep the GDP growing, we have seen a decline in the quality of our lives.

The GDP was invented for very good reasons. At the end of the Great Depression, politicians in the United States realized that they had poor statistics on factors like employment and average annual income. In order to measure rates of recovery, they had to have an instant snapshot of how bad things really were, and whether various economic policies were working. Then, during the Second World War, governments needed to be able to measure every aspect of our economies so they could figure out how much could be produced for the war effort. After the war, it was found that because he lacked such tools, Hitler had greatly underestimated his own economy's ability to produce. For all these purposes, an economic indicator was very useful. The GDP measures the actual cash transactions that take place during the course of a given year, an important statistic for a country to know.

There was never any suggestion then, however, that the GDP should be used for measuring the state of general prosperity of a country. It can't reflect pockets of misery, education levels, health and pollution problems. It can't gauge prosperity and well-being. Hazel Henderson says we need other forms of measurement that are multi-factorial and reflect more clearly the fact that we lead very complex lives. We can't reduce the joy our children bring us, our hobbies, our job satisfaction, the health of our environment and our neighbourhood safety down to a single number, especially a number that keeps track only of cash transactions.

Many different organizations, including the United Nations, have been developing alternatives to the GDP. Even a former president of the World Bank, Barber Conable, admits, "Current calculations ignore the degradation of the natural resource base and view the sales of non-renewable resources entirely as income. A better way must be found to measure the prosperity and progress of mankind."

Musings like Conable's appear to fall on deaf ears. Despite the efforts of Hazel Henderson and people like her, conventional economics continues to dominate our political thinking, especially under the two Bush presidential administrations. Again and again, politicians point to the GDP as a measure of the success of their policies. They get away with it because economics has all the trappings of a sophisticated, quantitative science. Articles in the field are filled with arcane terms, complex formulae and computer analyses, all of which give the appearance of mathematical precision. But Henderson says, "It's all complete rubbish. Economists use mathematics to cover up very, very simple propositions, which would be much better addressed in the mother tongue."

I once sat in on a university course in economics and was appalled when the professor said, early in the first lecture, "Sharing, co-operating and caring are emotional and irrational acts." He told the class that economists begin with the understanding that only acting in pure self-interest is rational; that's why modern economics rests on this principle. When I mentioned this to Henderson, she responded, "It's terrible. If the textbooks say the only kind of rational behaviour is maximizing self-interest in competition with all other comers, then that's a recipe for a society that is going to end up as a behavioural sink."

We Can't Grow That Fast

We seem to be moving towards the idea that rates of growth of money in the bank are autonomous, and that the natural world can accommodate this. Or if it can't, we can adjust it. That seems to me the wrong way to go.

— HERMAN DALY, ECONOMIST

Hazel Henderson could be thought of as an outsider, but there are respected scholars even within the economic mainstream who are critical of their own discipline. Herman Daly was a senior economist at the World Bank for six years and now teaches at the University of Maryland's School of Public Affairs. When we asked him what an economy is, he replied, "I would say that it's a community of people who, through division of labour and exchange, live a life in some mutuality or interdependence. And they depend not only on each other, but also on the natural world, the natural community, from which they extract materials and services and to which they have to return wastes. There are two levels of community: the natural community, which must be respected for the economy to be sustainable, and the human community, in its mutual interdependence and division of labour, where relations of justice and dependence have to be respected as well."

Daly is one of a growing number of economists who point out that nature sustains the economy, not vice versa. Everything we make comes from the Earth — plastic, paper, cars, computer disks — everything. As Daly says, "You sacrifice some of the natural system when you convert it into man-made things. And you have to balance those costs and benefits. In the past, we haven't focused on that. We've thought that if the economy grows, we'll be fine and, oh, well, yes, resources might become a little bit scarce relative to other things, but no problem, technology will solve that."

Not only is endless growth seen as possible, it's believed to be necessary. But growth in the economy and growth in nature are not the same. Money grows fast — so fast, in fact, that nothing in nature can keep up with it. For example, trees in British Columbia increase in size by about 2 to 3 percent a year. That means that if a volume of wood equivalent to 2 to 3 percent of a forest is logged annually, the equivalent of an entire forest can be cut over a period of twenty-five to thirty-five years and still leave an intact forest standing! In other words, trees could be harvested that way forever. Unfortunately, that would make no economic sense. It makes better economic sense to clearcut all the trees at once and invest the money. Returns would yield 10 to 30 percent a year! What CEO would keep his or her job with only a 2 to 3 percent return on investment?

Since money grows faster than trees, economic thinking demands that we trash the forest. "I once saw a very nice cartoon," Herman Daly says. "A fisherman is standing with a puzzled look on his face. On the one side, the fish in the sea and in his net are saying, 'Slow down! We can't grow that fast.' And on the other side, the banker is saying, 'Speed up! We have to grow faster!' And so, there you are. The fisherman is caught between the rate of growth of money and the natural, biological growth rates of species."

One doesn't have to be a forester or a scientist to know that our present rates and methods of forest extraction are unsustainable because they are driven by a corporate mandate to maximize profits. If we want "forests forever," as an ad for the forest company MacMillan-Bloedel once boasted, we should be working to match technology with the capacities of the planet. Instead, we stubbornly refuse to acknowledge that there are any limits to natural growth; we're trying to put nature on steroids. "We're striving to do the opposite of what we should be doing," explains Daly. "We're taking as given, or autonomous, the rate of growth of money in the bank and, by golly, we'll just redesign the genetic structure of everything we depend upon so that it grows as fast as money in the bank. So we increase agricultural growth rates and design new animal husbandry, using techniques of cloning and everything else, in order to get that growth rate up. And that strikes me as a dangerous thing."

In the past, local experiments with speeding up natural growth rates sometimes worked — such as when we learned to cultivate prairie land for grain — and sometimes didn't — such as when we desertified northern Africa, Spain and Greece by intensive agricultural practices. This saga continues to take place across much of the world; one recent example we found that illustrates the complexity of the natural world's limits to growth in the face of our constant expansion is in North Carolina.

Eaten Alive

We have a new thing that's come up in our rivers, Pfiesteria — *single-celled animals that have lain for thousands of years at the bottom of our rivers. But our rivers are becoming what they're calling nutrient-rich — I love that [phrase] — full of hog and chicken manure. We've woken up these animals, the* Pfiesteria, *and they are actually eating the fish alive. And feeding on the fishermen.*
— HOWARD LYMAN, HUMANE SOCIETY OF THE UNITED STATES

It sounds like science fiction, but it's actually happening. Beginning in the early 1990s, fishermen in North Carolina, along Chesapeake Bay, started to notice some terrible things. Fish had begun to die by the millions, and the dead and dying fish were covered with gaping, bleeding wounds. In 1991, North Carolina alone lost a billion fish; the piles of carcasses had to be bulldozed off the beaches. To add to their worries, some fishermen began to find open sores on their own bodies. They were told to wear gloves to avoid infection. Then water skiers in the bay found their legs were being attacked by a microscopic creature and eaten away. In 1995, after another 15 million fish died, researchers discovered that about 40 percent of the people who'd been exposed to the fish — the fishermen or those living nearby — reported memory loss and nervous-system problems. Others developed severe respiratory conditions that resisted every treatment. All these problems are now known to be associated with an estuary microorganism called *Pfiesteria*.

Where did this horrible bug come from? *Pfiesteria* is a dinoflagellate, an animal-like protozoan so small that thousands would fit on the head of a pin. It is found in nature all over the world. Normally, it's benign, sitting in the mud of estuaries like thousands of other tiny creatures we know little about, eating bacteria and algae. But under certain conditions, it can become toxic. At the moment, researchers like Dr. Joanne Burkholder, a professor of aquatic sciences at North Carolina State, believe that *Pfiesteria* use their toxins to immobilize fish. Then they transform themselves into amoeba-like creatures that attach to the fish and feast on the remains. We don't know what triggers this action.

"We have a very short-term database," explains Burkholder. "But some of the information, [which] we regard as provocative and somewhat disturbing, about cousins of *Pfiesteria,* other toxic dinoflagellates and toxic algae, is that they seem to be increasing in the world's coastal areas. In fact, over the past fifteen or twenty years, massive microbial blooms involving something like twenty-two species have increased to ones involving sixty or more species. So that's a sign to me of coastal water-quality degradation. Some of these species seem to be very highly correlated with nutrient over-enrichment and other pollution problems in coastal zones. *Pfiesteria* is one of those species. It has been highly associated in poorly flushed, quiet waters with nutrient over-enrichment from human sewage, swine wastes and other problems."

"Over-enrichment" and "nutrient-rich" are terms invented by various industries to describe water with too much nitrogen in it — in short, water that is full of feces. And that's just what the rivers and oceanfronts around Chesapeake Bay are loaded with, especially on the North

Carolina side. North Carolina and the Netherlands share the somewhat dubious distinction of being the most intensive producers of hogs in the world, followed not far behind in Canada by Quebec. In one small county of North Carolina alone, there are 2.3 million pigs — and until recently they were protected by state legislation that forbids any zoning against the industry. Methods of intensive hog farming produce tens of thousands of pigs in every barn, where they are so tightly enclosed that they can barely turn around or lie down. And the extremely sulphur- and nitrogen-rich wastes they produce are disposed of in exactly the same way they were when the area was home to more normal-sized operations and produced only a few hundred piglets a year. The manure is put in lagoons, then sprayed on the surrounding farm land as "fertilizer." It's also dumped directly into watercourses. Since there's far too much for the land to absorb, the nitrogen and ammonia drain directly into the water table and out into the estuaries, where they trigger changes in *Pfiesteria*.

Hog farming has been an ecological disaster for Chesapeake Bay, the second-largest estuary on the U.S. mainland. It sits behind the Outer Banks, where the water is relatively quiet and poorly flushed. As Joanne Burkholder says, it is exactly these kinds of places that are the most productive in terms of natural systems. "It's an extremely important place as a fish nursery ground. According to a recent Sea Grant document, this bay provides 50 percent of the total surface area of nursery grounds used by fish from Maine to Florida. They come to our nursery to breed or get born, and then they disperse north and south." But because of the poor flushing, it's a bad place to dump wastes. "Pollution," Burkholder says, "stays in the estuary, where it has time to cause problems like major algal blooms, which provide material to feed the *Pfiesteria* when the microorganisms are not killing fish. *Pfiesteria* likes these conditions, likes a lot of algae and other things that are associated with nutrient pollution."

Human and other crop wastes are also at fault in this situation. But the hog fecal output is the worst pollutant, the trigger for plankton blooms. How dangerous to people is exposure to *Pfiesteria*? Burkholder says there isn't enough scientific, clinical evaluation in the field, but the effects researchers know about with certainty are the open sores, serious cognitive impairment and short-term memory loss, as well as chronic respiratory problems already afflicting residents.

This completely unforeseen consequence of large-scale hog farming brings us back to the point made in the cartoon referred to by Herman Daly. We're trying to increase profits in the meat industry, but on a scale that outstrips nature's capacity to deal with the wastes. This effort to produce cheaper pork has created problems so horrific that they seem hallucinatory: people with serious brain damage or flesh invaded by microorganisms; fish eaten alive by the millions in the waters where they've come home to breed. Our drive for quick profits ignores any understanding of the interconnected natural systems on which the whole enterprise depends.

Joanne Burkholder says, "It's a subtle thing. These connections are difficult to make. In *Pfiesteria*'s case, who would have known that a little animal-like creature could be strongly stimulated by human and swine wastes? Who even knew that it was there until eight years ago? Who would ever have considered that diseased fish could translate into serious learning disabilities and memory loss for people? These kinds of effects come back to haunt us if we don't develop longer-range vision about what we're doing."

In Poland, Brazil, Mexico and Canada, more economic, ecological and social connections are being made for thousands of rural residents. There has been a massive influx of ILOs (industrial livestock operations) such as North Carolina's mega-hog farms into these countries over the past few years. They're looking for lax regulations, since both Europe and the United States have begun to pass laws to limit their operations. The already large hog industry in Quebec, for example, having seriously polluted eight major rivers in its original areas, is attempting, against desperate citizen and municipal resistance, to expand into the prosperous and idyllic dairy and apple country farther west, as well as throughout very unsuitable landscapes of the Maritimes. In the Canadian Prairies, there is a wholesale invasion of these enormous, water-consumptive and polluting installations, leaving a trail of vastly lowered property values, distraught, divided communities, and water and soil contamination behind them. Their products, however, don't feed the Canadians, Poles or Mexicans who are absorbing these costs. The meat, laced with the hormones, antibiotics and pesticides needed to offset the unhealthy conditions of extreme crowding, is exported to other countries, especially to Asia, where its subsidized prices undercut local producers and put them out of business. This rather insane way of treating our countrysides reflects the constant pressure our current economic model creates on more sustainable production methods. It is simply set up to reward long-term resource destruction for short-term gain.

Corporate Morality

Their proposition is to exploit, as much as possible, both people and resources. That's what they do.
— HARRY GLAESBECK, RETIRED PROFESSOR OF CORPORATE LAW

Until recently in the United States and in most provinces of Canada, the various levels of governments haven't protected their rural citizens from the smells, pollution and rapacious behaviour of ILOs. In Canada especially, they have usually worked in collusion with these industries, providing such lavish tax breaks, subsidies and "insurance" schemes that a given hog or cattle farm doesn't have to make any actual profit off the meat or milk it produces in order to prosper. In fact, governments these days are usually not very good checks on business.

The kind of large corporations that include these ILOs in their many interests not only support politicians financially, but they also know how to put on pressure on both politicians and voters to get what they want. A Republican congresswoman in North Carolina, Cindy Watson, dared to attempt to place some controls on the hog industry that was destroying her constituents' land, air and water. She became the target of a media campaign. Farmers for Fairness, a group created and funded by the hog industry, spent $2.6 million on television and other ads to discredit her for her opposition to their use of nature for experiments in pork supply and demand. And they won when she lost her bid for re-election.

This is just one example of the now well-recognized gains that big business has made in the political process. How did this happen? Most analysts agree it's because of a few legal decisions made over the past hundred years. Harry Glaesbeck, who taught corporate law at Osgoode Hall Law School of York University in Toronto for more than twenty years, says, "A corporation is really an amazing kind of creation of the legal imagination . . . . Any individual in Canada who is eighteen, sane, not bankrupt, can go to a registrar and register a corporation, just as you would register a dog licence. It doesn't have to have any objectives. There is not even a requirement that there should be any capital in it."

Once that corporation is formed, it becomes a legal entity *with the rights of a person*. As Glaesbeck says, this person "is instantly matured. There is no incubation period. It is immortal. It will not die unless it voluntarily dissolves itself, or if it's bankrupted and loses all its assets. It is completely unseeable. It's untouchable . . . . Individuals who manage the corporation and who actually own it are very hard to get at. It's not impossible, but it's very difficult. It creates a shield between people who take risks and the obligations that arise out of those risks."

Jane Ann Morris is a corporate anthropologist who works with a group called the Program on Corporations, Law and Democracy. She says that especially in the United States, but to some degree all over the world, "Corporations have written themselves into the democratic process in the place of human persons." She thinks it's important to remember that corporations are not people, yet we've accorded them virtually all the rights that people have. Add to those rights their wealth and immortality, and it's no wonder they can be pretty hard to deal with.

It wasn't always so. In the nineteenth century, laws in many American states limited corporations' powers, forced them to hire local workers or gave the state the right to review their behaviour annually and to terminate their existence. Early corporate charters did not relieve the stockholders of liability beyond the extent of their investment, as they do now. And corporations were often not allowed to hold stock in another corporation. (This stipulation alone would solve a lot of today's worries about mega-monopolies in banking or the media.) States also retained the power to terminate the existence of a corporation if it violated state laws. But in the late 1800s, corporations began to press for rights that would allow them to appeal state cases to the federal court, or to have the kind of protection and due process enjoyed by human citizens, even con-stitutionally guaranteed protection against "unreasonable" search and seizure. All these privileges were accorded, despite violent protests, and since then corporations have been building on them, and on a global scale.

Jane Ann Morris says, "Now, let's put all this together. Corporations have some of the rights of natural persons to equal protection and due process. The concept of property is expanded to include ideas and future earnings. You can see how the noose is tightening, as corporations are getting themselves more rights and reducing the rights of humans." But how does corporate acquisition of rights reduce those rights for humans? "One example of the way that humans' rights — living, breathing, flesh-and-blood humans' rights — are reduced is that corporations have been getting increasing rights to restrict what human beings can do on corporate property," she explains. "Think of things like malls and shopping centres and the workplace. In all those places, we're bombarded by corporate speech, yet the rights of humans to free speech are severely restricted." Think also of mandatory drug testing by companies. Think especially of commercial secrets, the kind of secrets that threatened the lives and vision of tens of thousands at Bhopal, because Union Carbide could not be compelled to disclose the composition of its private property, the ingredients of the chemical fertilizers that had been leaked from its factory onto the surrounding population. It had the legal right to do that. And that corporate legal right superseded even the moral obligation to save human lives.

Harry Glaesbeck goes so far as to say, "Corporations are criminogenic institutions, as we academics like to say. That is, the pressure on them to make profits is enormous. So in order to make profits, especially within a competitive situation, they have to mould the environment. They try to mould the environment to their needs, rather than the other way around. Now, they do that in a variety of ways. They participate in politics so that they get laws which are very lax. Then it's even easier to mould the environment. If they don't succeed in that, they'll break the laws or they'll try to have them not enforced."

We tend to assume that corporate interests coincide with human interests, and that can be true. Corporations employ people and manufacture the goods and services that we want. But they have also become a threat to humans, because our needs and theirs simply do not always coincide, and we are finding it increasingly difficult to control these ever bigger and more powerful multinational corporate entities when *they* want things that are bad for *us*. "It is beyond doubt that of Fortune 500's major corporations, 40 percent do not commit a crime annually. Sixty percent do," Glaesbeck asserts. "Now, some of the crimes are nothing. They're just small regulatory offences, not filling in the paperwork, stalling — I don't want to exaggerate. But some of the crimes are monstrous. They're like Union Carbide at Bhopal. They're like the asbestos boys. . . or the Dalkon Shield manufacturers, or the Ford Pinto. They made these things for money. . . hiding the fact that they later learned they were dangerous from the public and the consumers at all costs, and paying a paltry amount of damages afterwards."

Corporations easily bully governments by threatening to deprive even democratic nations of their wealth. If we try too hard to control them, they say they'll leave and take their jobs with them. And as Glaesbeck says, "They're aggressive people. They're expected to aggressively go out and compete and fight and smash their opposition. So they're naturally risk-takers, especially as the risk is not theirs. It's the risk of other people's money."

We do hear about "good" corporations, of course, ones that fund softball teams and women's shelters and progressive documentaries. But these activities usually provide tax write-offs, and generally are justified only if they improve the corporation's bottom line. As Glaesbeck says, "If they can show that altruism incurs better profit, then of course they're allowed to do it. It's defensible. But there is no such thing as genuine altruism in the corporate sector under our present system." The way we've set up corporations, even a majority vote of stockholders cannot demand that a corporation's policies reflect the public good or preserve the environment for future use. That's because profit is the one and only motive. It's up to government and it's up to people to protect the public interest. Corporations are simply not allowed to.

I remember an occasion years ago when I was battling the clearcutting practices of a major Canadian forest company and encountered the CEO. "Dave," he said, "my job is to maximize profit for my shareholders. And I'm good at it. If you disagree with the way we log, your quarrel is with government, because we do everything within the legal limits." He was honest and correct. If performing to serve the public good — say, to log forests sustainably so that future generations will have them, or to refuse to dump wastes into a nearby river or lake — in any way infringes on a company's profits, and it probably will, then shareholders can complain. That particular CEO won't last. What is more, as Glaesbeck points out, "How do the managers know the public good? They're not elected. They're unaccountable."

We always have to remember how corporations, those bulwarks of the economy, work. "The reason that they exploit the environment, dump wastes in the air and water, is because natural resources are considered to be owned by nobody; so they can be taken and exploited with impunity," Glaesbeck says. "When we put out regulations to stop that, corporations try to soften the regulations, soften the monitoring. And if none of that works, they break the law. Or they go elsewhere, where they're allowed to do what they want. They're always threatening to go elsewhere, where they can exploit people and resources to a greater extent. The consequence of that is we all lower our standards continuously. A balance between sustainable ecology and sustainable human life, on the one hand, and the unfettered drive for profit, on the other, is just an oxymoron."

Humans need to have control not just over environmental and justice issues, but also over the economy. Jane Ann Morris says, "Right now, some corporations are starting to say that they're socially responsible. And I think there are some genuinely well-intentioned people both inside and outside those corporations who want to be better than average and do good. But that doesn't help much if it doesn't change the locus of control. Instead of a corporation, think of a king. I don't want a king who's in a good mood or who's nice today. We don't want to be subject to arbitrary authority. Our problem today is not about a particular corporate harm, a sweatshop or a toxic dump, it's about the fact that we're no longer in control of the show. We need to understand that corporate, economic power has discombobulated our democracy. And we need to repair it or make it new."

One of the biggest obstacles people face today is the fact that many, if not all, Western governments collude with private business interests in a large variety of ways. They can even publicly admit this because of the old saw about supporting a growing and competitive economy. But they really do it because of the way their elective processes are currently set up. It has become so expensive to run for office that candidates know they can't get government jobs in the first place if they don't make deals with businesses. This is a terrific loophole in the democratic process that could be tightened if the laws that govern campaign contributions were altered. Fortunately, there are movements doing just that around the world, and in Canada and the United States, increasing numbers of city, state, provincial and even federal governments are starting, reluctantly, to make it a little harder for corporations to pull all the political strings.

Our job as citizens is to make sure these loopholes constantly get tighter. We can get a much better handle on private interest influences on the legislative process if doing so becomes clearly illegal. In the United States, it's a lot easier to get involved in such issues with new, Web-based movements like MoveOn (at www.MoveOn.org; see also www.progressiveportal.org, www.truemajority.org and www.commoncause.org, among many others) that make it possible to contribute to the elections of progressive politicians at the click of a computer. MoveOn alone raised more than $2 million to help elect four new U.S. senators and five new House of Representatives members in 2000. It launched an online Democratic primary, helped mobilize the millions of people worldwide that protested the Iraq war, and engaged normal citizens in discussions of what kind of a political system they really wanted. In Canada there is Democracy Watch (www.democracywatch.org), which also keeps track of who gets what money in that country's political system. Another good source and sounding board in Canada on these issues is www.Straightgoods.com; the Council of Canadians is useful too.

Greed Lock

> *Exponential growth . . . won't work. It works to a point, and you can have a big bash and a big binge, you can spend all your global capital, and then it's all over. A sophomore dropout can figure this out. Why can't all the bright people in global institutions figure it out?*
>
> — DAVID BROWER, FOUNDER,
> FRIENDS OF THE EARTH AND EARTH ISLAND INSTITUTE

The modern global, corporate-dominated economy didn't always exist. When we came through the global depression of the 1930s, what got the economy going again was a war — the Second World War. By the end of that war, the American economy had become the dominant economy in the world. Fuelled by both altruism and self-interest, the United States — both the government and eager corporations like Coca-Cola, Boeing and General Motors — exported its largesse and American ideas about progress and development around the world.

David Korten has a Ph.D. from the Stanford Business School. He taught at Harvard and spent most of his career in overseas development work. Like so many other young American professionals, he wanted to go out and help the rest of the world. He says, "I graduated from Stanford in 1959 and in that year made a decision that I would devote my life to development work, to bringing American management methods and economic theory abroad, so all the rest of the world could live properly — you know, like we do. At the time I was very conservative and concerned about the spread of communism and revolution. There was an underlying sense that if we didn't do something to end poverty, it would be a threat to our American way of life. But working within the system back then, through UN agencies and things like the Rockefeller Foundation, there was a very strong sense that we were doing this out of our altruism and our concern to help other people live better lives."

With youthful idealism and energy, Korten plunged into Third World development work. But over and over again, in places like Ethiopia, Central America and Southeast Asia, he kept noticing that the kind of prosperity and progress he expected from U.S. aid and modern business methods didn't seem to be materializing. He says, "As you participate in this process, you notice that a lot of things don't seem to work very well. But you assume it must be working better someplace else, because if it were working this badly everywhere, then somebody would surely notice it and do something about it."

What Korten began to notice was that the major beneficiaries of American aid were an elite minority within the developing countries. Large numbers of poor people remained mired in poverty. And just as alarmingly, the natural environment they depended on for their livelihoods was being rapidly destroyed by the industrial development that was enriching that tiny elite. He says, "Behind the facades of development, more and more people are being displaced. There's more and more deep misery. The forests are being stripped away. You go out snorkelling in the tropical coral reefs, and the coral reefs aren't there anymore. It looks like a barren landscape under the water instead of rich gardens of coral and fish along the coast. And then it gradually begins to dawn on you that something is really more deeply wrong."

These are just impressions, of course, of travellers who leave the hotel districts to explore the slums or the countrysides. But there are ways to determine whether the impressions are true. Korten tells us how: "If you start looking at the statistics and the studies, you realize these problems are not just local. They're everyplace. What really became a shock for me in Southern countries was continual, deepening poverty and inequality, environmental devastation, and a breakdown of the social fabric. That's when I realized that something deeply systemic is wrong. It's not that we don't have just the right fine-tuning on our concept of development. Our whole concept of progress is wrong, in that most of our dominant institutions were creations of a mindset that focused purely on growth and production and the monetary value of the products produced by the economy."

Not only has this mindset failed in the Third World, but by the 1980s all the problems in the developing world were also present in the developed world. They were, however, masked by our greater overall wealth. The paradox is that while the economy has continued to grow, and we have, in fact, made great advances in global medical care, more and more people are becoming truly poor, and overall human suffering has not diminished.

How do we know this? Certainly not from the statistics presented by international agencies like the World Bank, the United Nations and the UN Development Program (UNDP). These agencies report encouraging trends. They claim, for example, that great progress has been made in reducing poverty over the course of the past century, especially in the last decade or two, when the International Monetary Fund (IMF) and other international agencies began advising poor countries on how to manage their finances. Their measuring stick, the Human Poverty Index (HPI), indicates that only around 11 percent of Mexico's population is now "poor." Yet a report in the *Washington Post* states that real income in Mexico fell drastically between 1982 and 1992, infant deaths due to malnutrition *tripled,* the real minimum wage lost more than half its value, and the percentage of the population living in poverty increased from just under half to two-thirds of Mexico's 87 million people. Figures compiled by UNESCO tell a similar story. When the world's population is ranked by wealth and subdivided into five groups of equal number, "in 1990, the richest group received 59 times more than the poorest. Despite growth of almost 3 percent per annum worldwide of per capita GNP over the past three decades, this is almost double the ratio of 30:1 in 1960. In that year, the richest 20 percent already controlled over 70 percent of global GNP. In 1990, this had climbed to 83 percent. In 1960, the poorest 20 percent had to get by with only 2.3 percent, which by 1990 had fallen to 1.4 percent." The report goes on to state that "these figures conceal the true scale of injustice, since they are based on comparisons of average per capita incomes of rich and poor countries, and do not take into account the wide disparities between rich and poor people in each of these countries."

Michel Chossudovsky, an economist teaching at the University of Ottawa, is a specialist on the macro, or world, economy. One of his recent books is entitled *The Globalisation of Poverty.* He maintains that David Korten wasn't imagining things when he noted a general grinding down of social and environmental conditions in developing nations. Chossudovsky has observed a global decline in living standards that is a direct result of classical "development" policies. He says the UN agencies that deal with development, as well as international economic entities like the IMF and the World Bank, are completely committed to vindicating the free-market system. That is, they fully support the idea of continuous economic growth and the belief that natural systems are secondary to purely "economic" concerns, like industrial development and free trade. Chossudovsky points out what everyone knows: that the majority of people in the former Soviet Union, for example, are worse off now than they were in the 1960s. He also points out that in very recent years, despite UNDP reports that fail to note such things, "local famines have erupted in sub-Saharan Africa, South Asia and parts of Latin America; millions of schools and health clinics have closed down; hundreds of millions of children have been denied the right to primary education; and all over the Third World, Eastern Europe and the Balkans, there has been a resurgence of infectious diseases, including tuberculosis, malaria and cholera."

In large part, the conditions leading to the degradation of human health and poverty are the undermining of local communities and their air, water and soil, which are the real basis of well-being. David Korten puts it all together. He says that government agencies contend that things are getting better for everyone because, since 1950, "there's been roughly a five and a half times increase in total global economic output. And our policies are still driven by the presumption that the solution to poverty is growth. The solution to the environmental problem is also growth, because then we generate the resources with which we can be rich enough to afford to clean up the environment. And, of course, as we do that, what we do is we continue to increase the burden on the ecosystem. We break it down so that the very foundation of all wealth, of all of the total productive system, is being diminished. We're destroying the natural capital. But the other thing that we do is intensify the competition for that resource base between the rich and the poor. And, of course, the rich win that competition."

Unfortunately, when the economic advisers to our governments and to global bodies like the United Nations see increased social deterioration, more poverty and greater insecurity, they seem to have only one way to respond. As Korten says, "The worse the crisis gets, the more the people who are making the decisions become determined to press ahead with exactly the policies that are creating the destruction."

Trickling Up

The doctrine is that if the horse is fed amply with oats,
some will pass through to the road for the sparrows.
— JOHN KENNETH GALBRAITH, ECONOMIST

It's interesting to ask ourselves what money and wealth really are. Money is a construct, an agreement among people. Wealth is what we can actually use: food; shelter; clothes; transport; access to clean air, water and light — those last three increase the rent on an apartment, because everyone knows their true value. "When the money system is working properly, it's essentially a mechanism to facilitate transactions," David Korten says. "But in what some people refer to as finance capitalism, people are essentially creating money out of nothing. People think of it as creating wealth, but actually what they're doing is capturing wealth; they are creating claims on the wealth of the rest of society."

This gets us to the age-old question of whether wealth is created or merely taken from one place to another. The answer is both. Sometimes it's created, and sometimes it's just stolen. Korten says that the system currently favours people who are taking wealth, rather than people, like farmers or craftsmen, who create it. Of course, some say that it doesn't matter if more than 450 billionaires in the world have sequestered almost 80 percent of the world's wealth for their own use. As they do use it, we are told, it will "trickle down" and support us all. But Korten says, "It doesn't 'trickle down.' It's totally a process of 'trickle-up.' It trickles down to a few people, the rich guys' lawyers and accountants and pool attendants. But in terms of trickling down to ordinary people, that's not the name of the process, particularly since the reality is that we are dealing with a finite, natural wealth base. It is not a growing pie. It is a fixed pie in that regard. And so *to keep expanding the wealth of the very wealthy, you basically have to push more and more people off the basic source of sustenance* [italics ours]. And that's pretty much the story of our times."

What Korten says about the developing world is equally true within the richest country on the planet. The journalist John Cassidy wrote, in a 1995 *New Yorker* magazine article, "Until recently, it was an empirical law of American politics that the majority of citizens. . . received steadily rising earnings." And for the post-war boom years this was true. From 1945 to 1973, "the rich got richer, but almost everybody else got richer with them and at roughly the same pace." Cassidy says that in those years "the annual growth rate of family income was between 2.4 and 3 percent regardless of where the family stood in the income distribution." If the distribution of incomes is divided into fifths, or quintiles, then the incomes (in constant, inflation-adjusted dollars) in each quintile doubled. But since 1973, says Cassidy, "The bottom two-fifths of American families saw their incomes fall." Even with more working mothers, the average incomes of the middle quintile barely rose, from $36,556 in 1973 to $37,056 in 1993.

In contrast, the top third increased by 7.9 percent over the same period. Incomes of the richest 5 percent grew from $137,482 in 1973 to $177,518 in 1993, an increase of 29.1 percent, and the top 1 percent skyrocketed from $323,942 to $576,553, a 78 percent rise! Twenty-five years ago, the average CEO of a large corporation earned forty times as much as the average employee. Now the differential is nearly a thousand! Today the top 5 percent of American households receive 20 percent of the nation's income. Unless the figures are broken down this way, the extreme concentration of wealth among a diminishing minority of the world's population is hidden by simply averaging income.

It's a Mad, Mad World

There's this enormous irony that the so-called proponents of the free-market support when it comes to reducing public welfare programs. We must have a free-market, level-playing-field situation for young teenage mothers. But for the big industrial players, the arguments get turned on their head, and suddenly we absolutely have to have these government handouts and support payments and logging and agriculture and building subsidies.

— ALAN DURNING, HEAD OF NORTHWEST
ENVIRONMENT WATCH

Economic issues are constantly in the media and our consciousness. Governments are slashing social services, including medical care, welfare and education, to "reduce debt and stimulate the economy." With debt and deficit dealt with, we're told, economies will grow and we will be able to afford things like social security, better health care and environmental protection. But, as seen across Canada, even after the budgets were balanced, even after there were all kinds of surpluses, we still didn't seem to have enough money to pay decent salaries to nurses, finance our children's educations or make hospitals run properly, to say nothing of protecting wildlife and controlling pollution.

Alan Durning thinks there's plenty of money to pay for all of this, but right now it's being misdirected because of our wrong-headed approach to economics. He used to work for the Worldwatch Institute and now he's head of Northwest Environment Watch in Seattle. He's got a surprising list of ways we destroy our natural surroundings, and how we use our tax dollars to do it. He lives in the homeland of the Pacific salmon, as do I, and as Durning says, "You hear people in public life all the time saying that we may not be able to afford to save the salmon. It just costs too many billions of dollars to protect them. But, in fact, exactly the opposite is true."

Salmon are almost mythical creatures in West Coast society. They're born far inland in clear forest streams; they travel to the ocean, mature and then somehow, after years, they find their way back to their natal river, and fight their way upstream to spawn and die back in the headwaters where they started. Because of the immense distances they travel and their competition with humans for forest, farmlands, estuaries, oceans, rivers and so on, it seems as if it's very difficult to protect them. But Durning says that protection wouldn't be so hard if, at every stage along the way, we weren't spending taxpayer dollars to harm them.

Here's the full picture as he paints it. "In the headwaters, hundreds of millions of tax dollars are paid to build roads so loggers can go into publicly owned forests and cut them down. From the roads, soil washes into the salmon streams and ruins salmon habitat. A little bit farther downstream, we find big public subsidies for ranchers to run cattle on land that is also publicly owned, and that also erodes into the salmon streams. A little farther downstream, we find hydro-electric dams and irrigation dams built with enormous public subsidies that run to billions of dollars just in the Northwest region alone. Then, a little farther downstream, we come to the aluminum smelters that are using the subsidized electricity from the dams and emitting pollution into the water, emitting global warming chemicals into the atmosphere. A little farther downstream, we find suburban developments along the sides of the rivers, all of which are subsidized with local tax dollars, because developers don't pay the full costs of providing roads and sewers and so on. So people's property-tax dollars are going to subsidize local development. A little farther downstream from that, we find the U.S. Army Corps of Engineers dredging out the channel of the rivers so that ocean-going ships can come in and out of the river and put into port. And again it's tax dollars, tens of millions of tax dollars, paying for the dredging of the river."

If that wasn't enough, Durning continues, "And then, finally, we come to where the fish hatcheries are — tens of millions of . . . tax dollars going to provide fish hatcheries that are supposed to mitigate the effects of the loss of salmon upstream, where they are affected by the logging, farming, development, damming and dredging. But, in fact, we now know that the fish hatcheries are weakening the genetic stock of the wild fish. Then, because of all of this upstream damage, all wrought at our expense, the federal government has to put in disaster-relief aid to fishing communities, welfare cheques for unemployed fishermen, and flood relief when the denuded shores can't contain the river and it overflows onto the subdivisions. And then we spend even more money to try to repair habitat or do research on how to help the fish survive or to stop flooding. So it's not that we can't afford to save the fish. It's probably that we can't afford to keep on killing them."

This story illustrates one of the many ironies of our current economic system: we give with one hand and take with the other, impoverishing ourselves as we go. A few developers and loggers get rich while rivers are dammed or forests are clearcut and lost to future generations. And the salmon, a resource that once returned annually by the tens of millions, also need ancient forests and pristine watersheds, all of which are being destroyed. We're losing the fish not because we can't afford to save them, but because we're actually paying private corporations to rub them out and to sequester that wealth and our taxes for themselves. "Billions and billions of dollars out of government treasuries or in tax write-offs go to stimulate and encourage the pollution and resource depletion and habitat destruction that we're all so concerned about," asserts Durning. "We're paying to have it all destroyed."

Taxing the Wrong Thing

> *Automobiles' private benefits are enormous and well understood, yet their abundance makes them the source of a disturbing share of social problems. They are the cause of more environmental harm than any other artifact of everyday life on the continent.*
>
> — ALAN DURNING

What we've done to the salmon is a classic example of the misdirection of public money, but it's not the only one. All over the world, we are paying billions of dollars in perverse subsidies to industries that cause social and environmental breakdown. Alan Durning has many more examples. "Most of the resource-extraction industries are fairly heavily subsidized, [as are] the mining industry, the oil and natural-gas industries, the timber industries and to some extent agriculture," he says. "But just as significant are subsidies to urban real-estate development and particularly to suburban real-estate development. Subsidies to cars — to driving and burning fossil fuels — [are] typically about a hundred dollars a month, per car, in North America. Those subsidies range from the free parking provided all over many cities to the price of gasoline, which doesn't include the full medical costs of caring for all the people who are run over by automobiles or the $20 to $40 billion spent by the American military each year in order to project power to the Middle East to keep the shipping lines open and the oil fields providing the gasoline." And, of course, it is now dwarfed by the expenses of wars in Iraq and Afghanistan.

As Durning points out, "The roads that you drive on are subsidized. They're paid for not only out of gas taxes, but also out of local property taxes and sometimes income and sales taxes. Just about everything that you do in the car, you're paying part of the cost and somebody else is paying another chunk of it."

Once upon a time, a great deal of freight — to say nothing of people — was moved around this continent on trains. Most of that freight is now being transported by trucks, which are far less efficient and hence less environmentally friendly than trains. It's commonly assumed that trains just cost too much, that they're not competitive because they have to be subsidized. Yet trucks are heavily subsidized, too, and somehow mainstream economists consider that perfectly all right. "The huge subsidization of the highway infrastructure has resulted in a long shift away from rail and onto the roads for the movement of freight," explains Durning. "Now, in the United States at least, train companies own their own railbed. The truckers don't own the roadbed, right? It's paid for through tax dollars and subsidized. So the rail system has a real hard time competing. If we simply eliminated the subsidies to road transportation, that would shift some of the freight back onto the rails, and make that a much more viable venture economically."

So-called perverse subsidies are the tax dollars that support an activity that creates a social problem that we then have to pay tax dollars to correct. Durning says we could make everyone's life easier and happier if we stepped back from our current arcane and absurd economic rules and rewarded people for doing things that make society better. For instance, taxation of payrolls could be stopped to encourage employers to hire people and give them careers. Instead, taxes could be levied on polluting activities. The industries would not be any worse off, but society would fare far better. As Durning says, "We're taxed for working, for saving, for investing. And we're subsidized to pollute, to deplete natural resources and to consume natural habitats. If we actually eliminated all of these subsidies, then we would find incredible improvements in environmental quality, and in economic performance as well, because all of these subsidies are big drains on the true economic vitality of our continent." This kind of sea-change in the allocation of government funding isn't even a pipe dream. It's a reality for most of Europe.

In the European Union, perverse subsidies are being turned on their heads. For example, in Germany, government aid that used to help German farmers purchase chemical fertilizers and pesticides is gradually being phased out. Now farmers are awarded tax breaks and social help for converting their fields and stock to organic methods. This is Germany's way of making sure it gets a serious, long-term handle on problems like mad cow disease and water pollution. The Germans are well on their way to their current goal; 20 percent of all their agriculture will be organic by 2005.

Sweden, Denmark and Holland have all made major changes in some of their most negative financial policies. Financial analysts are well aware that taxing payrolls and income discourages positive social actions like employment and savings. So these countries have lowered payroll taxes, getting that same revenue from new taxes on real social negatives such as waste incineration and pesticide and electricity use. In the United States, social-security payments that an employer has to make on behalf of each employee increased by 1,325 percent between 1970 and 1992. Naturally, this tax is a disincentive to hiring people. In Spain, the government was able to cut its hated payroll taxes by the simple expedient of raising taxes on fossil fuels, making employment more attractive while discouraging pollution. In fact, in eleven European countries, for every percent that a given industry is taxed for pollution emissions, that company will have to pay 1 percent less payroll tax. The workers still get their benefits, but the company is motivated not to fire people, but to clean up its emissions (see *Good News for a Change*, pages 301–09). The kinds of perverse subsidies that economically force taxpayers to destroy resources like clean water, pristine forests and beautiful salmon could easily be legislated away in the same manner. It's such a no-brainer it's hard to understand how these perverse subsidies happened in the first place.

We have to look at history to see how we got ourselves through this particular looking-glass. Back when these subsidies were enacted, we saw forests as impediments to farming and rivers as roads to fill with ships. We didn't yet understand how they provided us with our most basic needs: fresh water, the millions of fish we scooped up and used for fertilizer, our very climate. We paid people to set up industrial infrastructures because we thought that was where the public good lay. But now it's time to say that if goods like timber, upscale housing and metals are really worth having, then the activities that exploit them shouldn't have to be subsidized. As Alan Durning says, "These subsidies reflect political power. They reflect organized constituencies that, over many decades, have sufficiently ingratiated themselves with the lawmakers that they got subsidies written into the books. The public has yet to be sufficiently informed and organized to get the subsidies out of the books. That's all it is."

Sharing the Pie

An economy is a little like a human child. When it's born, it has a tremendous appetite for raw materials, and it clamours for them. The more raw materials it consumes, the more waste and pollution it creates and the more its appetite grows. This persists, but only until the child matures. Then, all of a sudden, the phase of physical growth levels out for good. That doesn't mean the growth of that person has come to an end. On the contrary. The richest and most protracted phase of growth is just beginning — mental, intellectual, emotional, spiritual growth. Somehow, we have to get our economies out of this adolescent phase into a more mature adult phase. It won't be easy, but we can do it. We have to do it, because the growth economy we've had over the last few decades simply cannot continue.

— NORMAN MYERS

Herman Daly, a highly respected mainstream economist, is one of a growing number of thinkers who believe we have to reassess the entire purpose of economics. He says, "I think that economics should be about sufficiency as much as it's about efficiency. But it's not." Daly and others are beginning to challenge the idea of limitless growth, for the very good reason that nothing else on this planet is able to grow without limits — except cancers, and even cancer has limits set by the death of its host. That there are limits to growth is a simple law of hard sciences, like math and physics, but it hasn't penetrated soft ones, like economics. Where did this idea, which has no echo in the natural world or in human experience, come from? Daly says, "We built the modern economy around the idea of growth, I believe at least partly, in order to avoid facing up to the problem of sharing. If you don't continue to grow and you still have poverty, then you have to redistribute. You have to share in order to cure poverty. How do you cure poverty without sharing? Well, the only way we've been able to come up with is by growing."

How does growing avoid sharing? Well, it involves that old assumption that the rising wealth of the elite will somehow lift up everyone else. There will be that famous "trickle-down" to the poor. "Number one," says Daly, "that generally doesn't happen. But even if it did, it's still a substitution of growth for sharing, to avoid the moral problem of sharing. We find moral problems too difficult. So we convert them into technical problems of just growing faster. Then we won't have to deal with them."

The problem with continuous growth is that, as the science of physics tells us, we live in a closed system with respect to matter. We can't have a systematic deterioration of the productivity of nature — of species diversity, for example — and still maintain the ecosphere's abundance. And in order to have a chance of acting sustainably, we need a society that is just, equitable and distributes resources fairly. Without such a society, some people will take far too much while others become so desperate that they will chop down the last trees or use up the last drop of water.

Paul Hawken, a businessman and author of several classic books including *The Ecology of Commerce,* developed a concept he calls "natural capitalism." This notion maintains that we need to see nature as the true capital on which our lives and economy depend. And if we learn to value nature, our real wealth, we'll take better care of it. He says that our economic system "works for practically no one, except maybe the 1 percent at the very top. Our system wastes the environment. It wastes people. And it's very, very expensive. We need a radical change in how we relate to resources and people and the environment."

It's plain that if we are to survive on Earth, we have to make some fundamental changes in our economic value system. Growing forever and not sharing haven't worked, and we're now stretching the limits of the natural envelope that sustains us all. As Herman Daly says, "We got away with that approach, growing and not sharing, for a long time because we were living in a relatively empty world with abundant resources and plenty of space. Now that we're in a full world with much tighter limits, I don't think we can get away with that strategy any longer. And maybe, in some way, that'll force us to face up to the moral issue at long last."

Conventional economics suggests that with steady growth, life will get better and better. But even in industrialized countries, rising GDP and more material consumption have been accompanied by family and community breakdown, pollution, atmosphere change, depleted resources, violence, alienation and drug abuse, downsides to the affluent life that are rarely mentioned when a community debates whether or not to allow a new industry into its borders. It is time to look for different goals and other ways to assess progress. As an example of how we can change, Hazel Henderson points to Jacksonville, Florida, which has had its own indicators of quality of life and progress since the 1980s. "The citizens get together, and if they're doing well on their school and education indicators, for example, they'll say, 'Okay. Let's move the goalpost now. We're ready to achieve higher standards here.' Then they do the same with water quality. They have a feedback component from the citizens, a huge kind of forum, and the media picks this up. This is a living, breathing process. It's part of a new system of collective decision making."

If Jacksonville can do these things, so can many other cities. Henderson says, "For 98 percent of our experience, we have lived in nomadic tribes of twenty-five or so, as gatherers and occasional hunters. We got quite good at living in villages. We had to develop these kinds of feedback systems to run our affairs. And then we moved into towns and cities, and we now are living in mega-cities, and they're new to us. We have absolutely no experience of how to manage our affairs in these big arenas. These well-being indicators are a rudimentary form of social intelligence. I see them as real social learning and cultural evolution, which is really what has to happen right now."

We have made changes before. In fact, that's what humans are best at — adapting to new conditions and change. And we've had the paradigm of growth economics and corporate power for only about a hundred years, which is a brief moment of our evolutionary past. As Hazel Henderson says, "We're talking about cultural development and full development of the human being, not just economic development. Basically, haven't we all been talking about the evolution of the human species?"

I once flew into a village of 200 people in a remote part of British Columbia. I was met at the dock by most of the villagers, and that night went to the community centre for a feast. Everyone was there, and the tables were overflowing with delicacies from the sea, aromatic fruits and wild vegetables. After dinner, the tribal leader began his speech to me by saying, "We are poor. We need development." He then went on to explain why it was necessary to allow forest companies to log. When it was my turn to speak, I began by saying, "I live in a city called Vancouver, which is highly developed. In the one block where I live, there are probably three times as many people as there are in this village, yet I know only ten or fifteen of my neighbours. At night, I lock the car and my house because we've been broken into several times. My children can't play in a nearby park at night for fear of deviants or drug users. We buy our water in bottles for more than the price of a similar amount of gasoline. We have to stay inside when there's a smog alert because the air is so bad it harms our health. I could never put on a feast like the one I've just eaten here. To me, you are far richer in community and resources than anyone is in Vancouver."

On my way out of that place, I realized that it is our idea of economics and progress that actually impoverishes people. Those village folk had a rich, productive ocean and forest and a vibrant community. Yet they believed they were poor, and people like me seemed rich in comparison. Our current use of economics forces us all to destroy the very things that sustain us. Of course, those people in the villages want many of the things that we take for granted in cities: books, doctors, even TV and running shoes. But must they come at the cost of vibrant societies and rich ecosystems? We need new ways to measure wealth, ones that put value back onto community, air, water and the richness of life around us. Some people have already started finding out how to do this. They are shifting their priorities and setting up institutions and agencies to pursue that kind of wealth. We've done that before, when we founded democracies to strive for the common, not the particular, good. We can do it again.

CHAPTER 8

GLOBALIZATION BLUES

*Economic globalization involves arguably the most
fundamental redesign of the planet's political and economic
arrangements since at least the Industrial Revolution. Yet the
profound implications of these fundamental changes have
barely been exposed to serious public scrutiny or debate.*

— JERRY MANDER, HEAD OF THE INTERNATIONAL
FORUM ON GLOBALIZATION

Obviously, the global village anticipated by Marshall McLuhan is now a
reality. We are plugged in electronically to the remotest parts of the planet. Through that network, consumer goods, information, culture, politics, even sports and children's entertainment, are being transported
around the world. But the force driving this takeover of the world is not
just military might, as it was for the great political empires of the past.
Today, power is no longer the exclusive prerogative of the nation-state.
Now it is increasingly exercised by private corporations, both on their
own and in a kind of unholy alliance with nation-states. The change has
been revolutionary.

John Cavanaugh is the author of a book entitled *Global Dreams* and
the co-director of the Institute for Policy Studies in Washington, D.C.
The institute is a think tank that examines the way national and international political and social policies affect people's lives around the world.
Cavanaugh reminds us that of the one hundred biggest economies in the
world right now, fifty-one are private companies. "That means that the
big companies like Mitsubishi, Mitsui, the giant Japanese trading companies, for example, are bigger than Denmark, Thailand or Indonesia," says

Cavanaugh. "That means that Toyota or Royal Dutch Shell is bigger than Norway. Exxon is bigger than Finland. Wal-Mart is bigger than Poland." Not only are these corporations bigger and more powerful than countries that have hundreds of years of independent history, they are also "economic entities of a size and power that are unequalled in the history of the world." And like any powerful economic entity, such as the Church in the Middle Ages and the Renaissance, they have infiltrated their members into the highest echelons of government, so that governments are increasingly responsive to them as much as, if not more than, the citizens they govern.

There have been big companies in the past, mostly beginning in the nineteenth century. But the fact that now there are so many and they are so huge reflects the current demand for constant growth. Since individual countries and their consumer markets are finite, the only way for these corporations to keep growing is to go beyond national borders. Take a look in bars in Australia, South Africa or Thailand and you'll find Heineken, Guinness or Tuborg beer, all originally coming from small European countries — the Netherlands, Ireland and Denmark. As Cavanaugh says, "They all, over twenty-five years ago, saturated their local market and, by nature of the corporate need to expand or die, went overseas . . . . In the past twenty-five years, most large companies have gone global. The extreme case would be a company like Coca-Cola. It's over one hundred years old. Americans are saturated with Coke; you'd be hard pressed to get them to drink any more of the stuff. But the Chinese market isn't saturated with Coke. Nor is the Indonesian market." Today, whether you make cars, software or running shoes, you'll eventually have to go global to stay competitive and profitable.

These companies amassed their enormous wealth so quickly by expanding into new territories just as aggressively as Alexander the Great and the British did when they claimed their empires. This corporate empire-building went largely unnoticed, however, because their power was measured in terms of wealth, not military might. Cavanaugh says, "For much of the Cold War period, 1945 to 1990, the growing power of these global corporations was obscured by the Cold War itself, and by the dominant military and economic force of the United States in global markets and in world politics. With the end of the Cold War and with the shift from what we might call the age of national security to the age of globalization, more and more people started to become aware of the fact that the true economic power in the world is no longer the United States, per se, but a phalanx of 200 large companies whose sales are the equivalent of over one-quarter of global economic activity. It is these entities that more and more determine the working conditions, the health conditions and the environmental conditions of people around the world; in this regard, they have more power than the military might of the United States by itself."

The idea of a global economy is usually portrayed as a natural outgrowth of modern progress. Globalization has been depicted not only as positive, but as inevitable. Jerry Mander, head of both the International Forum on Globalization and the Foundation for Deep Ecology in San Francisco, says, "Globalization tends to be described as something that evolved, as though it were some sort of force of nature. But . . . it's not an accident, and it's not a natural evolutionary process. It is a designed system, set up for the specific purpose of enabling corporations to make the rules of economic activity globally."

Globalization was deliberately designed by a particular group of men quite a while ago, according to Mander. "This global economic design was in the minds of economists in the 1920s and 1930s, but it came to fruition . . . at the end of the Second World War in the United States during meetings at Bretton Woods, where the leading economists, corporate CEOs and government leaders got together to make new rules. They were coming out of the Great Depression and the Second World War. There had been tremendous devastation, and they were trying to figure out how to design a world economy in the wake of all this horror and destruction."

These people weren't trying to seize power from democratically elected governments, and in many cases weren't trying to amass wealth for themselves. "I think they thought of themselves as 'do-gooders,' " Mander says. "They were trying to think of what kind of system would prevent such terrible things from happening again; what kind of system would provide the things people needed; and how they could help spread the wealth, so to speak. So they designed a system. The best one that they could come up with was one that accelerated economic development everywhere in the world. They believed that economic development was the final answer to people's problems, so they actively tried to accelerate free trade. They broke all the rules that countries used to control economic activity, and they put real power in the hands of large corporations via such institutions as the General Agreement on Tariffs and Trade, the World Bank and the International Monetary Fund."

When news of these new global institutions first hit the media, most people were confused and a little bored. These new acronyms seemed like arcane entities that had little to do with people's daily lives. But their impact on ordinary lives has been immense. As Mander says, "Since then, the GATT, NAFTA, APEC [Asia-Pacific Economic Community], the European Union and all these mega-institutions have essentially integrated all economic activity and put all countries under the same economic system. Now they create rules that all countries must follow. The purpose of these rules is to free up corporate and financial activity so that industry can be completely unfettered, so it can find and use up more resources, locate the cheapest labour and operate globally without any interference. The theory is that this has general benefit for us all."

Winners and Losers

> *Under a system of global free trade, the losers will, of course, be those people who become unemployed as a result of production being moved to low-cost areas.... The winners will be the companies that move their production offshore... that benefit from paying lower salaries at home... that invest where labour is cheapest and that, as a result, receive larger dividends. But they will be like the winners of a poker game on the* Titanic. *The wounds inflicted on their societies will be too deep, and brutal consequences will follow.*
>
> — JAMES GOLDSMITH, PARLIAMENTARY REPRESENTATIVE
> FOR BRITAIN AT THE EU

In the aftermath of the Second World War, the new monetary twins, the World Bank and the IMF, were impressively successful in aiding the recovery of the economies of war-ravaged countries. With Europe and Japan back on their economic feet, these financial institutions turned to the Third World, whose poverty, it was believed, could also be conquered by greater trade and economic growth. The wealth created would trickle down to the benefit of all sectors of society and the global community. The Organization for Economic Co-operation and Development (OECD), which is made up of twenty-nine of the world's richest industrialized countries, is a kind of club, an economic think tank of people who consider themselves to be in the best position to give financial advice to others. The OECD was founded after the Second World War to monitor and encourage trade in the interests of peace and stability. Today it's one of the leading voices supporting economic globalization. As recently as 1999, OECD secretary-general Donald Johnston, a Canadian by birth, was claiming that since all these new organizations have taken over the business of global trade, we're all much better off. "The numbers are certainly very impressive," he asserts. "The trade in North America has increased dramatically from 1993 to 1996, an increase of something like 43 percent. Mexico has now replaced Japan as the third-largest trading partner with the United States. Unemployment in all of the countries has actually decreased, and direct investment has increased. So I would say that NAFTA has been a great success. And, in fact, one could say, looking at it from a regional point of view, that this is hopefully what this so-called globalization would bring to an ever-increasing universe."

To flourish, the globalization movement, like the rest of the economy, needs to grow. So whenever there's a new union, such as the European Union, NAFTA or APEC, the OECD applauds and calls for even bigger economic organizations. As Johnston says, "In other words, these regional agreements should be looked at as stepping stones to a more global arrangement — so that there will not be excluded countries, so that there won't be anyone on the margin."

The new free trade regime is certainly making sure nobody is left out. Following the predominant economic theory, it has to expand ceaselessly not only in terms of organizations to enforce and monitor its dictums, but to keep finding brand-new areas to govern. So the original GATT, established in 1948 but vastly expanded in the 1980s and 1990s, spawned the WTO in 1995. The European Economic Union keeps finding new republics to join its trade consortium; countries such as Turkey, Poland and Bulgaria are coming in hard on the heels of Greece and Spain. NAFTA, the trade agreement between Mexico, the United States and Canada, is now having an oversize child in the shape of the FTAA, which hopes to bring in economic powerhouses such as Brazil and Chile. And it's not just the exchange of hard goods, like corn or pork bellies, that's being monitored and regulated on a global level.

Thanks to the WTO, we now have Trade-Related Intellectual Property Rights (TRIPS), binding global agreements that help companies like Syngenta, Novartis and Monsanto to find patentable green gold in the rain forests of the world and also make sure nobody is planting their seeds without paying royalties. Most governments on Earth, including Canada's, have signed onto Trade-Related Investment Measures (TRIMs), which dictate what governments are allowed to do in terms of regulating foreign investment, that is, whether they can demand that a foreign company employ a certain number of its own people or pay state taxes, even whether it can set standards for its own purchases, such as ecological sustainability, that might in any way "discriminate" against a cheaper product. There's the Agreement on the Application of Sanitary and Phytosanitary Standards (SPS Agreement), which "constrains government policies relating to food safety and plant and animal health" — in other words, it makes sure there aren't too many trade-disrupting prohibitions on pesticides or GM foods. Six other major agreements under the powerful aegis of the WTO include the Agreement on Agriculture. The AOA sets rules on the international food trade and thereby can restrict domestic agriculture policy, such as government legislation that supports farmers or the maintenance of emergency food stocks to make certain citizens have an adequate food supply. There's also the General Agreement on Trade in Services (GATS), which makes sure labour can't point to unfair advantages such as high wages and benefits or more accessible education in one country over another.

It would be one thing if these global regulations were being set up to assure citizens of the new global economy that animals would be treated as humanely in Mexico as in Britain, or that food standards in India would be gradually brought up to Germany's. Proponents of globalization envisage one world so closely linked by economics that there would never be a reason to go to war. Why would there be ethnic or religious hatreds when we have so much in common? A global economic vision even foresees a future in which humanity would have ways to unite, to combine our labour and aspirations, to work for global goals. For anyone who wants global sanctions against dumping toxins into the air or the seas and endangering species, globalization could be a very good thing. For a number of reasons, consolidating into larger entities is not necessarily bad. But the first problem is to judge whether the current method of economic union is working, and if humanity in general is benefitting. If the answer is yes, it's a lot easier to jump on board.

Despite the glowing report card that Donald Johnston and many others give NAFTA, escalating numbers of critics believe that it's been a disaster, and that further economic unions have to be very carefully assessed. Maude Barlow is one of the critics. She's the chairperson of the Council of Canadians, a grass-roots organization that led the battle against NAFTA and its predecessor, the FTA. She claims globalization has had very different results. "What free trade did was improve the health of about 200 corporations that work across the North American border. And it improved the investment opportunities for some of the people who are able to invest in those companies. So, yes, some people did just fine. But it was not fine for the 58 percent increase in child poverty. It is not good for all the people who had their hospitals closed down and their schools underfunded because we suddenly have no more public funding for social programs in this country. It's not good for the seniors who've seen $8 billion come out of the pensions in the past five years."

Jerry Mander says that the theory that free trade benefits the general population has had twenty-five years to prove itself. Initially, during the particular conditions of post-war recovery, it looked good. But he says it is not living up to that early promise. "It's already clear that it doesn't promote the general welfare; that poor countries are getting, relatively speaking, poorer as compared to rich countries. The poverty gap is widening within the rich countries too. The people who have really benefitted by this expanded trade are the corporate heads and very few others. They talked about it being a rising tide that will lift all boats, but as it turns out it really lifts only yachts. Worse than that, this accelerated trade, this freeing-up of all resources, the 'no controls' on development (which means no environmental controls, no labour standards, no social programs) have brought us to the brink of an environmental breakdown faster than anything before has. We're seeing global warming. We're seeing ozone holes. We're seeing the loss of species and habitat. We're seeing pollution and terrible over-exploitation of the oceans. And all of this because of the theoretical benefits that will come to the general human population from this free-trade system of unlimited, global industrial development."

At least as far as environmental issues go, anyone who has looked at the statistics since the start of globalization will be inclined to agree with Mander. Maude Barlow says, "Globalization is not good for the environment. We have seen cuts of two-thirds support for natural-resource protection at the federal level [in Canada], and between a half and two-thirds in most of the provinces across the country. It's not good for the animals. It's not good for the forest. It's not good for our water protection. It's good for the mining companies. It's good for some of the forestry companies. It's good for American transnationals and their subsidiaries here. Yes, it's good for them."

Barlow and the other critics contend there's no money to pay nurses or radiologists or to maintain art, music and sports programs in schools, despite newly balanced budgets, because of the naked grab for public money by the private sector. Because they no longer pay taxes at all commensurate with earnings, corporations are having the wealth that once went into public coffers rerouted back to them. So when it comes to assessing globalization, Barlow says, "This system has created winners and losers. And the analysis that says that free trade is working is an analysis coming from the winners."

What's Good for General Motors Is Good for Us All

> *Corporations, by their nature, do not function as democratic organizations, yet it is they who have stepped into the vacuum created by failed political institutions and taken up the daily work of politics. Their tremendous financial resources, the diversity of their interests, their squads of talented professionals — all these assets and [more] are now relentlessly focused on the politics of governing.*
>
> — WILLIAM GREIDER, POLITICAL JOURNALIST

When Canada signed NAFTA, many people were worried about its effect on political sovereignty, national culture and our ability to protect the environment. But we were assured that a separate environmental department, the North American Commission on Environmental Cooperation (NACEC), would harmonize the environmental laws of the three signatories, Canada, the United States and Mexico. The assumption was that Mexico's laws would be brought up to our level. But that's not what has happened. After environmentalists struggled for decades to get better standards, trade agreements like NAFTA have pushed these standards down. Maude Barlow says governments no longer have the power to establish independent regulations, and "a lot of us don't have problems with that, if we're talking about higher regulatory standards. But what happens is that these trade agreements take precedence over the international environmental agreements. Multilateral Environmental Agreements, or MEAs, are overwhelmed by trade agreements like NAFTA, because trade agreements have enforcement measures that the environmental agreements do not. The untold story to date about the Canada-U.S. Free Trade Agreement, the GATT . . . NAFTA, the World Trade Organization, and now the new FTAA, is that they all try to lock in environmental standards where they are now, and they actively try to roll them back in some of the countries where they have attained 'too high' a standard to suit the activities of the transnational corporations. They do try to establish international environmental arenas where global standards are set, like the Biodiversity Convention or the Codex Alimentarius of the United Nations, but these are increasingly influenced — and even staffed — by the corporations. And therefore, when they talk about 'standards,' they're not standards that most of us would consider acceptable."

Barlow thinks this situation threatens not just environmental standards, but also our right to have a say in how countries and corporations are allowed to behave. She says, "We're basically entering a period that some of us are calling 'corporate rule,' where transnational corporations are using governments as fronts to set the dominant standards around everything from social programs to health-and-safety standards to the environment. And they are using international agreements to cement these new rights in international law, so they are able to supersede those of more democratically run nation-states."

Barlow's statements sound a little exaggerated. After all, surely Canada and the United States and Australia and the United Kingdom — the wealthy democracies — would never give up the right to decide what kind of laws and regulations their citizens will live under. But what we take for granted — sovereignty over the basic rules that govern a nation — is in the second stage of a profound attack. How do companies acquire the ability to nullify the regulations of nation-states? As John Cavanaugh says, it's been a gradual process: "A hundred years ago, companies were local and served primarily local markets. They employed local people, were the pillars of the local tax base and cared, necessarily, about the health of the community, because that's where they got their sustenance. In most countries, those companies created national markets by the 1930s or 1940s. In 1952, the chairman of General Motors, which has long been the largest U.S. industrial company, testified before the U.S. Congress and uttered the now-famous words 'What is good for General Motors is good for the United States. And vice versa.' "

Cavanaugh says that back in the 1950s, you could legitimately make that argument. General Motors, like many other companies — think of Sony in Japan or Volvo in Sweden — was still perceived as rooted in a national place and was a major local employer. It was also a leading force for racial integration in the United States and brought many black Americans into the unionized workforce. GM paid millions of dollars in taxes back when corporate earnings were still taxed significantly. So it made sense for U.S. policy to make life easier for the company. Today, however, what's good for U.S.- or Canadian-founded corporations like General Motors, Bombardier, Monsanto or Coca-Cola is not necessarily good for the citizens of their founding countries. "General Motors thinks globally," Cavanaugh explains. "It doesn't care about unemployment in its 'home' country. It cares about making cars cheaply and profitably using a global production base. I think fifty years ago you could say the interests of large Canadian corporations were linked to the Canadian economy, and the interests of large U.S. or British companies were linked to the U.S. or British economy. They now no longer are, and that's creating enormous turmoil, social tensions, environmental problems. It's created a mess."

Jerry Mander concurs. He says all these corporations went global partly because "the high economic growth rate that became the norm throughout the rapid development of the 1960s was not sustainable. Corporations began to run into shortages of resources; they began to max out their markets, so they couldn't keep selling at high profits to masses of people. This combination of declining resources and used-up markets meant that in order to keep growing so wildly, corporations had to be able to get at the resources that were still supposedly 'locked up' in countries that were protecting them from serious exploitation through trade or environmental or labour barriers. And the corporations also needed to get at the cheap labour markets, which in theory would produce goods less expensively, give them bigger profits and enable them to sell to people they hadn't been able to get at before. This corporate desire for expansion was building up, and then when the Soviet Union broke down there was no opposition to any of the prerogatives that the corporations wanted to establish. There was no resistance to this process at all. And so, since then, it's been a free-for-all."

The role of government has always been to regulate and balance the various sectors of society in order to protect citizens from the excesses of, say, the military or other specific interests. One of the most common questions people ask is, If MacMillan-Bloedel's or Monsanto's or Shell's interests aren't really linked anymore to Canada, the United States or Holland, why do politicians in these countries still bend over backwards to make policies that favour them? Why would national politicians give up their own powers and responsibilities to private corporations? John Cavanaugh says, "There's a certain amount of corporate blackmail that goes on . . . . For instance, corporations can say to governments, 'Look, we still provide some jobs here. We still pay some taxes here. If you don't continue to give us the sort of rules that we want, the sort of corporate favours that we ask for, we can pick up our production and move it overseas.' And so, to a certain extent, governments are in a horrible bind, trying to create the incentives that keep corporations at least partially loyal to their national roots. It's a horrible game that governments play. And ultimately I think it's a game that they lose."

A big part of our problem is that politicians are elected for only a few years at a time. In that time, they have to provide benefits that can be seen quickly, like construction jobs, new factories and shopping malls. Since social and environmental problems take years to be revealed and just as long to be averted, different politicians will be around to take the credit or the blame when they become critical. So people elected to office find themselves naturally concentrating on short-term quick fixes, like helping a pulp-and-paper company open a new mill. They know they'll be long gone by the time the effects of deforestation or dioxin contamination from that mill are felt by the public.

Most democratic governments have another built-in problem

that explains why national politicians and transnational corporations have grown so cozy. "Politicians in most of the world are elected into office in increasingly expensive campaigns with corporate money," says Cavanaugh. "It's not in the interests of government leaders to make it clear to the public that the interests of large corporations are now diverging substantially from the interests of countries."

On top of all this, we now have an enormously powerful global body, the WTO, that has the power to punish countries that dissent from its policies with crushing economic sanctions and boycotts — the kind of thing that brought South Africa to its knees, but in a very different context. This worries a lot of people, for the simple reason that the WTO, as Maude Barlow and Tony Clarke write in their booklet on the subject, *Making the Links*, "is set up to serve the interests of big business and to promote economic globalization in a world increasingly dominated by transnational corporations." The coziness that exists between powerful oil companies like Halliburton and Unocal and George W. Bush's administration has long been noted, as has the undue influence on bodies such as the FDA and the Canadian Food Inspection Agency by corporations like Monsanto. The U.S.-based Pharmaceutical Research and Manufacturers Association (sometimes dubbed "Big Pharma") spent almost $200 million to elect Republicans in the November 2000 presidential election. This was, in fact, "the most money ever spent by any corporate sector on a presidential election in American history." These representatives have dutifully worked ever since to protect their patrons' patent monopolies. Barlow and Clarke note that the Business Roundtable (or BRT), representing 200 of the United States' largest corporations, even helped fund the failed 2003 ministerial meeting of the WTO in Cancún.

Globalization means opening up markets, lowering trade barriers, becoming more competitive. Transnational corporations want the same rules to apply wherever they go, so that they can operate globally the way they did in their countries of origin. They demand what's called a "level playing field." But getting that level playing field involves a lot of multi-level dealings. It means jobs from northern hemisphere industrialized countries going to the South. It also means that people in the South will never acquire unions, minimum wage and workers' compensation. As Cavanaugh says, "We live in a world of very unequal nations. Rich nations, poor nations. Countries like Canada with relatively strong union structures, versus countries with no independent unions, like China. Countries that have strong social welfare states and programs, again like Canada, versus much of the world, where that doesn't exist. This creates a beautiful atmosphere for large companies based in Canada or the United States or Western Europe to move production into poorer countries . . . like Mexico or Indonesia or China, where they can exploit workers. They can ignore environmental standards. They can press production costs way, way down, yet still get very high levels of quality, very high levels of productivity."

Cavanaugh continues, "When northern workers try to get better wages or hours or working conditions, the corporations have enormous bargaining power to say, 'Look, we, the Ford Motor Company, or we, General Electric, can make these products at a tenth the cost somewhere else and still get the same levels of productivity and quality. Why should we be paying you this much? Why should we be paying you these lavish health-care benefits?' That dynamic is one that in the United States, for example, has led to a stagnation or decline in real wages for 60 to 70 percent of the workforce over the last generation. In other words, there's been an enormous shift in power from workers to corporations with this advent of a global economy."

Under Our Thumb

As a result of GATT and the WTO, Third World countries are under obligation to accept all investments from abroad; give "national treatment" to any foreign corporation that establishes itself within its borders . . . eliminate tariffs and import quotas on all goods, including agricultural produce; and abolish . . . [any] regulations to protect labour, health or the environment that might conceivably increase corporate costs. Conditions more favourable to the immediate interests of transnational corporations can scarcely be imagined.

— EDWARD GOLDSMITH,
PUBLISHER OF *THE ECOLOGIST*

One of the oldest of the organizations set up to monitor a new global environment is the IMF, founded in 1944. In a pattern that has been repeated again and again, the good intentions it started out with were transformed into something quite different. Michel Chossudovsky, an economist mentioned in chapter 7, says the IMF, like all the other global institutions, was founded for all the best reasons. It was set up with a pooled fund from several nations to stabilize the rate of exchange for member countries, so they wouldn't be subject to the disastrous inflation and currency devaluations that were experienced during the Depression, and which contributed to the start of the Second World War. But over the years its role shifted almost 180 degrees. "In the late seventies," Chossudovsky says, "the IMF started to police country-level economic reforms of debtor nations, imposing very harsh conditions on . . . Third World countries that had become indebted both to governments and to private banks. They were unable to service these debts. Then the IMF came in, more or less representing the interests of creditors, and said, 'We will help you to reschedule your debt. We will give you certain facilities. We will give you money. But you have to accept and implement a certain number of reforms.' Essentially what happened is that the creditors of the state, using the IMF as their bureaucracy, started to impose a whole series of economic policy measures. And in many regards, these were very drastic reforms."

We have to remember that Third World countries got into such terrible debt in the first place largely because organizations like the IMF and private banks and corporations urged them to. It was like being encouraged to use a new credit card to buy expensive goods before you have a steady job. A line of credit was offered to these poor countries so they could build infrastructures like roads, modernize and accommodate the factories of all the corporations that wanted to exploit their new markets. They were urged to grab this opportunity to take part in the same measures that had lifted Italy, Japan and Germany out of post-war poverty. And most countries did. Throughout the 1960s and 1970s, in Africa, South America and Asia, governments were encouraged to make their resources available to transnational exploitation and grow cash crops like coffee and shrimp in order to "participate in world markets." These new industries were funded with generous loans from the World Bank and private corporations. But by the 1980s, the debt load of most developing countries had become so enormous that they were struggling just to meet interest payments. The profits generated by the economic activity went directly into the coffers of the multinational corporations, or into the pockets of a tiny cadre of their local friends, while the general economies of these countries stagnated.

At this point, the IMF came in to help. Their solution was called "structural adjustment," or "economic rationalization," and involved massive cutbacks in public health, education, environmental protection and social services, as well as the dismantling of food and farm subsidies. With missionary-like zeal, the IMF imposed a kind of laissez-faire capitalism on countries that had neither basic development nor a long history of citizen involvement, and where riches were already concentrated in private hands and thus unavailable to most people. Michel Chossudovsky thinks several of the most horrible human catastrophes of recent years were the direct result of this economic policy. He has lived and worked in many parts of the world and continues to work for the Third World Network, a highly effective group of academics and activists that tries to protect the poor of the world.

He says that if we look at apparently political or ethnic tragedies around the world, we often find that they have been triggered by an economic crisis, and that the IMF was frequently there first. In Yugoslavia, for example, internal quarrels and schisms led to tremendous instability and culminated in the ethnic cleansing in Kosovo. Chossudovsky says, "In 1990, the IMF . . . intervened in Yugoslavia and implemented a very drastic austerity program, which was based on cutting transfer payments from the Belgrade government. It was a federal system, so there were transfers to the separate republics or provinces, just as there are here. It's as if in Canada the federal government said, 'From now on, the provinces are left entirely to their own devices.' Well, obviously, under those conditions, the federal system collapses. And that is exactly what happened in Yugoslavia." The IMF actions were meant to redirect payments to the creditors of the Yugoslavian state — that is, to rich North American and European banks — but the result was that the federal fiscal structure was destroyed, and within eighteen months Croatia and Slovenia had seceded.

Chossudovsky adds, "Another very important policy was implemented in 1989 and 1990 in Yugoslavia. Called a bankruptcy program, it was run under the jurisdiction of the World Bank. Similar types of bankruptcy programs have been adopted in virtually all these debtor countries." The World Bank would come in and pinpoint various sectors of the national economy, regardless of whether the industries within those sectors were state enterprises or were managed by the workers. And, Chossudovsky says, the World Bank would decide "this one, this one and this one have to close down . . . . Ultimately what happened is that virtually the entire industrial sector in Yugoslavia was closed down. This was not a spontaneous, market-driven process of bankruptcy; it was a bankruptcy program imposed by Western creditors so they could get their money back."

The same kinds of bankruptcy programs have been implemented in Indonesia, Thailand, Korea, many African countries and Russia. Chossudovsky says that after most enterprises were declared bankrupt, Western companies came in and bought them for a song. General Electric, for instance, purchased for $330,000 a company that had equipped the entire civilian aircraft industry in the Soviet Union. "This wholesale privatization of an entire industrial system is dramatic because, in fact, it's a whole national economy that has been sold. In other words, this is a process of recolonization. There's no justification for these programs. They are destructive. And ultimately they lead these countries down the drain."

And as we have known for at least 300 years, when a population is impoverished and desperate, it often turns on its neighbours. When people don't really understand what these international agencies have done, why an industry has collapsed and jobs and a future have disappeared, it's easy to believe that old enemies — Serbs, ethnic Albanians, Tutsis, Catholics or Jews — are responsible.

The Global Diner

Trade liberalization was imposed on Mexico [by NAFTA], and specifically, Mexico had to import corn surpluses from the United States. That destabilized grain producers throughout Mexico, especially in the southern region of Chiapas, where small farmers were literally destroyed. [The IMF and the World Bank] say these are political problems; but, in fact, if

*you follow through these reforms, you see how an insurgency in
southern Mexico was triggered [by] an economic process which
impoverished people.*

— MICHEL CHOSSUDOVSKY

Of course, even the IMF and the World Bank can't impose their will
without a country's political co-operation or obeisance. How do they get
it? Michel Chossudovsky thinks it's a combination of co-option of the
powerful through both money and ideology, and plain old arm-twisting.
He says it goes something like this: "If you don't do as we say, you'll be
left out of the global trade club. You'll be isolated and have no access to
modern markets and the riches that we have." And then, of course,
there's the biggest threat of all: "You'll starve."

Proponents of globalization point to increased trade figures to
prove that countries that co-operate with their reforms are better off
now in terms of food trade than they were before. But there are many
critics who take issue with that. Vandana Shiva, the physicist and author,
says, "As for the myth that through trade there will be more food security...you just have to look at the Indian experience in 1998. We had
more trade in food, but we had a declining consumption of food
domestically. Let me illustrate this with concrete facts. In June of that
year, we were told by the World Bank, 'Your food stocks are high.
You're spending too much money holding your stocks. Sell your
wheat.' So India obeyed and sold 2 million tonnes of wheat. In
December, obviously we were short of wheat because we had sold the
wheat that we were holding in stock for our own use, and India had to
import wheat at a much higher price. As far as trade figures are concerned, there's been a tremendous growth. After all, 4 million tonnes of
increased trade in wheat took place. And the pundits of free trade say,
'This is 4 million tonnes more of wheat!' But it's the same wheat going
out and the same wheat coming in. And in the meantime, more people,
more peasants, have become poorer and are eating less food. So if you
look at the economy of food, people are worse off. But if you look at
the economy of trade, you say, 'Four million tonnes more wheat!'
Finally, if you look at the discussions at the WTO and the discussions
undertaken by global agribusiness, every single one of them says, 'More
trade means more food.' No. More trade can mean *less* food."

Shiva adds, "Trade liberalization is, in fact, creating a situation where more and more people are moving out of food production, more and more peasants are being made unviable. Agriculture is being pushed into growing shrimps and flowers for exports, not food. Add all that up together, you don't have more food. You have more food traded, but you don't have more food grown. You have less food grown and less food security overall."

Chossudovsky concurs. "I would say it's the most serious economic crisis in modern history. What has happened in Southeast Asia, for instance — Indonesia, Korea, Thailand — is abrupt and rapid impoverishment of large sectors of the population . . . . What we are now seeing is that all the achievements of the post-war period in the developing countries — in other words, the post-colonial economic- and social-development process — are being dismantled by this new system of globalization."

The IMF and the World Bank are still giving out advice based on policies that have already proven to be failures. In Mozambique, the minimum wage was set by the IMF at seventeen dollars a month, after a series of devaluations. "You can imagine what kind of standard of living you have at seventeen dollars a month," says Chossudovsky. "But the IMF said to the Mozambique government, which wanted to increase wages, 'No. That would be unsustainable. You won't be competitive on the world market with wages of twenty-three dollars or twenty-five dollars a month.' In other words, they intervened, blocking the government's desire to raise these wages."

Somalia followed World Bank reforms for an entire decade. The reforms started in the early 1980s, and before the end of the decade public-sector wages had fallen to three dollars a month. The country collapsed and civil war ensued. Says Chossudovsky, "The World Bank said, 'Well, perhaps we've gone too far. We will now negotiate a program with the government that is intended to raise the levels of wages in the public sector. However, this program will have to be financed by firing 25 percent of the Somalian civil service.' The whole exercise was absurd, because by this time the whole civilian administration had collapsed and the country was at war. But the IMF was still talking about austerity and budgetary restraint."

The austere measures imposed by the IMF and the World Bank on developing countries are familiar to those of us in richer countries like

Canada. The civil service has been stripped down, hospitals closed, entire research departments terminated and schools underfunded. In fact, our social problems, which have been made worse by severe economic policies, are the same as the Third World's. It's just that our greater wealth has helped partially mask the destructiveness of these policies.

This largely U.S.-led program to force other countries to "live up to their fiscal obligations" by repaying corporate loans at any cost, with resulting domestic political crises, may be about to backfire on its originators. George W. Bush's reckless expenditures on war, security and oil exploration inspired the IMF, in January 2004, to roll up its sleeves and warn the Americans that they could face the same sanctions and "structural adjustments" as have so many other, weaker countries. In other words, under the free trade system, even the world's most powerful nation is no longer in sole charge of its economy.

Evolutionary Endgame

> *Champions of corporate libertarianism gleefully greeted the disintegration of the Soviet empire . . . as a victory of the free market . . . Francis Fukuyama proclaimed that the long path of human evolution was reaching its ultimate conclusion, a universal, global consumer society. He called it the end of history.*
>
> — DAVID KORTEN, *WHEN CORPORATIONS RULE THE WORLD*

Many well-intentioned people have argued that the setbacks and turmoil experienced by countries in Africa or the former Soviet Union as they move to a capitalist economy and embrace free trade are just difficult pills to swallow while they evolve slowly but surely towards prosperity in a new and more modern world. After all, these people say, we went through it, too, and look at us now! Indeed, the United States and Canada, Europe and Japan, and many other leaders of the globalization offensive, did experience social unrest and widespread poverty when they first embraced capitalism and industrialized in the nineteenth century. There was terrible air and water pollution throughout the British Isles and the American East. Belching factories filled the horizon with their wastes as they provided the ill-paying jobs that foreigners and rural migrants came to the city to pursue, just as they do in the Third World today. But free marketers believe, or at least say they do, that "the Market," like some beneficent, all-knowing god, will eventually regulate these temporarily painful evolutionary steps for the best in its own mysterious, unfathomable way, the same way it did for us. That is, the enormous profits accruing to millionaire factory owners will, over the next generation or so, trickle down to the working masses, and they'll end up like us, with houses, nicely mown lawns, family pets, TVs and hobbies. And, presumably, job benefits, schools and health regulations. It's a nice thought, and as Herman Daly comments in chapter 7, it gets you neatly out of having to share anything right now.

Economic and business analyst David Korten says free-market supporters especially like to cite the recent history of the fall of the Soviet Union to bolster their positions. Capitalism, they say, has proved itself far better than communism or socialism, so the more of it, the better. The only trouble is, if we look at the history of the nineteenth and early twentieth centuries, that's not what happened. "Contrary to the . . . claims of corporate libertarians," Korten points out, "the West did not prosper in the post–World War II period by rejecting the state in favour of the market. Rather, it prospered by rejecting extremist ideologies of both Right and Left in favor of democratic pluralism: a system of governance based on a pragmatic, institutional balance among the forces of government, market, and civil society." In fact, the first round of what was then termed "laissez-faire capitalism," rampant throughout the nineteenth century and which now goes under the name of "free trade," was, following the Great Depression and the First and Second World Wars, radically modified and placed in a regulatory straitjacket. The misery and rage this economic system caused among the majority of the world's populations inspired social upheaval, wars, colonial rebellions and depressions, and the state had to respond. Capital, as always, continued to call for its complete freedom, but socialist theories inspired citizens to demand, and fear of socialist gains eventually got them, the job benefits, school support, health and environmental regulations, and financial and hiring constraints that most modern countries still enjoy today — in short, everything that's being thrown out under the WTO.

Korten says of the United States: "A relatively egalitarian income distribution [following the Second World War] created an enormous mass market, which in turn drove aggressive industrial expansion. America was certainly far from socialist, but neither was it truly capitalist . . . . It was the America of democratic pluralism and equality that defeated communism, not 'free-market' America." As for the Soviet Union, it "embraced an ideological extremism so strongly statist that the market and the private ownership of property were virtually eliminated. The same ideology resulted in eliminating the civic sector's essential public oversight role. This left only a hegemonic and unaccountable state."

The key word is: *unaccountable.* We have made business corporations as all-powerful as the old Soviet state, and now we're entrusting them with governing us. Korten points out that the reasons our current free-market institutions are increasing economic inequality, poverty and environmental damage and failing to cope with hunger and social unrest are "virtually identical to the reasons why Marxist economies failed." Both concentrate power in "unaccountable, centralized institutions"; both create systems "that destroy the living systems of the earth in the name of economic progress"; both depend on "mega-institutions" and extremely complex, high-maintenance technologies, instead of simple, local oversight; and both take such a narrow, materialistic view of human needs that the "sense of spiritual connection to the earth and to the community of life essential to maintaining the moral fabric of society is undermined."

David Korten is no socialist. He likes the idea of market governance a lot more than the people who claim to support it, and quotes Adam Smith, author of the globalization Bible *The Wealth of Nations.* Smith wrote in the eighteenth century, and "his vision of an ideal market economy was one composed of small farmers and artisans, a circumstance in which owner, manager and worker are commonly one and the same. Communism vested property rights in a distant state and denied the people any means of holding the state accountable for the exercise of its rights. Capitalism persistently transfers property rights to giant corporations and financial institutions that are largely unaccountable, even to their owners." There is a third way. Korten describes what has been called a "community enterprise economy," composed "primarily, but not exclusively, of family businesses, small-scale co-ops, worker-owned firms and neighbourhood and municipal corporations." He outlines many other ideas as well: how to tax so as to benefit the largest number of people and

not unduly curtail enterprise; how to restore money creation to governments (money creation has devolved, although most people don't realize it, entirely into the hands of private banks); how to control rapacious corporations and share intellectual property. The whole recipe is published in Korten's *When Corporations Rule the World*. "There are signs of hope," he writes, "even in the growing excesses of the corporate world, because the more obvious and arrogant the excess, the faster the spiritual and political awakening of the world's people unfolds."

Developing the Poor

> *The massive efforts to develop the Third World in the years since World War II were not motivated by purely philanthropic considerations, but by the need to bring the Third World into the orbit of the Western trading system in order to create an ever-expanding market for our goods and services and a source of cheap labor and raw materials for our industries.*
> — EDWARD GOLDSMITH, "DEVELOPMENT AS COLONIALISM," *THE CASE AGAINST THE GLOBAL ECONOMY*

There's something else that most analysts don't discuss that's responsible for the great wealth of the countries of the North and the West, both before and after the Second World War. They all were able to gain control over resource bases of true wealth, either within their own borders, like Canada and the United States or, as in Europe's case, abroad in former colonies in Africa, South America and Asia. Goods like fruit, palm oil, furs, fish and timber rolled out of these countries and ended up in the kitchens of Europe. Riches such as ivory, gold, silver, tin, oil and gas built opera houses, hospitals, colleges and new roads; they also funded the institutions of business and science that have kept the North in the driver's seat ever since. So the question free marketers need to answer is: even if you're right about wealth trickling down, where does the Third World get *its* Third World resources from, that is, its source of tangible, material wealth? Where are the diamonds, goldmines, timber, elephant tusks, fish and food that will pour into Uganda and Bangladesh from *somewhere else* as their newly industrialized economies supposedly give their citizens the incomes to purchase them?

As any analysis of human population increases and their demands on the finite material wealth of the earth will show, material riches are no longer available as they were at the beginning of the twentieth century. The Ugandans' and Bangladeshis' own material wealth is being siphoned off by the North a good deal faster than it can be renewed, as are India's, Peru's, Iraq's and even Canada's. Free marketers make noises about the wealth that lies in pure business ideas, the Internet, new products, and new needs such as manufacturing cell phones and manning computer centres, or in the intangibles of the modern economy that will somehow, rather mysteriously, lift the average Indian or Chinese peasant out of her hut and into a high-rise, but as we discussed in chapter 1, that idea has already been discredited. Wealth has to be tangible for humans to want it and pay for it over any length of time. Electronic products on the Internet, high-tech and biotech bubbles on the stock market, may provide sustenance for some brokers for a few (likely a very few) years, finding markets as trinkets and amusements for the rich. They might even help a few taxi drivers and restaurant workers to keep their heads above water if they also run one of the South's many barely functioning Internet exchanges. But that's not what will feed, clothe and house billions of dispossessed farmers and peasants in the Third World, and it is ridiculous to imagine that it could. What will feed them and make them healthy is the stark material reality that has always fed humans: good, available soil, seed and livestock; clean water, clean air, usable materials like wood and metal; decent prices for their crafts and finished products.

Like many others, Alan Durning, in his book *How Much Is Enough?* points out that countries in Central America grow export crops such as bananas, coffee and sugar "on more than one-fifth of their available cropland." Export cattle ranches in Latin America and southern Africa have replaced vast amounts of rain forest and wildlife habitat. Since that land is owned not by the people themselves, but by a very few elite families or, in the case, say, of the Dominican Republic, almost wholly by foreign corporations, the lion's share of the money those exports generate doesn't do local citizens any good at all. You might say it gives them a job, but very often the land they work on for truly miserable amounts of pay used to belong to them and was confiscated by their government and given to rich landowners or corporations. WTO agreements have not only legitimized such behaviour, they have actively encouraged it.

As for those of us who live in the global North, we're living in a dream world if we think all the true wealth of the globe will be ours forever. The forests and soils that are being depleted so heedlessly now will eventually affect everyone. Durning notes that "Japan imports 70 percent of its corn, wheat and barley, 95 percent of its soybeans, and more than 50 percent of its wood, much of it from the rapidly vanishing rain forests of Borneo." In the sweet, liberal heartland of the Netherlands, in order to provide European consumers with their high-fat diet of meat and milk, "millions of pigs and cows are fattened on palm-kernel cake from deforested lands in Malaysia, cassava from deforested regions of Thailand, and soybeans from pesticide-dosed expanses in southern Brazil." It's pretty obvious that this kind of resource use can't continue, even in the North. How is it supposed to spread to the South?

Save the Dolphins

> *Under the new system, many decisions that affect billions of*
> *people are no longer being made by local and national*
> *governments but instead, if challenged by any WTO member*
> *nation, are deferred to a group of unelected bureaucrats*
> *sitting behind closed doors in Geneva. [They] decide whether*
> *or not people in California can prevent the destruction of their*
> *last virgin forests or determine if carcinogenic pesticides can be*
> *banned from their food.... Moreover, once these secret*
> *tribunals issue their edicts, no external appeals are possible....*
> *A country must make its laws conform or else face perpetual*
> *trade sanctions.*
>
> — RALPH NADER AND LORI WALLACH OF PUBLIC CITIZEN

A uniquely Canadian story illustrates how globalization, as currently practised, not only impoverishes and undermines developing nations, but also endangers the most comfortable and complacent countries in the industrialized world as well. Ethyl Corporation is an American company that introduced lead into gasoline decades ago as a means of increasing octane to get better combustion. But lead proved to be a toxic neurological pollutant, particularly affecting growing children, and was eventually banned from use in gasoline. Ethyl Corporation then switched to a manganese-based compound called MMT, even though studies have shown manganese to be a neurotoxin that can attack the brain and induce tremors similar to those in Parkinson's disease. Car manufacturers complain that MMT gums up engines, and many refineries have refused to work with it. Several American states have banned it, and it is generally not in use south of the border. In Canada, however, MMT has been allowed in gasoline for the past twenty-five years. Eventually, objections to its use mounted, and in June 1997, after a parliamentary debate, the Canadian government banned the compound.

Ethyl Corporation saw the ban as restraint of trade and invoked Chapter 11 of NAFTA, claiming that Canada, as a signatory to this agreement, couldn't ban its product. Ethyl sued the Canadian government for $350 million for restraint of trade and for daring to discuss the subject in Parliament, thereby damaging "the reputation of the corporation." Instead of fighting the claim, the Canadian government settled out of court in June 1998. Canada paid Ethyl a penalty of $19.2 million, rescinded its ban on MMT and issued a formal statement that there was no evidence after all that MMT was hazardous.

Analysts believe there were two reasons for the government's about-face: first, it was afraid it would lose if the matter went to the trade tribunal, because *NAFTA does indeed have the power to override national health laws;* and second, Canada, as a committed partner of the United States and other global powers in the trade revolution, didn't want to appear uncooperative. So Canadian children will continue to be exposed to an unwanted additive that is largely banned or unused in its producer's home country. And Canada certainly could have used that $19.2 million for hospitals or schools.

For Canadians, this is a vivid example of how the rules of free trade can override the sovereignty of national governments and the most basic interests of citizens. The Ethyl case has even greater implications for everyone's future. Maude Barlow, whose Council of Canadians fought NAFTA and supported the now-rescinded ban on MMT, says, "What we are so fearful about, and the terrible precedent that the Ethyl case has set, is that governments are going to negotiate with these corporations before they get to the trade tribunals, so that nothing becomes public. We call it the chill effect. And what will happen is that . . . governments will no longer even contemplate certain legislation because they will be told by the industry sector, 'Well, we've had our auditors and our accountants do an assessment, and we think it'll cost you a billion dollars to stop us. You'll have to pay us off if you try to legislate against us.' " As we noted in chapter 1, many governments, rather than face that kind of payment or the humiliation of the public knowing about it, are either backing off legislation or, if there's sufficient pressure in the community, paying off the corporation when no one's looking. "So you'll find taxpayers in Canada and other countries," says Barlow, "*paying* transnational corporations for their governments to have the right to implement legislation that the people voted them in for."

The MMT case demonstrates the vast implications global trade has on the environment. Environmental regulations have been enacted in a helter-skelter way from country to country. Some populations demand strict controls on pesticides and toxic-waste dumping, for example, while others allow corporations to do whatever they want. When trade is liberalized, corporate demands often come into conflict with the more stringent local regulations.

Today, in the early twenty-first century, the WTO is arbiter of all trade disputes and makes the important decisions about global environmental standards. Maude Barlow points out that laws in individual countries have to fall in line with whatever the WTO says. "Under the WTO, some very important American environmental laws have already been knocked down — for example, the law that banned the sale of tuna caught in drift nets that kill dolphins, the famous 'dolphin-free tuna.' Mexico challenged that ruling at the WTO successfully, although few people know it. So you can't protect dolphins that way anymore. There was a more recent case of [a challenge to] a law to protect rare sea turtles from drift-net fishing that catches shrimp in Southeast Asia. Again, that's been knocked down through the World Trade Organization, which is why so many protesters dressed as turtles during the anti-WTO demonstrations in Seattle."

Why would the WTO say that Americans or Canadians, for example, no longer have the right to ban ecologically destructive practices? Barlow says, "The way it works is that they use what's called the principle of the product as you see it. In other words, a tomato is a tomato is a tomato. Tuna is tuna is tuna. It doesn't matter how it was grown or caught, or by what humane or inhumane practices, including labour practices." That means that it doesn't matter what terrible things were used — slavery, inhumane animal trapping, banned toxins — in the production or the sale or the growing of the product. "What happens is that if it's still a piece of tuna or a tomato in the end," says Barlow, "that's all you're allowed to look at."

The proponents of free trade argue that the environment is protected in the global economy. The case for dolphin-free tuna might have been lost, but they point out that the WTO considers the individual merits of each case brought to it. In principle, this is supposed to protect the interests of local regions and countries on a case-by-case basis. So what is the WTO's track record on environmental issues? Elizabeth May of the Sierra Club of Canada says, "When you look at the incredible power of the World Trade Organization, you realize that it's not just now and then that they strike down an environmental provision. There has never been a domestic environmental protection regulation that's gone before the World Trade Organization panel on a challenge and survived. Every single time an environmental regulation has been challenged in a trade dispute before the WTO, the environmental regulation has been struck down." Except for a French ban on the importation of asbestos, she's correct. This wasn't an environmental issue, but a human health one; it came up just after the Seattle protests, which didn't hurt its chances. It remains the only one.

Who are these people who decide whether anyone on Earth has a right to protect dolphins, avoid dioxins or hire more research scientists? May says that when you look at the composition of the trade panels, you realize why environmental regulations are not high on their priority list. "None of the people who sit as judge and jury at the WTO has any knowledge of environmental or health issues. All they know is the doctrine of trade liberalization. It's a very effective cadre of people who have a very complete and reassuring ideology that trade liberalization will lead us to a better world. For example, I heard them speaking at the WTO First Ministerial Conference, and one of the delegates said, 'You know, someday, if we go on down this road of trade liberalization, telecommunications and globalization, someday every single village in Africa will have its own telephone. And that will mean survival!' I was thinking, 'What are they going to do with a telephone? Order out for *water*?' What is this priority that every single village must have a telephone before every single village has drinkable water? They're certifiably nuts! But that's the governing ideology. Honestly, I've met some of these people. They're not evil. They really believe in the mantra that trade liberalization will lead us to a better world. It could. But not the way they're doing it."

Some people will argue that the globalization juggernaut must be stopped in its tracks. But May believes that there is a way to bring a global world economy into balance with the environment and with human rights. "I think the best way to redress this imbalance between the rights of economic entities and the unspoken rights of all living beings, like forests and the creatures of the oceans, is to have real global clout for all those rights that have been abandoned at the nation-state level," she says. "We need a global institution that is as powerful as the WTO, but that's a world environment organization. We need global standards that can be enforced at least as powerfully as the WTO enforces its pronouncements. This is the irony of globalization and free trade. People like me — environmentalists — are not necessarily against either globalization or free trade. But we're desperately against the way it's being forced upon us now. There's no logical reason why you couldn't construct a system where expanding trade and expanding globalization mean that every country has to meet the highest standards of anyone else within that trading system. And then we'd all really be better off."

Invisible Government

> *The proposed FTAA rules for Market Access include measures*
> *that would . . . restrict the sovereign right of governments to*
> *limit exports, including in times of austerity or catastrophe.*
> — MAUDE BARLOW AND TONY CLARKE, *MAKING THE LINKS*

There are some signs of a reaction to the usurpation of democratic rights by trade bodies. For example, the European Union is lining up against the United States and the WTO, insisting on its right to regulate the quality of food its citizens eat. Europe is also stubbornly defending its right to subsidize farmers and maintain social security. The higher health, employment and other standards maintained by the EU no doubt reflect its lonely status as the only large trade body that has elected and therefore accountable officials. One reason the U.S. and Canadian governments are so keen to start the FTAA, an even bigger trade organization than NAFTA, is so they can put the EU in its place.

Cracks in the free-trade monolith are multiplying, however. The collapse of the WTO ministerial in Cancún was a tremendous blow to its supporters, and the new alliance between developing countries like Brazil and China that took form there shows no signs of weakening. Mexico, the poster child of NAFTA, with its supposed economic boom in the *maquiladora* industrial sites that sprang up along the U.S. border when NAFTA came into effect, is seeing the huge majority of those jobs, dangerous, ill-paid and insecure although they were, literally go south — to Sri Lanka, Burma and the Far East, where people will work for even less. The flooding of Mexican markets with U.S. crops also displaced more than a million small farmers, who have literally lost their ancestral lands because of NAFTA — and they know it.

In their analysis *Making the Links,* Maude Barlow and Tony Clarke add that "the toughening of Brazil's stance with the election of the Lula government could at least result in a slowdown of the FTAA negotiations." Various moves by the United States to protect its domestic economy (the latest Farm Bill, jumping on any chance to close its borders to Canadian food products, steel tariff hikes) may also "signal some loss of confidence in the free trade model" in its very heart. Barlow and Clarke point out that conditions are becoming sufficiently desperate in countries such as Haiti, Guatemala and Brazil that their governments are getting worried and are demanding greater priority be given to the serious gap between rich and poor countries that is only widening under these free trade regimes.

Despite these developments, there is still plenty to worry about. The European water giants, Suez, Vivendi Universal and RWE Thames, are waiting for the FTAA to help them pry open public water utilities to make profits for themselves. The United States wants Mexico to divest itself of what few agricultural protections it has left, and if it and other poor countries resist, President Bush has announced the new "Millennium Challenge Account." This is a fund administered by the U.S. National Security Council, of all things, to help poor countries "increase their capacity" to participate in trade talks, and has all the earmarks of a rather menacing bribe veiled as aid. Certainly, as Barlow and Clarke write, countries "that did not support the U.S. and the U.K. in their invasion of Iraq might expect some form of retaliation through trade . . . keeping the border closed to Canadian beef so long after the single case of mad cow disease, and the sudden U.S. tariff hikes on Canadian wheat, are indicators of U.S. trade retaliation against Canada for its opposition to the invasion. Other U.S. trading partners . . . may find themselves in a similar situation."

We might say that countries fighting out their differences on the trade floor instead of the battlefield is at least a progressive, less violent development in human history, until we reflect that trade and economic sanctions can often end, not just in individual losses and misery, but in actual starvation and war in the countries that are targeted. Moreover, the people who suffer most aren't soldiers and businessmen, but usually children and the most defenceless members of society, and the gutting of their resources destroys their chances for future recovery. Barlow and Clarke conclude: "Little wonder that a growing number of citizen's organizations, labour unions, environmental groups and mass movements of peasants want to abolish the WTO and the FTAA altogether. For many of these groups, the most dangerous threat of this 'two-headed monster' is the assault on democracy itself. Under these free trade regimes, the 'rights' and 'freedoms' of corporations are enshrined while the rights and freedoms of peoples are largely trampled on . . . . Simply put, this is government of, by and for transnational corporations. As such, it threatens to become a form of tyranny against people and their democratic rights."

We're All Brown Now

Chrétien wanted to look good in front of the military dictators of China and Indonesia. There they don't see or hear protests, and he promised them the same when they came here. It's ironic, because people always say, "Oh, if you trade with them, then they'll get our human-rights standards." But it appears that if we trade with them, we'll get theirs.

— DENIS PORTER, STUDENT AND APEC
DEMONSTRATION PEPPER-SPRAY VICTIM

We were sitting in a United Nations meeting, talking about the extra-ordinary powers that corporations now have to restrict the freedom of everyone, not just the poor, and not just those in countries of the southern hemisphere. The three women we were talking with were from the Philippines, Malaysia and India. One of them smiled at our outrage and said, "Yes, we're all brown now." She meant that these new global rules give corporations the right to take away due process, discourage dissent and even provoke violence on the coddled elite of the northern countries. We're finally beginning to understand, firsthand, what it means to be colonized.

A Canadian boycott of Daishowa, a giant Japanese timber company, a few years ago, illustrates the extent to which transnational companies can override our sovereignty. Back in 1989, this multinational was given logging rights to a huge area of Alberta at bargain-basement prices. The area included the traditional territory of the Lubicon Cree, and the Lubicon have been fighting to save their forest ever since. To show their support, three students in Toronto, Kevin Thomas, Ed Bionchi and Steven Kenda, got together and organized a boycott of Daishowa prod-ucts. Among other things, Daishowa made paper bags that were bought by big chains like Roots, A&W and Kentucky Fried Chicken. The stu-dents were pretty successful in persuading a growing number of com-panies, as well as individual customers, to support the boycott. Daishowa stayed away from the Lubicon lands for several years. When they eventually went to the courts, however, they targeted the so-called Friends of the Lubicon. They accused the three students of "political guerrilla warfare and economic terrorism," for, as Kevin Thomas puts it, "leafleting in front of stores where Daishowa sold its bags. We hand-ed out flyers letting people know who Daishowa was, whether the com-pany they were patronizing bought from Daishowa and what the situation was with the clearcut. That was the extent of it."

For these acts of "terrorism," the three students were sued by Daishowa for $12 million in damages; the company also obtained a blanket injunction outlawing any boycotts against Daishowa products across Canada. For several years, the three activists lived with the suit hanging over their heads. When other groups heard about the case, they also began to back away from boycotts. It was astonishing. A for-eign corporation was given huge amounts of Canadian territory and multi-million-dollar incentives to extract the most valuable and irreplaceable resources. They were given tax dollars to build themselves a huge, subsidized pulp mill to make the clearcuts even more profitable. And now they were using Canadian courts to sue Canadian citizens and deprive them of their right to complain about it all. The company was even able to forbid the defendants from talking publicly about their lawsuit, on the grounds that it might cost Daishowa more business.

Karen Wristen was the staff lawyer with the Sierra Club Legal Defence fund, which defended the three students. She says, "The issue was freedom of speech versus corporate profit. There's just no other way to look at it. The significance for the Canadian public at large is enormous. The effect of thinking that if you go out and advocate for change in any way, and if you make statements about corporations and their activities in the world, you could be subject to a lawsuit where you're going to have to pay damages of $12 million — that's going to have an incredibly chilling effect on people. They're not going to step forward to advocate. And we see it happening right across Canada. [These cases] don't make sense as commercial lawsuits. None of these people can ever pay the millions of dollars they're being sued for. So you have to look beyond that and wonder what is the real reason. Well, the real reason may very well be to shut down this kind of legitimate public debate."

Fortunately, this story has a happy ending: the Ontario court found in favour of the defendants and their right to boycott. But the process took more than two years, and not everyone can find a lawyer willing to work pro bono. The headaches and constant pressure were enormous for the defendants, but not for a corporation like Daishowa, for whom the costs were only a drop in the bucket. And consumers could have lost one of the only tools they have available to defend themselves from corporate excess — the right to know who makes what and the right not to buy it!

The courts aren't always so reasonable. In Quebec, a similar advocacy group tried to help people in Guyana who alleged they were poisoned when effluent from Cambior, a Canadian gold-mining company, contaminated their river and estuary. The allegation was that the spill killed thousands of fish and made many people sick. When the Guyanese couldn't get legal standing to pursue their claim to hold the company liable for damages in their own country, a Canadian advocacy group headed by Dermot Travis tried to get it in Canada. They were immediately hit with a Strategic Lawsuit Against Public Protest (SLAPP) suit, which prevented them from discussing the situation with the press; they were also sued for millions. Once again, the defendants were accused of terrorism. But a Quebec court ruled the company could not be sued in Canada, not even when Cambior's Guyanese lawyer admitted there was no way for people to address the situation legally in their homeland. Cambior was never sued or made to defend its actions on the merits of what occurred in Guyana. The effect of the courts' decisions was to grant the company what amounts to complete immunity from liability, both in the country where it was founded and in the country where it practises its activities. The implications are pretty obvious.

Today, in the post 9/11 world, the idea of being seen as a "terrorist" of any stripe could mean weeks in prison with no right to lawyers or outside communication. How can people being branded with this word feel as free to advocate for their democratic rights, for the safety of their food, for the beauty and preservation of their land, with such laws in place and with global corporations poised to exploit them? Globalization and the security laws following the actual terrorist attacks aren't achieving their primary goals: to make the world more unified, monolithic, and therefore safer. On the contrary, most analysts point to cross-border trade in commodities like enormous, shared power grids and industrially produced food as an opportunity for incredibly dangerous terrorist activity. It's a lot easier to poison the vertically integrated, mixed-source food supply of North America with toxins or genetically engineered diseases than it is to affect the independent, locally sourced food still available in many of the countries of Europe and South America. The economic and democratic price we would have to pay to provide security to these gigantic, multinational food, energy and production systems is beyond imagining.

Many activist groups like the Council of Canadians are suggesting that societies should take a cue from the computer world and think about decentralizing; it would be a lot harder to bring down every PC in the world than it would be to attack an old-fashioned mainframe. Mad cow disease, hoof-and-mouth, SARS — all these crises highlight the extreme dangers inherent in constant, cross-border exchange of commodities and living things. Add genetically engineered organisms to the mix, and there's a real heyday for terrorism, and for the kinds of constant international food, disease and energy crises we're beginning to experience since the implementation of globalization, to say nothing about the steady erosion of our democratic rights and freedoms. These purely pragmatic reasons are one key element of an international turn to local sourcing — and local control — for any group's most essential commodities.

Monoculturing the Planet

> *Ideas, knowledge, art, hospitality, travel — these are the things*
> *which should of their nature be international. But let goods be*
> *homespun whenever it is reasonably and conveniently possible;*
> *and above all, let finance be primarily national.*
> — JOHN MAYNARD KEYNES, ECONOMIST

Globalization as it is now practised erodes the ability of governments and citizens to freely make decisions about their countries' welfare. It's destroying the public sector itself, the social glue that has kept our societies together. In 1997 alone, $167 billion worth of public assets around the world were transferred to the private sector — prisons, hospitals, roads and schools. And the process is escalating. Maude Barlow says, "The essence of globalization is the assault on everything left standing in the public sector, everywhere in the world — from social programs to public education to the protection of the environment. That includes the notion that this river belongs to the people who live on it and fish it; the notion that the air that we breathe is common property; the notion that knowledge is our common heritage, that the genetic inheritance of the developing world belongs to the people who cultivate it. All these fundamental beliefs are under profound assault."

Barlow backs up statements like this with examples that should strike a chord of recognition among Canadians. In chapter 1, we mentioned the effects of global trade restrictions on the everyday lives of people in the First World, even in terms of such mundane things as car insurance. It is rulings from NAFTA, the WTO and the IMF that help explain why, in so many cases, politicians like Canada's former prime minister Jean Chrétien and Ontario's former premier Bob Rae, who were elected to protect social services and clean air and water, seemed to switch priorities once in office. Barlow says, "Bob Rae promised to bring in public automobile insurance when he was elected in Ontario.

And within two years of being elected, he reneged on that promise under the threat of a billion-dollar lawsuit from the American private automobile-insurance industry. They said, 'If you do that, we've got rights in your country now because of NAFTA. We'll sue if you don't run things the way we want.' "

David Korten believes that the very nature of the corporate system, which not only encourages greed but actually demands it, is affecting our ability both to govern ourselves and to control private interests. Corporations are compelled by the drive for profit to squeeze out even the small percentage they'd lose if a national government put limits on their activities. That's why they expend so much energy trying to control both governments and people — to make sure they don't miss out on a single dime.

"These corporations . . . are under enormous pressure to maximize their profits," Korten asserts. "The kind of contemporary experience with profits that we're experiencing creates an expectation that if you have money invested, you should rightfully be getting a return of 20, 25, 30 percent, and that if you're not, then somebody's ripping you off. Now, there's no way that you can get that kind of return from a real, productive, long-term investment if, in fact, it's a responsible investment that's producing useful goods and services, you're paying a decent wage, paying your share of taxes, being responsible about the environment, and so forth. It would probably be closer to historical returns of somewhere between 5 to 7 percent."

The inflated investment profits we've come to expect are based on taking public resources — labour, state subsidies or a clean environment, for example — and turning them into private profits. CEOs are under staggering pressure to increase the returns they're giving the shareholder every single quarter; in other words, every three months they have to show not just the profit they had the previous quarter or year, but an even bigger one. This has become the norm, and in the past few years the biggest corporations have been increasing their profits at a rate of 20 percent each and every year. Korten asks us to put ourselves in the place of one of these CEOs: "Now, you're running a multi-billion-dollar corporation. How do you get that kind of increase in profits every year? Well, you'd better find a lot of people to fire. You'd better contract out more of your production to sweatshops. You'd better find places to dump your environmental waste more cheaply. You've got to press the government and buy some politicians to get tax breaks and public subsidies and so forth." Obviously, you'd also better make sure you can control any protests from the citizens by using public relations, lawsuits or government back-up.

Big business has always wanted to grow at rates of 20 percent a quarter, but for several generations previous to the present, it wasn't allowed to. Korten's system of "democratic pluralism" kept it in check, helped it, in fact, to lead a much longer and more stable life. "This was a system," he says, "in which you had a balance of corporate power, state power and union power, to represent worker interests. That system was enormously productive, enormously successful. It had very, very little in common with the system that emerged, triumphant, from the fall of communism. In the new system, government was being discarded and the labour unions were being destroyed, so there are no checks and balances within it."

Korten continues, "As long as our minds are trapped in the idea that the only choices are the communist system or the capitalist system, we're dead, because they're both failed systems." The advocates of capitalism would have us believe that there is no alternative, but that's not true. We do have other choices. "One of the things that we might actually try would be a market economy, because again we're told that we have a market economy, but that's like being told that we have democracy," says Korten. "We do truly have a capitalist system in the sense of rule by big capital. But that's very different from a market system." Korten says that if we look at what that patron saint of the free-market economy, Adam Smith, actually wrote about it, we'd see that one of its most critical tenets is "the assumption that the costs of production are *internalized* by the producing firm and reflected by the prices of its products. Now that, of course, is one of the most egregious areas in which market theory is violated by our present system."

Perverse Subsidies

Why should companies receive tax credits for their research and development when they are simultaneously shrinking their [local] employment? Why should governments pick up the tab for cleaning up social [and environmental] problems that

*were generated by private employers who failed to observe
minimal obligations to their workers and communities? . . . As
it stands now, in the name of fostering prosperity, workers and
taxpayers are helping to finance enterprises that do not
reciprocate the loyalty.*

— WILLIAM GREIDER

Environmentalists and social critics say the problem with corporate
power as it exists today is that it keeps *externalizing* its costs by making
society pay for its wastes and getting its resources for free or at the
expense of the state. If industry really had to pay the true market rate for
resources, labour and waste disposal, things would change rapidly. In his
book, *The Tyranny of the Bottom Line,* Ralph Estes estimates that in the
United States alone, the externalized costs of industry to society — that
is, costs that should be borne by the private sector but that are borne by
the public — amount to $2.6 trillion every year. As David Korten says,
"That is one incredible corporate subsidy! If you begin to look at the
extent of externalization of costs . . . it becomes very clear that these
mammoth institutions could not survive without massive subsidies. So if
we began to apply just a single principle of a true market economy —
that the producer must internalize costs — that in itself would result in a
massive break-up of these enormous mega-corporations. And it would
very quickly move us in the direction that an awful lot of the alternative-
development thinkers are committed to: namely, economies in which
markets and capital are more localized. You might call it real people's
capitalism. And the more we localize markets, the closer we move to the
more traditional village concept, where people begin to learn to live
within the limits of their own ecosystem."

The mechanics with which we could gain control of the situation
are not beyond our grasp, even now. What is needed are real regulations and controls at the global level. That's not impossible. Indeed,
we've been doing it in North America since the 1930s. And today, the
same lightning computer accounting and information networks that
have made business global could obviously be used to control it. What
if profits above 7 percent could trigger a worldwide legislative body to
investigate a given company to see how, if it were really paying its taxes
and paying decent wages and internalizing its pollution, it could

generate such profits? If it was found to be profiting through tax eva-
sion or dumping wastes, the company could be fined to such a degree
that continuing to behave like a buccaneer would cost it more than it
could make. This should be at least as easy as our current methods of
pursuing individual taxpayers who appear to have too much money for
their stated incomes.

The global bodies are already set up; the OECD or even the World
Trade Organization could do it. But the proponents of a "free market"
argue that global economic transactions by multinational corporations
must be completely unregulated so that their capital can "flow freely."
They also insist that, to get debts under control, national economies
must be regulated and controlled by austerity programs imposed by
these global authorities. It's this idea of debt that puts countries at the
mercy of global institutions like the World Bank in the first place.
Third World countries got into debt in an effort to catch up with us.
But First World countries like Canada and the United States, which
were fully industrialized and economically robust after the Second
World War, ended up borrowing heavily too. That's because they not
only gave away public resources to private industries, but stopped taxing them as well.

We are constantly told the reason countries like ours got in debt is
that we "overspent" on social programs. We allowed ourselves the luxury of health-care systems, good schools and community centres.
Michel Chossudovsky has an opposing theory. "How do we get indebted?" he asks. "First of all, we get indebted through the handouts we
give to corporations. That's been ongoing since the Reagan-Thatcher
era. We give out generous tax write-offs and state subsidies and grants.
But there's another phenomenon, which is not always understood.
Since the early 1980s, we've seen the development, on a massive
scale, of offshore banking — the development of tax havens. All
the Canadian chartered banks have offices in the tax havens. All the
major Western banks have branch offices in the Cayman Islands, in
Switzerland, in Luxembourg. And the amount of wealth held in
these offshore havens is staggering. It's equivalent to something like
25 percent of world income."

A lot of us wouldn't mind paying taxes in exchange for what we
used to get: dependable research and regulatory agencies; schools and

health care; protected ecosystems; and laws that respond to local needs and desires. But now most of us feel that we're not getting much for our money, and that we pay far more than we ever did before. Governments in the United States and Canada are striving to balance their budgets; some even boast of surpluses. But budgets are balanced at the cost of hospitals and schools, animals and trees, water and soil. And the sacrifice was unnecessary, claims Michel Chossudovsky. "What has happened is that corporations and rich individuals have transferred large amounts of money to these tax havens. And this money, of course, has escaped tax. It's been estimated that in the United States, the capital flight to the offshore havens, in terms of undeclared profits, is equivalent to the U.S. budget deficit; so if we were able to appropriate even a fraction of this capital flight, we wouldn't have had these budget deficits in the 1990s." Even Canada's current prime minister, Paul Martin, is accused of using these tax havens to shelter his shipping company from paying proper taxes to the country he now heads.

One suggestion to redress this imbalance is to impose a small tax — a fraction of 1 percent — on every single international financial transaction. The tax would help keep track of these exchanges and could fund useful reforms. The Sierra Club is one of the many citizens' and environmental groups around the world that support this idea, which is known as the Tobin Tax, after James Tobin, the Nobel Prize–winning economist who proposed it. Elizabeth May says the tax would provide a way of tracing the "hot" money that has no nationality, that pays no taxes and accepts no internalization of costs. The only major criticism of such transaction taxes comes from supporters of globalization, who claim they would be too difficult to collect. However, it's not difficult to trace computer transactions. Supporters say all that's needed is the political will. And that's where citizens of all countries can get involved. "People can understand what it's about, how it could work," says May. "Maybe it's not the best tax, maybe there's another way of organizing it. But the notion that the global economy should be wholly unregulated is just bogus."

In recent years, investors making what amount to bets on the stability of currencies have pushed societies into economic chaos in countries like Mexico, Korea and Brazil. In Canada, the dollar dropped to close to half its previous value because currency speculators decided to bet against it in order to make some money. And that could happen again at any time. "It took the stock-market crash of the 1920s in the U.S. to bring about the kind of constraints and regulation of the stock market that were necessary to protect all of society and to keep the whole system from collapsing," asserts May. "Yet the globalization cheerleaders seem to be living in some sort of dream world. They don't see that having unregulated currency transactions at the level of over a trillion American dollars a day whizzing around the world is inherently unstable and dangerous."

We're Not Powerless, We're Just Lazy

> *Little by little, we have lost our sense of mutual aid and co-operation. Two-thirds of us give no time to community activities. Fewer than half of all adult Americans... regard the idea of sacrifice for others as a positive virtue. Over 70 percent of us do not even know the people next door.*
>
> — DAVID MORRIS, COMMUNITY ACTIVIST

So what can the average person do? Elizabeth May says, "It may sound too goody-two shoes for words, but write Paul Martin or any other high-level politician. Say, 'Come on. We know what's going on. We think this is an important issue. We want to see Canada in the lead at the G-8, discussing and promoting the need to do this. We want to see our representative at the IMF change our tune.' I mean, changing the positions Canada takes globally will happen only when they realize that votes might be lost or won over these issues. Our governments have decided that the environment doesn't lose them votes, doesn't win them votes, so it's safe to lift the ban on a neurotoxic gas additive and issue a bogus statement written in advance by the lawyers for the manufacturer of the chemical," asserts May. "It's outrageous. But if the public didn't just put the newspaper down and think, 'Well, isn't that awful,' but actually started organizing, started contacting their newspapers, writing their MPs, contacting the ministers, they might change their tune."

Even our representative at the OECD, Donald Johnston, makes it clear that politicians and civil servants alike wait for input from us. When they don't get it, they feel they can go ahead and make all the decisions themselves. Johnston says, "Once I called a town hall meeting in

Montreal because there were so many important issues on the constitution and health care and so on . . . and three people showed up . . . . Having been active in the political process as an elected representative, as a member of government and also as a party official, I can tell you that people in Canada, and I think elsewhere as well, really should spend a lot more time involving themselves in the fundamental political process. And they don't do it. They should be there at the conventions of the NDP, Conservatives, Liberals, adopting resolutions that deal with these issues. Then they can deal with global institutions and [agreements like the FTAA]. They can deal with health care. The fact of the matter is that there are very few Canadians involved in that process."

Citizens have to get involved because the corporations and the people who believe so fervently in globalized free trade do get together and they are organized. They're coming up with new international bodies, new ways of making sure their agendas are not disturbed by diverse laws, national traditions or citizens' movements. A citizens' movement is, in fact, exactly what destroyed one piece of the projected trade regime in 1998. What was called the MAI or MIA, the Multilateral Investment Agreement, was such a naked grab for corporate power that citizens' outrage, when its provisions were leaked onto the Internet, caused its supporters to withdraw it. But it's far from gone. As Elizabeth May explains, "The MAI extended the definition of investment dramatically. It included culture. It included natural resources. It included every kind of so-called financial transaction you can imagine. So from the day that the MAI would be signed — and eventually they want most countries in the world to sign it — any government could be sued by any corporation that felt that any piece of legislation they were implementing or even starting to enforce in a more organized way than they had done in the past would . . . cost the corporation money. They could sue for financial compensation."

Even though its withdrawal was seen as a victory for democratic input into this highly secretive process, many of the MAI precepts are already enshrined in NAFTA and at the WTO and are being expanded in new agreements like the FTAA. May says, "The Multilateral Agreement on Investment was just one form of a concept that's already established in Chapter 11 of NAFTA. That concept is that transnational corporations are considered equal to nation-states in international law, which then supersedes the constitution of individual nation-states. Then they are given tools with which to sue if their interests are in any way hurt . . . . In other words, no one can sue corporations back, and that includes environmental groups. The MAI is not only not dead, it's here. It's living. It's already present." Maude Barlow and Tony Clarke's analysis of the proposed Rules and Disciplines of the FTAA, for example, demonstrates that the MAI's enshrining of corporate rights above government's is simply coming in the back door. Astonishingly, especially when dealing with countries such as Ecuador, Peru and Mexico, where famine is never far away, there are no protections for national food security programs, which "may be considered trade barriers that must be removed. The necessity of land reform is [also] not recognized or protected." These small, struggling countries must "eliminate tariffs on all imports within a maximum of ten years, *despite needs*," thus opening their borders to heavily subsidized and mass-produced American and Canadian goods. Conversely, they aren't allowed to limit their exports of needed commodities, *"including in times of austerity or catastrophe."* Suppose Canada had a tremendously cold winter. It wouldn't be allowed legally to keep some of its own natural gas and hydro power to keep its own citizens warm. Suppose Ecuador had a crop failure. It couldn't keep any of its corn for itself. Ecuador would have to sell the corn, then buy more on the open market at normal prices, if it could afford to do so given its problems at home.

Countries signing this agreement, that is, almost all of the western hemisphere, aren't allowed to use any deterrents, such as export taxes, to discourage the sale of their forests, fish, coral reefs "at prices less than substitution costs," as if such things could be replaced. Environmental or social regulations "that are more restrictive of trade than necessary" are prohibited. Sovereign countries won't be able to limit the establishment of private service and utility companies coming in to make money off their public health, education and water systems; foreign private companies can challenge any government service "monopoly" by demanding the same treatment, so public utilities, postal services, schools, more prisons and health regulation bodies could all be vulnerable to corporate challenges demanding they be run for profit (assuming there is a profit to be made by operating such services decently and humanely). Governments would even be, amazingly, "prohibited from applying performance requirements on foreign corporations to ensure they meet social and environmental responsibilities." So there would be no way to make sure people will even be paid minimum wage, let alone that their towns don't become vast toxic waste dumps.

These provisions can be read on the Internet, and they're beyond shocking. Their reach goes beyond federal and state to municipal governments, so forget about even those little urban oases like Portland, Oregon, or Hudson, Quebec, which have made environmentally sustainable by-laws. Most egregious of all, like the MAI, the FTAA would allow foreign corporations "to demand financial compensation for government regulations that affect their profits." So if an American or Brazilian timber consortium, for example, was *not* allowed to clearcut some wide section of crown land in British Columbia because of government protections, say, of wildlife habitat or Native rights, it could sue the Canadian government for all the years of its projected lost profits, and Canadian taxpayers would have to pay. It's as blatant and crude as that.

This brave new world was fashioned with the best of intentions. It was based on the idea that a single world built on co-operation is a good thing, and that trade is the unifying principle that will promote both peace and prosperity. But like any single idea in a complex world, the New World Order of Free Trade has subverted some of our most commonsense observations about human society. The flawed but nonetheless struggling democratic governments we have painstakingly established over the past 300 years are now seen as being disruptive to a more important social good — the free movement of capital. In this new world, it is believed, the rich will care for the poor voluntarily, not because some government makes them. And because the rich will have so much wealth, they will inevitably spread it around.

Maude Barlow believes that Canada is not just a place where getting rich is all that matters. She doesn't believe Canadians think that the rich will help others voluntarily. She doesn't want "Survival of the Fittest" to be the country's national slogan; and she's backed up by dozens of recent polls and studies on Canadian social attitudes, especially after 9/11. She says, "This is not the Canadian way. We built a fundamentally different society here, one based on the concept of universality, not the notion of deserving poor — you know, the widows' and orphans' fund. Instead, we built a system that said 'We are a family.' Any of us can lose a job or lose our health or lose a child or a spouse. Any of us, and all of us, at one time will need our family. And we'll need the safety net our family provides. Because we live next to the largest superpower in the world, we know that we have to share for survival. And that, in my opinion, is the Canadian narrative. That's the soul of our history. And when we share, we do it by establishing systems of interdependence — like our child payments, our social and cultural programs, our universal health care, our pensions."

The new economic ethos that encourages the amassing of wealth at great speed and all costs is being supported by the impulse to give corporations legal rights that are not even held by countries or people. "I can't find words strong enough to say how this betrays one hundred years of jurisprudence in our country," says Barlow. "This removes the rule of law. We're dealing with international agreements that have given transnational corporations — that is, private capital — fundamental rights in international law for the very first time. There's never been anything like it anywhere in the history of jurisprudence."

Imagine

The real voyage of discovery lies not in seeking new lands
but in seeing with new eyes.

— MARCEL PROUST

It's a frightening situation, but David Korten has some reassuring thoughts. "To me, the fundamental principle that we as citizens, as ordinary people, have to bear in mind is that these seemingly overwhelmingly powerful institutions have only one source of power. It's the power that we as people yield to them. They're all human institutions. And in a sense, they exist only in our minds, and they draw power only from the extent to which we accept them as legitimate and yield to them as legitimate institutions. If we withdraw that legitimacy and withdraw that power, they almost literally no longer exist. It's like the fall of the Soviet Union. When people finally got their act together and said, 'This institution is not serving us. It's not legitimate,' it collapsed in an instant."

The first efforts to pass the MAI were stopped in their tracks by groups of citizens that circulated its provisions, wrote governments and organized teach-ins all over the world. As a result, a lot more people are aware of what's at stake in the global casino these days. And citizen input has gotten a lot more physical. Seventy thousand people took to the streets in Seattle to disrupt the WTO meetings in December 1999, and since then not a single WTO ministerial or global ministers' assembly, or NAFTA, IMF, World Bank or G-8 gathering, has been able to operate except under siege in whatever country the get-together is held. Following the biggest anti-globalization demonstrations of all time, in Italy in 2002, where half a million protesters had gathered, the next WTO ministerial eroded what little international credibility it had left by hiding out in Qatar, a country that forbids any form of free speech or public gathering. Even then the prime ministers and presidents slept on cruise ships off the coast so they wouldn't have to see anyone who didn't like what they were doing. Holding the next ministerial in Cancún was an effort to get back some global dignity, but that meeting, as we mentioned in chapter 1, ended in a shambles, brought about as much by the stolid and dignified protestors outside the meeting gates as by the coalition of poor countries within.

"There's no question in my mind," says Korten, "that we're seeing a massive shift in consciousness among people all around the world. The system has really overstepped itself with this press towards globalization. The consequences are breaking out in ways that people can see, and they can see the connections between the trade agreements and the fact that more and more people are losing their jobs, and that the labour unions are having a harder time maintaining wages, and the egregious environmental abuses, and so forth." Korten jokes about something we've all noticed. "If you went back, say, [seven or eight] years ago, and you tried to get a group of people to talk about the GATT and trade policy, you'd have a very small meeting. People's eyes would glaze over. But now . . . you can get a pretty good turnout and people don't fall asleep. In fact, they're on the edges of their chairs. And the more they learn about it, the more incensed they become."

John Cavanaugh of the Institute for Policy Studies says, "The key here is that governments have been getting all of the pressure from the corporations. Corporations have been giving them the money to get elected, and they've been saying, 'In return, we want new rules. We want a World Trade Organization. We want a NAFTA. We now want NAFTA expansion to the rest of the world.' Governments have not been getting enough of a backlash from citizens' groups saying, 'Absolutely not. You can't give this away.' Governments, after all, respond to pressure, and they've been responding overwhelmingly to corporate pressure and corporate money. And now in Canada, in the United States and in many parts of the world, people are standing up and saying, 'This incredibly accelerated push towards corporate globalization is not in our interests, and we're not going to let you get away with it.' "

Even Donald Johnston, who as secretary-general of the OECD fervently believes in the benefits of globalization, says, "Here at the OECD, I brought with me the concept I had of government, which is that what one has to do is maximize economic growth . . . because without that you cannot have social progress. But you also have to have social cohesion and social stability. So the benefits of economic growth have to be transferred, in an equitable way, to populations at large. And the agency for doing that . . . is good governance. And it's these three together, essentially, which make societies progress. And when any one of them breaks down, countries and societies come to a halt. That domestic triangle is now international. We have to ensure that the benefits of the liberalization-of-investment agenda . . . are equitably distributed on a global basis. And that has not happened yet."

Maude Barlow argues that not only do the current global institutions *not* ensure the balance Johnston was still hoping for when we interviewed him in 1999, they actively work against it. She says, "We have to stop these international agreements, and we have to build international systems and codes and processes that will create truly international standards. I'm not a nationalist in that sense. I would easily give away a great deal of sovereignty in this country for higher economic, social, environmental standards around the world for everyone. That's got to be the goal. There's a great deal of work being done now on ways that could happen — how governments could work to cut off tax havens for corporations; how they could work together to establish higher standards in terms of these global agreements; how you could build a Multilateral Agreement on Investment that gave citizens, governments and NGOs, and not just corporations, fundamental rights."

We have to imagine the kind of world we want and then work to create it. That's what active citizens have done in the past, and that's how we got universal suffrage, free public schooling and national health care.

If we could do it nationally, we should be able to do it globally. Barlow is trying. She says, "What if we said that we would revoke the licences of corporations that pollute or that use child labour? There are a thousand ways that we, as citizens of the world, can start to talk about how we are going to bring the rule of law to global capital."

What suffers silently, and in terms of our future, most tragically in the drive to acquire riches as quickly as possible is nature: the trees we chop down and the soil and oceans we mine for quick returns. The unsustainable loss of the natural systems that are so vital to our lives is directly related to economics and globalization. "If we don't bring the rule of law to global capital," says Barlow, "we cannot save the environment. We cannot save the Earth. We know the environmental realities of global warming, for example. But unless we address the systemic undermining of our democratic ability to do anything about that through these international trade agreements, through globalization, through privatization, through deregulation — unless we address that, we're sitting in a leaking boat, taking water out with a little thimble. And we're going to drown."

CHAPTER 9
THE OTHER WORLD

Shell has waged an ecological war in Ogoni since 1958.
An ecological war is . . . omnicidal in its effect. Human life,
flora, fauna, the air and finally the land itself die . . . .
Generally it is supported by all the traditional instruments
ancillary to warfare — propaganda, money and deceit.
Victory is assessed by profits, and in this sense Shell's victory
in Ogoni has been total.

— KEN SARO-WIWA, NOBEL PRIZE–WINNING AUTHOR,
EXECUTED BY THE NIGERIAN GOVERNMENT IN 1995

Despite all our modern, sophisticated electronic devices, we still perceive the world through our sense organs and our personal experiences in our local surroundings. It is natural, therefore, for us to focus on our own local problems. It's difficult to identify with people who live far away and quite differently from us. We worry about our taxes, the decline in the quality of our schools or health care, and local problems like air pollution, urban sprawl or computer breakdowns. But people in many other countries of the world are worried, too — about us. They have to be, because our demands, and especially those of the companies that are responsible for providing us with the goods that are part of our comfortable, modern lives, are ravaging their countries and their futures.

The enormous wealth of resource- and expertise-rich nations enables us to live very well while mitigating the extra costs of worsening air, water, soil and social conditions that are a direct result of our lifestyles. But increasing numbers of people in the industrialized countries are questioning the value of a quantity- rather than a quality-driven lifestyle. And they are finding that some of the societies in the Third World that we think of as poor shelter values and customs from which we could derive enormous benefits.

The Niger River Delta in Nigeria is a typical area of the "undeveloped" tropics that is afflicted by the lifestyle of people in industrialized countries. The delta is incredibly rich in biodiversity, one of the largest and most productive wetlands in the world. Its mangroves and lowland rain forests are nurseries for thousands of species of tropical animals and plants. Around 7 million people from twelve different tribes live there, making their livings through subsistence farming and fishing. But there's another presence in the area. One of the largest and richest multinational corporations in the world, Royal Dutch Shell, the international parent of Shell Oil, has been drilling in the Niger Delta for the past forty-five years. The company has extracted an enormous quantity of oil and gas to feed the fuel demands of countries in Europe and North America that make up what is called the First World. It has made vast profits, much of which have been externalized (that is, hardly any of that money has gone towards land rights, taxes, equitable wages, compensation or waste management). Shell has been able to keep such costs largely nonexistent by allying itself with successive governments and forming a mutually beneficial relationship.

Project Underground is a small but effective California-based human-rights organization that works with communities threatened by mining and oil operations. Its spokesman, Danny Kennedy, told us, "As long as Shell Oil has been in Nigeria, they've had this revolving-door arrangement with Nigerian dictatorships, one after another." Today, now that Nigeria is ruled by at least a titular democracy, that relationship still hasn't changed very much. "It's a constant transfer of power," Kennedy said, "from the upper echelons of government — and vice versa. Shell has extracted huge wealth from the nation — in terms of dollars, something like 30 billion. And they have done very little in terms of basic, compassionate, human welfare considerations, community relations practices, let alone environmental management, which a normal resource company might engage in."

But how is this a problem for the rest of the world? At first glance, it

seems to be a problem only between Nigeria and Shell. Oronto Douglas
is an environmental human-rights lawyer who works with the Environmental Rights Action (ERA) group of Nigeria. He says of Shell, "They
control government and control our very lives, our very existence. So I
wonder, how did this come to be? . . . Why have we Nigerians sold our
conscience, our humanity, to multinational corporations? Why can't we
take back what rightfully belongs to us? Part of the answer, I think, is that
the governments in the North are allowing the multinationals to do what
they are doing."

Douglas is a member of the Ijaw tribal group from the Niger
River Delta. In 1995, Ken Saro-Wiwa, the Nobel Prize–winning
author, was executed by the Nigerian military dictatorship for
protesting Shell Oil's activities. Douglas was Saro-Wiwa's lawyer. Just
for showing up to defend his client, he was arrested, flogged with a
six-millimetre electric cable, and tortured. He still bears the scars. He
says that behind the torture and deaths of more than 2,000 of his
countrymen and -women stands an unregulated corporation that
provides the rest of the world with fuel.

Ken Saro-Wiwa was a member of the Ogoni, one of the smaller
tribal groups in the area. He was also an internationally known author.
Beginning in the early 1990s, Saro-Wiwa found himself speaking out
against, and then rallying his people to protest, the presence of the big
oil companies. In 1993, he organized one of the largest peaceful
protests ever seen on Earth. Three hundred thousand Ogoni (out of a
population of only 500,000) turned out to tell Shell Oil to get out.
Using non-violent tactics, they blocked access to the oil-flow stations
and pipelines, as well as the few roads and airports entering the delta.
Shell backed off and let the Nigerian military in.

At least 2,000 Ogoni have been killed since 1993, yet as Danny
Kennedy says, "The Ogoni still haven't broken. They still haven't let
Shell come back in, and as a result their land is actually healing in some
small measure. We were able to conduct the first scientific analysis of the
background pollution in the Niger Delta in February of 1997 . . . . We
took samples, the first in Ogoni. We found that in an apparently normal
stream, the levels of background hydrocarbon contamination were over
300 times what the European Union considers a safe public health

standard! Elsewhere in the delta outside Ogoni, where Shell has continued to operate . . . we found background hydrocarbon levels *over 700 times* what the European Union considers healthy."

What does this mean to the health of the people and the ecosystems of the area? Environmentalists keep repeating a simple fact: gas, oil and their derivatives — fossil fuels — are highly toxic substances. That's why they're so dangerous in car exhaust. In Alberta, about 4 percent of the gas released by oil drilling is flared (i.e., burned), a level that's considered dangerous and unacceptable by many countries and is illegal in the United States. In Nigeria, more than 70 percent is flared. As a result, toxins are released directly into the air and rain down onto vegetation, the soil and bodies of water. Danny Kennedy reports that partially burnt hydrocarbons permeate the soils and the waters of the whole region. "The yam yield has been reduced, and that's the staple food. Everyone from the delta talks about how the gardens are less productive. Obviously, they would be. People's skin gets covered with blemishes because their bodies are reacting to these toxins." Still, this situation can be remedied. Standard industry practice is to reinject the gas down the drill hole instead of burning it. But, of course, that costs a little more money and has to be monitored.

Shell is not the only offending corporation in Nigeria. There are many others, including Chevron, and they're often as culpable. But Shell is getting most of the heat because it was the first to drill in the region, it has a documented involvement with the Nigerian military and it produced more than 50 percent of the highly repressive former Nigerian government's revenues. One would assume Shell has a lot of influence with any government, but what has the company got to say about the ecological problems associated with oil development? Basically, that it's an internal problem for the country, and the company wouldn't dream of interfering.

There are judicial proceedings against Shell, a worldwide boycott and general bad press. When we were preparing our radio series, we invited the company to discuss the issues at length, corresponding with them for weeks. But in the end, Shell refused the opportunity to present its position. The company did send us some glossy public-relations brochures on how ecologically damaging practices are slated to be "phased out" in another decade. In earlier brochures, the date for the phasing-out was supposed to be 1998. Shell also claims there's no proof that its activities cause environmental damage. However, until a recent and superficial World Bank study, Shell itself had done all of the environmental assessments of its own activities.

Death and the Bottom Line

Shell paid field allowances to the operatives of a squad called Kill and Go, the mobile police squad of River States, which comes into villages and does what its name says. They bought ammunition and other munitions for the military actions in the River State. All of this is on record. They've acknowledged all of it.

— DANNY KENNEDY, PROJECT UNDERGROUND

Shell's behaviour, although particularly well-known, does not stand alone and, in general, merely points out a basic truth about global business: when an unregulated company can accumulate wealth without internalizing any of the costs, it will do so. In countries that either cannot or will not protect their citizens, corporations will be tempted to ignore every precept of morality. Unregulated businesses driven to maximize profit were simply not created to take responsibility for protecting people or nature.

We operate under a system that talks about spreading wealth and helping the poor, and that urges one model of development to do so. But that same system is encouraged, even forced, to find every possible loophole, every imaginable shortcut, every devious way to save money in order to return larger profits. In countries that for whatever reasons do not protect labour or the environment and do not impose stringent health standards, corporate logic encourages private enterprises to take advantage of the situation, and to fight any upgrading of standards that will hurt that lush and unfettered bottom line. Shell Oil is one of the most flagrant examples of this behaviour, and it's been richly rewarded by this system; it is the single most profitable company on the planet, according to *Forbes* magazine.

Danny Kennedy says that even when its own shareholders voted in unprecedented numbers — 12 percent of all the stockholders — for more adherence to Shell's own stated but unenforced human-rights policies, the company's response was merely to increase its public-relations efforts. Meanwhile, Shell has admitted that throughout the 1980s and 1990s it purchased weapons for the notoriously violent Nigerian military, and maintained a private "security force" of its own. This is a chilling view of how the future will look if corporate powers are allowed to operate without restrictions and without outside controls.

We interviewed Oronto Douglas at a teach-in on corporate rule at the University of Toronto. He had to be smuggled out of Nigeria for the event because he cannot travel to and fro at will. His life is in constant danger simply because he tries to tell outsiders what is happening in his country. But he takes these chances because he's trying to get us to help, even a little.

Douglas has shown great courage in his fight for his people. Indeed, it is remarkable that Ogoni strategies remain staunchly non-violent, especially when one considers the violence and brutality of their opponents and the immensity of the forces arrayed against them. "What we are saying is that the West, the North, should join hands with us by fighting for these ideals — for us primarily, perhaps, but also for the world generally," says Douglas. "How can you do that? You can exercise your freedom that you hold so dear, that you fought for, and say, 'Look, Shell, you can't do that anymore.' People should boycott Shell. Cities should boycott Shell . . . . Governments should impose sanctions. And people should boycott any other multinational that is implicated in the ugly situation in the Niger Delta."

There has been an international boycott called against Shell by many NGOs since Ken Saro-Wiwa's death. Although many people still avoid filling up at Shell, there hasn't been much progress in the fight to tame these big oil companies, however; in fact, with the second Bush administration in power and in Iraq, these most powerful of powerful corporations, like Halliburton, are more influential than ever. But there's not enough oil even in Iraq to keep them happy. Outgoing President Clinton, quickly followed by George W. Bush in 2003, suddenly put

Nigeria on the must-visit list. That's because a great deal of oil has been discovered offshore in the Gulf of Guinea, maybe as much as has made Shell and Chevron so rich in the Delta. In a recent article, syndicated political analyst Gwynne Dyer posed some trenchant questions about what happens when a country gets lucky like that.

Dyer asked, "Why has Algeria been devastated by tyranny and civil war, while neighbouring Morocco is peaceful, relatively democratic, and no poorer? Why is Angola, once Portugal's richest African colony, a wasteland of poverty, violence and corruption . . . while Mozambique, its poor relative in colonial times, is now peaceful, fairly equal, and politically open? Why was Iraq under Saddam Hussein even more violent and repressive than Syria, its near twin that has also been ruled by the Baath party for over three decades?" His answer is simple: "All of these countries had fragile political structures, ethnically complex populations and difficult colonial pasts, but the ones that descended into a full-spectrum nightmare were the ones that struck it rich with oil." The same must be said of the deeply impoverished but also peaceful West African nations of Ivory Coast, Benin and Senegal, when compared to the history of culturally similar, neighbouring Nigeria. Oil is a great part of the reason why Nigeria is cursed by instability and violence, and now that oil has been found offshore as well as throughout the Niger Delta, bleeding into the territory of little São Tomé, an island in the Gulf of Guinea, more coups, violence and rebellion have landed in the region, right next to Air Force One.

As a global society, we may be learning something from the past. British Petroleum, the most progressive of the big oil companies, has launched an interesting initiative. It posted on the Internet its production-sharing agreement with Azerbaijan and the signing payments it made to Angola, in a bid to head off the usual charges of bribery to government officials. It's trying to avoid those dirty private purchases of public oil and mineral rights that have characterized the oil industry's past and have led to unholy alliances between corporations and repressive regimes, resulting in outrage and violence in the citizenry. "Mobil is facing a huge criminal case over alleged bribes in Kazakhstan, and the French company Elf faces similar action over payments in various African countries," Dyer says, and heaven only knows what deals are being made about who will govern Iraq. These legal demands for accountability reflect a new public understanding of the ravages of the oil economy that need to be pushed much further.

As for the development argument that says Nigeria can support its people only by mining its resources, Douglas responds, in a way that should make us all pause to reflect, "What we have is not just oil wealth. There are more riches in the delta. A lot of people don't know that we have wonderful forests, mangroves, tropical rain forests. There are other minerals, wonderful foods, crops, a lot of birds flying everywhere, fish. This is our wealth . . . . We should not be thinking of fifty years from now. We should be thinking of 2,000 years from now. That is an African philosophy; that is our perspective. If you come to my village today, people will tell you about past ancestors who fought to protect the community, ancestors who link us to the past in terms of religion and so on . . . . You go to a typical African community, they think both retrospectively and prospectively. They do not think only of themselves."

A friend who has lived in Africa once remarked that the reason famines there are so devastating is that no one starves until all the food is gone; then everyone starves, because they share. Africans often have values and a way of life whose positive aspects are completely unknown in the West. Oronto Douglas tried to express them when he said, "We do not own this planet. You will live and you will die, and some other person will come. Why should you make life difficult for those coming after us? Why? We think about what the planet has given us, and what we give the planet in return . . . . This struggle is not an oil struggle alone. It is a struggle that is all-embracing. It is holistic. The utilization of oil or any other wealth must be done with the consent of the people, with equality. As a people, we have mobilized to be the guardian of our environment and our resources so that the resources will be used for the benefit of all and not by a greedy elite, whether they are locally based, or nationally or internationally based. We will never accept that. We will not."

Recolonizing the World

We must find new lands from which we can easily obtain raw materials, and at the same time exploit the cheap slave labour

*that is available from the natives of the colonies. The colonies
would also provide a dumping ground for the surplus goods
produced in our factories.*

— CECIL RHODES, NINETEENTH-CENTURY
INDUSTRIALIST AND POLITICIAN

The people of the Niger Delta are fighting for their very existence,
and the struggle has brought out a heroism that's dramatic and overwhelming and far from anything most of us have ever faced. And yet
it's not atypical. Many of us have encountered similar situations in
poor countries of the Third World. On the one hand, there are
terrible tragedies — violence, environmental despoliation, injustice and
greed — that are both endemic and external. On the other, there are
large groups of citizens, and sometimes even their governments,
who have a very clear vision of a better world, who are tirelessly
fighting injustice, environmental destruction, commercialization and
corruption.

As a member of a rich, industrialized country, I brought to these
countries my preconceptions about their societies. The first of these is
based on our notion of "development," and holds that all people who
do not have telephones, shopping malls, cars or VCRs are "poor."
Especially since some of the specious analysis post-9/11, we have come
to believe not only that everyone *should* be like us, but also that they
want to be like us. So we take it as our responsibility to help them achieve
what we think is a universal human aspiration: to acquire what we have.
And it isn't just clothes and food and shelter anymore; our brand-name
consumer lifestyle is presented as a kind of heaven, a pinnacle of desire
that everyone *must* achieve for true happiness.

Today's drive to push Western development onto Third World countries is very much like our previous forays into new lands. Many centuries
ago, Europeans brought Christianity and Western medicine, as well as
diseases, to other countries. The devastation unleashed on these often
viable, balanced and healthy cultures and economies was incalculable. Yet
in general, the colonizers really thought they were helping. They thought
that European life, worship, dress, thinking and farming were all
supremely desirable, and that any other way of life was barbaric, primitive, heathen, dirty and sinful.

Now we're back — with a largely North American vision of economic progress and global development that looks at any other kind of society as underprivileged, inefficient, outmoded, dirty and poor. This time the conquerors are armed with a new religion: material gain and salvation for all through global trade. The first wave of colonialism forced everyone to become Christians, farmers or producers of wealth. Now we want everyone to watch TV, drink Coke, wear Nikes and consume the goods we offer. But a lot of the people we're trying to convince remember what happened the first time. The inducements we offer are not instantly seized upon by people in the Third World. Indeed, just as in the First World, billions of dollars are spent annually on advertising to convince people that they really want and need products they've never heard of.

It has been said that television softens up Third World cultures for consumerism the same way Christianity softened them up for colonization. Images of *Survivor, American Idol* and *Star Trek* are dispatched as harmless fun. Jerry Mander, the ex-adman who runs the International Forum on Corporate Rule, says, "About 80 percent of the world population now has television. Very little programming is produced locally; in most cases, it's from the United States and a few other developed, northern countries. So you have people in the tundra up in the Arctic and people living in grass houses in the South Pacific and people living in slums in Asia and South America seeing a bunch of white people in Dallas, say, standing around swimming pools drinking martinis and aspiring to nothing more than killing each other to take over each other's corporate activities and make more money. Or they're watching cartoons or MTV or the Nike ads. And what you have is a set of images that are homogenizing consciousness in the world." The incalculable cost of this has been the erosion of local, diverse values, cultures and communities.

Mander points out that the reason advertising is so successful is that it works — images have power. Our brains have evolved to take visual stimuli very seriously. Once we see images, we can't forget them, and their contextual emotions of happiness, fun, contentment and triumph become associated with the commodities they sell. "It's a tremendous invasion of cultural and economic values . . . " says Mander. "Since these images are so believable, all that goes with them appears to be attainable, achievable, easy, nice and good. It's got such a gloss and an attractiveness to it that everybody kind of wants to go for it. And it's presented to them as a real alternative to the way that they live . . . . There's no corresponding counter-force that tells people that this stuff that they're watching, this lifestyle, is producing alienation — drug abuse, violence, suicide, family violence and disempowerment — on a level they've never imagined. And it's also bringing a tremendous breakdown of the environment. The level of consumption that's presented in this kind of imagery is directly connected to the overuse of the resources of the planet and the terrible waste problems that cause global warming, ozone depletion and our current destruction of habitat. All of the tremendous problems that are bringing us to the brink of evolutionary breakdown — forever — are hooked directly to this set of images that look so attractive and harmless in the first place."

There are not enough resources on the planet to provide everyone with the lifestyle of *The Gilmore Girls* or *Will & Grace*. This is not only an impossible goal to achieve, however; by aspiring to it, we accelerate social and environmental degradation. As Mander says, "The sets of values that are really survivable and sustainable are being driven under completely, because these consumer values are replacing them. To have sustainability on this planet, we need cultural values that are consistent with less consumption, the development of local economies, less transport of goods, a more diverse economic base within regions and more equitable distribution. These are the things that lead to a future!"

For some, it may be reassuring to get off the plane in Manila and hear the same joke you saw last week on *The Simpsons*, to exchange hip talk with an Asian teenager about the latest supergroup and to meet people who have a clear picture of your country based on television images. It *is* easier to understand people when you share some of the same culture. But as the people of the Third World realize that although a global economy may make a pair of Nikes accessible, an education or decent drinking water is not an option, their enthusiasm is replaced by frustration and anger. In many parts of the Third World, people have already recognized the new colonization for what it is.

Jocelyn Dow is a businesswoman from Guyana, a poor South American country of only 6 million people, mostly ex-slaves from Africa and the rest of East Indian, Chinese and Native origin. She says, "The mysticism and the romancing of the idea of commodification should not fool a single Caribbean person; you have to remember that except for the indigenous people, we're all here because we were the actual commodity. We were the energy, the slaves and indentured labour." Buying and selling to send riches around the world looks better if you're not one of the things being bought and sold. "Caribbean society was designed for a world that wanted sugar," says Dow. "It was designed to produce sugar. There was a whole philosophical system, a whole religious system, to support the reason why this commodity and this system of production were not only justifiable, but desirable. And I always say that what you have to remember about this whole cult of the market is that for 400 years *we* were the market. We were the items. So we should not in any way be caught up in any world-view that talks about how wonderful the market is."

Fighting Back

There simply aren't enough policemen in India to deal with 600 million farmers. And that's something that's difficult [for companies] to understand. Our quarter of a million farmers in Canada can't conceive of that kind of direct action. But there are people who can.

— Brewster Kneen, food system expert

John Cavanaugh says one of the countries where citizen opposition to corporate globalization is the strongest is India. He says, "There is a strong history [in India] of people rising up and fighting for their rights. In that country, half a million people took to the streets to protest Cargill, a giant agribusiness conglomerate that is trying to take over their seed banks in an effort to gain further market share. Thousands went out into the streets against Kentucky Fried Chicken, who they saw as trying to homogenize consumer food tastes." In fact, at UN meetings, on the Internet or on television news, if you're watching for it, you'll see that the countries our culture is trying to colonize are fighting back hard. Cavanaugh mentions the Philippines, Thailand and Ecuador, all of which have made efforts to protect their markets and their crops from global homogenization and piracy.

But one of the most volatile battles is going on closer to home, and it's being fought by one of Canada's key trading partners. "More and more organized resistance against corporate globalization [is] coming from indigenous communities," says Cavanaugh. "In Chile, for example, 10 percent of the population are Mapuche Indians, and they have joined in an alliance to keep NAFTA from expanding to Chile. But by far the most impressive indigenous opposition has come from the Mexican state of Chiapas. In the southernmost state, where Mexico's most heavily concentrated indigenous population exists . . . there has been a massive resistance to NAFTA, and to the entire Mexican economic model that has come with NAFTA, which is disenfranchising people from their land."

As they did with Rwanda or Yugoslavia, members of the media portray the rebellion in Chiapas as an internal or ethnic conflict when, in fact, it reflects a global economic and trade crisis. In Chiapas, it has a lot to do with First World corporate and government desires to make money at the expense of the poor. "Much of the issue in Chiapas is land," says Cavanaugh. "With NAFTA came new Mexican laws that permitted what were formerly state lands, communal lands, to be sold; and the selling of land in indigenous Mexico is really the end of life and the end of culture. So there's been a massive resistance. It's been an armed resistance as well as an organized, citizens', non-armed resistance."

Not only was the traditional land base sold to agribusiness, as also happened in other developing nations, but both the World Bank and NAFTA tied agricultural reform to their loan and trade deals. That meant that there are no more subsidies to or protections for the indigenous corn growers, and that at the same time the market has been flooded with cheap U.S. grain — in this case, corn. These people — whose agricultural base, not to mention their entire culture and history, is centred around the growing of corn — were instantaneously deprived of a market, as well as their land. They were also left with no homes. One assumes the proponents of these economic measures expect the displaced peasants to migrate to the cities and find computer jobs, or else just disappear. What was not expected was that they would arm themselves and go on the Internet to publicize the situation.

What is most remarkable about the ongoing Chiapas rebellion is the outpouring of sympathy and support it inspired from middle-class people not only in the United States and Canada, but also in Mexico itself. In the fall of 1998, 1,001 Zapatistas, some of the Maya of Chiapas, started to retrace the famous entrance into Mexico City of Emiliano Zapata, a hero of the Mexican Revolution. They set off to walk barefoot hundreds of kilometres to the capital in order to publicize the plight of the Indians of Chiapas.

By the time they reached Mexico City, they had been joined by many thousands of other indigenous people, as well as by ordinary Mexicans of every social class. Together, they entered the city's central square, directly in front of the main cathedral and the president's mansion — the equivalent of massing in front of the White House. Hundreds of thousands of residents of Mexico City had turned out to greet them, waiting more than seven hours in an incredible crowd estimated at half a million. They were all there, as John Cavanaugh says, "to announce, very vocally, their opposition to NAFTA and to the entire free-trade model. It was an enormous statement in a country where the racism towards indigenous people is even greater than in the United States or Canada. It was an enormous statement of solidarity among the various segments of Mexican society that are adversely affected by these economic policies."

Eyewitnesses reported that thousands of middle-class Mexicans, most of whom had also been disenfranchised by the collapse of the peso, NAFTA policies and IMF austerity programs, burst into tears when the bleeding, exhausted Indians finally appeared. The crowd began to chant, "You are not alone! We are with you!" One can imagine how this was received — hundreds of thousands of people chanting and crying together. Cavanaugh says, "Among other things, Mexico City had recently elected a mayor who came out of the anti-free-trade camp, Gauatemo Cardinas. So you had this wonderful symbol of an indigenous movement against the dominant model of recolonization, marching into a Mexico City that was now ruled by a leading opposition figure. It was a sign that Mexican history has been changed forever."

The people who feel the effects most directly are showing the rest of us how to have a say in this new, globalized, monocultured world. "Mexico, I think, is perhaps two or three years ahead of most of the rest of the developing world in terms of the crises and the contradictions and the pain and the struggle of corporate globalization, partly because it was so singled out as the model of free trade," states Cavanaugh. "So much attention was put on it, so many financial resources were lavished on it. So the growth was faster, the inequalities were greater. And now the divisions and the pain are greater. But in Mexico, you can feel history being turned."

When the Student Is Ready, the Teacher Appears

We really would like to continue to exist in this world, but in
a manner that we ourselves define. A manner in which all
the others can respect our diversity. And maybe some others
would also like to learn from us, and adapt our values to
their communities.

— VICTORIA TAULI-CORPUS, IGAROT TRIBE MEMBER,
PHILIPPINES

Mexico? India? These are not countries many of us think of as having answers that could benefit *us*. The fact of the matter is that we think of the Third World as a place full of people we have to save. We never even think of them as being able to take good care of themselves without our help, our money and our technological expertise. We've felt this way for a long time. Way back in the sixteenth century, we were going to save their heathen souls and now we're bent on saving their economies, their businesses, their farms, their lives.

In the Third World, food and crops are always key issues of discussion. We're always talking about developing new technologies to save the poor of the world from starvation. Vandana Shiva has spent much of her life investigating the politics of modern economic development, especially in terms of food. She likes to quote Alfred Howard, a British agronomist who's been called the father of modern agriculture. He was sent to India at the beginning of the twentieth century as the imperial agriculturalist, and was supposed to teach the Indians how to farm in the modern European way. In his book *The Agricultural Testament,* he says, "I came equipped with sprays, because I was, of course, a modern agriculturalist. And very soon, I threw my spray cans away, because in the fields where I was working, there were no pests. At that point I decided that I had to learn from the peasants and the pests about what system keeps pests under control; because quite clearly, pesticide sprays do not."

Shiva says that methods for keeping pests under control are known and practised by many Third World farmers. "The first strategy is nutrition. If your crops are getting organic food, they are much more resistant to pest attacks. If they are fed with chemical fertilizers, the vulnerability to pests is much higher; there's data, endless data of comparisons available now, both from government and university studies. The second most reliable alternative is to have polycultures, or mixtures of crops, in the same field, so that your pests never literally have a field day; the entire field is not food for them. Somewhere along the way there are other insects, which are also their predators. In any case, the variation and diversity of the crops keeps the pests within a limited population. Even if some pests still occur, there are all kinds of ecological ways to control them."

We are told that high-yield modern farming methods developed by northern countries necessitate heavy use of petrochemicals (i.e., fertilizers and pesticides), and that's true — if you insist on growing huge fields of monocultures. The small land-holdings of the Third World have always been planted in a bewildering variety so that the farmer could feed his family, his cattle or his goats, and put some back into the soil. These farmers are consistently told by outside experts, and their own modernizing governments, that their methods are outdated and inefficient; that without monocultures and petrochemicals, their crops will fail. Yet as Shiva points out, "The very same companies that sell pesticides also pirate our knowledge of plant-based pesticides, without admitting that the fact that they are patenting this knowledge is proof that it works. Otherwise, they wouldn't be trying to commercialize it and trying to monopolize the market of ecological pest control. Every field on every farm in India that has not been destroyed by the Green Revolution is a biodiversity system of pest-control agents. Whether you're using the oil or leaves of the neem tree or the pungamia, they're right there. They've been replanted by the farmers over centuries because society had experimented and had already figured out that these species control pests."

There are pest repellents as well, as any good gardener knows — marigolds, niger, amaranth, pepper, even marijuana. They can be planted near susceptible plants like beans, and they will protect them from infestation. Growing monocultures in huge fields obviates such methods, and makes necessary the use of dangerous chemicals, heavy machinery and certain forms of land ownership. It results in the poor being thrown off their ancestral lands to make room for rich landowners or agribusiness, as happened in Chiapas. Shiva calls modern farming "a substitute for intelligence." She says, "I don't see why we should adopt a very distorted system of agriculture, which must be based upon pushing 98 percent of the farmers off the land. It has been disastrous for the North in many ways, and would be totally explosive for the Third World. It's a system in which the only way you can farm is by literally being in constant war with all the other species that grow or live around you, by being at war with your own soil, by being at war with nature's diversity itself."

Shiva is not alone in her assessment of the efficiency of traditional agriculture. In 1997, the World Bank carried out productivity studies on variously sized farms, as did the international Food and Agriculture Organization (FAO). "It turns out that the highest productivity is [on farms of] up to three acres," explains Shiva. "After that, there's a decline. After only three acres, you cannot increase productivity. You can manufacture data that seem to prove you're producing more, but you can't actually do so."

A recent study reported in *Scientific American* corroborates what Shiva told us. A chemical-industrial farm requires 300 units of energy — fertilizers, fuels and so on — to produce only 100 units of food. An organic polyculture farm requires only five units of energy to produce the same 100 units of food. It's only when looking at the "harvestable material" data so popular with industrialists that a chemical farm looks good. Those data do not include the costs of fertilizers and pesticides or the non-specific biomass, such as edible weeds, fodder, medicinal plants and green manure, all of which are eliminated in an herbicided monoculture. It's only when we start to look at the Earth as a finite system, and we try to figure out how many units of the sun's energy we have to expend to get our food, that we realize that it is close to insane to farm the way we do now, essentially converting petroleum, very inefficiently, into food.

Shiva has used modern methods to point up the failure of modern agriculture, and so have researchers from the North, especially over the past ten years. Marty Crouch was a well-known plant geneticist working on crop modification at Indiana University. She had gone far and fast in the field of molecular biology. Because of her prestige and expertise, Crouch had the opportunity to travel and teach others how to farm the modern way. She says, "I taught molecular genetics in China and New Zealand, and I worked in Pakistan. I got to see a lot of the other ways that agriculture is done around the world." The situation changed for

Crouch when she toured a Pakistani fertilizer plant, where, she says, "The guide said, 'We make our nitrogen fertilizer out of natural gas and this is what has allowed yields of rice and wheat in the region you're visiting to triple in the last decade.' So someone asked him how long the natural gas was going to last. He said, 'About ten more years.' They said, 'Can you afford to buy it from somewhere else?' He said, 'No.' "

While she was thinking that over, the same tour went by a valley where they'd been growing rice continuously for more than 2,000 years in a traditional way. Yields weren't as high as they were with the fertilizers, but production could continue forever. Crouch says she suddenly thought, " 'Hey! We're putting our lives, billions of people's lives, on the line with a technology that is destined to fail, and to fail utterly, on a grand scale. We're using ten times the energy to produce one calorie of food.' In the old methods, I realized, you'd take calories from the sun and you'd turn them into food calories, and you'd get a net increase. Now we have an actual decrease. Our agriculture is at war with nature and we're eating fossils. How long can we eat fossils?"

Having had this revelation, she transformed her career and began to work to revitalize traditional methods and to restore an energy balance that could be sustained indefinitely. She says many people in northern countries resist this approach because they think it's a return to the past, "and it scares them, deep in their bones. Because we've all heard these horror stories of the past: how life was brutish, hard and short; that the past was ignorance and disease and starvation. But I didn't have that memory."

Marty Crouch grew up with her grandmother, who was a traditional subsistence farmer in the Midwest. Neither of them thought that life was so bad. She has also lived in the Third World, and knows the disadvantages and the advantages firsthand. She says we in the North have an unconscious cultural mindset based on a desire to conquer, colonize and dominate. It's not only part of our culture; it's part of our food-getting process. She says, "In the West, we think of agriculture as a punishment. In the Judeo-Christian tradition of religion, we were kicked out of the Garden of Eden, and as punishment, forever more, we have to do this terrible, hard toil. We have to wrest poor little tubers and things out of the ground. Our entire relationship with our food is one [of] hard work and punishment. Crops are something we have to force the Earth to give us."

Crouch was especially impressed by what she saw on a recent trip to Peru. She says, "They have no such tradition. They go into their fields with a feeling of ritual; it's where they go to listen and to converse with the soil . . . with the weather and the plants and the animals. There isn't a big separation between the cultivated things and the wild things either. And they think of that as a very centre of diversity. And I sure would like to integrate more of that kind of thinking into the North American idea of agriculture."

Vandana Shiva says that when she talked to an eighty-three-year-old peasant farmer in India who has developed twenty varieties of rice, she asked him if he would try to patent them in order to control them and make money. "How could I own them?" he replied. "They are my friends. They are my partners. I have to be a good friend to them, and they will then be good to me. But without my treating them as friends, they will never feed us. We are in a partnership of species." In Hindi, there is even a word meaning a "partnership of species"; there is also a word to refer to the "earth family" — that is, all species together.

Shiva says, "There's now data coming out of India that show that the agricultural real yields, before colonialism destroyed the practices, were at least seven or eight times higher than the Green Revolution yields. These are not anecdotes. This is documented data from the tax records. So we had a very high productivity system, which is why Alfred Howard said, 'What I see in the fields of India is something as perennial and as long-lasting as the prairies or the ocean, because they have worked out their agricultural system to perfection.' These kinds of practices are part of the culture of many peasant farmers all over the world. These have to come into the thinking of a twenty-first-century enlightened humanity, so we, too, can live in a consciousness beyond war with other species."

The Meek Shall Inherit the Earth

Africans are more in harmony with the natural environment,
more directly dependent on the natural environment. They
don't have the illusion at this stage that artificial fixes can

solve the problems of life. The very poverty attributed to Africa makes us realize how closely dependent we are on natural life-support systems, and how delicately balanced those life-support systems are.

— TEWOLDE BERHAN GEBRA EGZIATHER,
MANAGER OF ETHIOPIA'S ENVIRONMENT
AND PROTECTION AUTHORITY

Immense pressure is being put on the Third World to conform to a certain vision of economics and of farming. And the last frontier for industrial farming, including pesticides, fertilizers, monocultures, genetically engineered organisms and the rest of that chemical package, is Africa. Because these industries are facing a shrinking global market as more and more countries, especially in Europe, adopt biosafety laws and limit chemical use, Africa has become the new market of choice. For example, the United States Agency for International Development (USAID), mentioned earlier, is spearheading a U.S. marketing campaign to introduce GE food into the developing world and is also attempting to water down legal efforts to control the technology. This is just one instance of how Western industrial farming techniques are currently invading Africa. From a business point of view, the industry simply has nowhere else to expand. But the reason Africa has been "left out" of the global move to industrialized farming over the past four decades is not just geographic, and it doesn't jibe well with our vision of northern countries as the great providers of food and technology.

Tewolde Berhan Gebra Egziather is trained as a plant ecologist. He is the manager of Ethiopia's Environment and Protection Authority, and he's also been the Ethiopian delegate and head of the African coalition at the Biodiversity Convention of the United Nations. He is a gentle, diplomatic and very knowledgeable man. He says, "There is a view that Africa is now the most backward continent, because it is 'forgotten.' In fact, it can to some extent afford to be forgotten, because it is the most self-sufficient of the continents. It does not need the rest of the world to maintain itself as much as the interlinked continents do. The reason is very simple: the rest of the world depends on a small number of crops, but Africa has a huge variety of food crops that are specific to Africa. That is why it has been 'forgotten,' why the agricultural research and development that has taken place in the past few decades has bypassed Africa."

Modern varieties of food plants have been introduced to the continent, but with so many local crops and crop types to choose from, they haven't done well. And they won't do well — unless farmers are forced to buy them through complicated programs tying aid to petrochemical use and "agricultural development." Tewolde Egziather says modern methods do not appeal to the average farmer partly because "Africans are very aware that their lives depend very much on the well-being of their own environment and of their life-support systems . . . . They don't get the foolish feeling that science and technology will solve all the problems of humanity."

Many Third World countries have learned the hard way that science and technology often fail. As we noted, the Green Revolution was a terrible experience for Thailand, Peru, India and many other countries. The initial high yields were not only illusory, but also left a legacy of disease and despoiled ecosystems, destabilization in land tenure and mass migration to the cities. Africans see these innovations as often well-intentioned, but they are increasingly viewed with reserve. Tewolde Egziather says, "I remember an American lecturer in agriculture in one of our colleges seeing our traditional plough, which has evolved over a long period not to disturb the soil unduly because, in fact, tropical soils are a lot more delicate than temperate soils. And he berated our technology, saying, 'What you need is a deep plough that will turn the soil over!' When that was tried, it destroyed the soils and their fertility. Such ill-advised but well-meaning efforts at . . . so-called "development" no doubt will continue to take place. But I hope that we are learning to catch up with our own mistakes and identify them as soon as we start committing them."

Often it was not just technologies, but entire Third World ways of life, including diets, that were denigrated as primitive or backwards. All over Asia and South America, for example, native staple varieties of millets, lentils, pulses, oilseeds, kasha, quinoa and hundreds of other seed-producing grasses have been replaced by industrially developed soy, corn and wheat, while wonderful, nutritious foods have been forgotten, even lost. Tewolde Egziather recalls the situation in Ethiopia fifty years ago. "When I went to school, we had European-style schools for the first time in Ethiopia, and I was one of the earliest to go. One statement we used to hear very often was 'What is this grass you eat?' — as if grain, all grain, weren't grass . . . . So our agricultural researchers for a long time very seriously tried to eliminate our main cereal crop just because they thought it was inferior. Now it turns out that it has superior qualities. In terms of minerals, it is one of the best grains you can grow. It has a high lysine content. It is now recognized as a special food. This [kind of thing] has happened again and again in many countries."

We're seeing the growth of a monoculture in economics, in culture, in food production and in diet all over the world. What happens if new diseases turn up, if genetic technology runs amok or if climate change demands new kinds of crops? Ethiopia and a few other places still have the diversity of wild and cultivated plants from which to find strains need-ed to flourish in new situations. But as Mexico's experience teaches us, they can easily be destroyed, and Ethiopia is being pressured to become just like everyone else. Even our much-touted seed banks are being pri-vatized all over the world as part of the economic rationalization mantra demanded by globalization. Collections that were amassed from farmers with public money are being turned over to private corporations that may sequester, misplace, sell or even destroy them with impunity. This last scenario is not far-fetched; companies often acquire other seed com-panies and seed banks to gain a monopoly on the market, not because they want to experiment with biodiversity.

When Tewolde Egziather talks about Africa being forgotten, he means that its countries are not as plugged into the Internet as other countries are. Africa is not a prime site of industrial development as South America and Asia are, nor is it a home for research and development like India and Thailand. "Although Africa is the second-largest continent," he says, "people think of it as one country. You can hear people refer to, for example, the Netherlands and Africa, as if they were comparable enti-ties. Africa is huge, but the population is still fairly small, relatively, so the level of demand on the environment has not been great enough to stimu-late the need for massive efforts to produce more on ever-decreasing land." Africa has the most mineral wealth of any continent; there is virtually no valuable mineral that is not found there. "That means that the world as a whole has a vested interest in taking what Africa can offer," asserts Tewolde Egziather. "And the way to do that is to keep the relatively small population disorganized. This means that the focus is not to treat

Africans as trading partners, but as a source of raw materials." This is a position familiar to many other countries, including Canada.

But Africa is both too large and too rich in minerals, land and biodiversity to remain forgotten in the twenty-first century, and Tewolde Egziather knows it. In the Third World, there are many people like him who see in their countries opportunities for change, instead of disadvantages; who see multiplicity and adaptability, instead of poverty. He says, "I don't think there is only one way to development . . . . If you were to go back to the time of the dinosaurs, and you were to see a little mammal running about, you would never think that this mammal would succeed the dinosaurs. Who can tell which is a dinosaur and which is a mammal in our time, in this era of really frightening possibilities? That does not imply that I view the African way as the mammalian way. It is simply to say that what is successful now, and what will be successful in the future . . . is not necessarily something linear that proceeds from what we know at this moment. Therefore, it is absolutely essential to keep our options open." Africa may have been forgotten or "left behind," but it has retained the possibility of trying other experiments. Other places, as Tewolde Egziather says, "have destroyed many of the bridges they would have had into other possibilities."

The Law of Social Diversity

> *A vibrant civic society is a precondition for having markets develop and operate in the way that they should.*
> — ADAM SMITH, EIGHTEENTH-CENTURY ECONOMIST

In the second chapter of this book, we discussed at length the importance of biological biodiversity to the survival of the planet — how a rich mix of species is absolutely crucial to the healthy maintenance of our atmosphere, soils, freshwater supplies and our own bodies. It is increasingly apparent that politics, culture and economics become unbalanced and inefficient without diversity as well. In his classic book *Making Democracy Work,* the sociologist Roy Putnam describes his efforts to figure out why the citizens of northern Italy are considered by both outsiders like the UN and Italians themselves to be so much happier, more prosperous and healthy than those living in southern

Italy. Putnam looked at every factor he could think of, including history, geographic position and industrialization, and he got some surprising answers.

What made the difference finally seemed to be the extent to which people belonged to civic organizations — voluntary groups like football and camera clubs, choirs and collectors' groups — which transcend religious, cultural or ethnic boundaries. Anthropologist Steve Rayner corroborates this. "It turns out there is an astonishingly strong correlation, about +92, between people's membership in these organizations within a region and the successful performance of the government in that region. Similarly, there's a negative correlation, about –75, between the performance of government and the systems of patronage that divided religious, cultural or ethnic groups employ. . . . So it seems to suggest that in fact there really is something very important about people being engaged with each other in things that don't appear to have any direct impact on government, or on the environment, or on any of the things that we're concerned about."

Jeremy Rifkin calls such voluntary organizations the Third Sector. He says that every non-communist country in the world has three sectors: the market, the government, and the Third Sector. "The Third Sector," explains Rifkin, "is every organization, every institution, every affiliation in a country that is not a corporation in the marketplace and not a government agency. It's much broader than non-government and non-profit organizations. It includes them, but it also includes cultural organizations, arts groups, sports clubs. It includes church and secular organizations, fraternal lodges, service organizations, civic and neighbourhood organizations. In a sense, it's all the institutions that make up the culture of a country. This sector is far bigger than the marketing or government sectors, and if you were to wake up in your country tomorrow morning and this Third Sector had disappeared, how long would the country survive?"

Currently, we're completely focused on government and the market. We take the Third Sector entirely for granted. But like the environment, which supports the economy, it's the invisible layer that supports all of society. "We call this the Third Sector," says Rifkin, "but it's misnamed, because it's the primary sector in every country. We need to go back and understand the lessons of cultural anthropology. . . . We need to understand that this Third Sector is the primary sector; it is the wellspring upon which the other two sectors depend. The other two sectors, market and government, are derivative of the Third Sector."

Analysts such as Barlow, Clarke and Michel Chossudovsky point out that under the rules of the proposed FTAA and many other free trade governing bodies, even the most benign and obviously non-profit aspects of this Third Sector of society would come under the scrutiny of a system that imposes a monetary value and market control on every imaginable human activity. Daycares, public parks, village libraries, even mainstream charitable organizations like the Girl Scouts, Lions Club or the Shriners, to say nothing of the NGOs that are currently fighting them, would come under the control of trade governing bodies. They could legally demand that all these groups be reorganized, as is already happening with health care and prisons, to show a monetary profit, and that any benefits that come to them as non-profits — tax breaks, government subsidies, the right to raise funds in the community — should be taken away.

We have to remember that in the history of democracies, institutions that deal in clearly non-marketable "goods" are nothing short of vital. There is a remarkable partnership between the civil groups of the Third Sector and the success and happiness of the people in a society. Why? Steve Rayner, the chief scientist at the Pacific Northwest National Laboratory in Washington, D.C., is an adviser to countries around the world that are trying to create healthier political and social systems. He wondered the same thing. "The hypothesis that I'm currently investigating," he says, "is that these multiple overlapping memberships . . . . give you a much richer toolkit, or repertoire of societal strategies for dealing with problems, than if you only have a hierarchy, or if you only have a market. Or indeed, if you only have voluntary associationalism. You need all three of these elements present in a sort of creative tension if you're going to have a sustainable kind of government."

It appears that as in the natural world, our social world requires symbiotic and complex relationships to produce a healthy system. In fact, the richer the system, the healthier it gets. And the simpler, the sicker. "Now why should that be?" Rayner asks. "Well, if you've got only one way of doing things, you've got only one way of doing them.

If you have only hierarchy, like in most governments and in business, and something goes wrong or you encounter a new situation that's not working, the only solution available to you is more hierarchy. If you have only a market and you encounter a new situation, all you can do is more tinkering with the prices. If you have only egalitarianism, or voluntary association, all you can do is say, 'Well, we must get more people involved and have more participation.' And yet, the definition of madness is surely doing the same thing over and over again and hoping the result will be different."

When we discussed biotechnology, we said it is impossible to predict what will happen when scientists introduce a new gene into an organism, because the interactions between all the various elements are so complex that they're transcalculational, meaning that the richness of a working system is mathematically extremely complex. This complexity also explains why we don't know how many species we can take out of an ecosystem before it collapses. Interestingly, social structures need mathematical complexity to work well too. Steve Rayner has tried to figure out why this is the case. He says that as soon as you have two elements in a social structure — say, government and market — "you have a slight increase in the number of strategies available. But still, it's either market or hierarchy, to use the conventional dichotomy in political science and economics. The fascinating thing is, if you introduce just one more element, a third element, the number of strategies or the number of elements that you can do and the number of ways you can combine them rises geometrically, not arithmetically. So that's just a very simple mathematical indicator of the fact that you have a huge increase in societal complexity and capacity for coming up with novel solutions."

If we should find ourselves suddenly invaded by Martians, overrun with genetically mutated crops or forced to cope with flash floods, global warming and ethnic wars, we'd obviously be better off having a lot of possible ways of dealing with the situation, rather than just one. "I think that's the key to civil society," says Rayner. "It keeps that whole repertoire in play all the time. And it exposes people to different ways of doing things so that when they encounter new situations, they can just switch strategies and experiment in ways that you can't if you have only one or even two options."

The Law of Requisite Variety

> *If we are just going to let markets all over the world*
> *commercialize and commodify everything, then what do we*
> *have left? The things we have been telling our children, the*
> *things that our ancestors have taught us in terms of how to*
> *value life — the integrity of creation, the integrity of all forms*
> *of life — all of these things will amount to nothing if we don't*
> *take a very active role in sustaining and developing our view*
> *of the world.*
>
> — VICTORIA TAULI-CORPUS, IGAROT TRIBE MEMBER

These days it is widely thought that if the Third World is to become democratically governed, like most of the North, it needs prosperity, the global
market and industrial development. But Steve Rayner demurs. "Money
doesn't buy you democracy, that's for sure. I mean, one of the things that
was always a great puzzle to me was . . . why, for example, did the former
British colonies in Africa collapse so soon after independence? After all, for
all the evils of colonialism, the British left those countries with nice, new
parliament buildings, with pluralistic political parties, electoral machines,
judges with all the robes and wigs and other paraphernalia of the modern
state, and even half-decent economies. And yet within about ten years,
most of them had become dictatorships and had a lot of environmental
degradation occurring, along with poorer economic and health circumstances. And it seems to me that the one thing that the British didn't leave
behind, and probably couldn't create, was a civil society; in other words,
those voluntary associations across the tribal, ethnic boundaries that
would have been the key ingredient in the stability of those societies."

Rayner has other examples. "Another interesting contrast is Pakistan
and India. India hasn't had a single military coup in fifty years.
Pakistan hasn't managed to get a single, civilian-elected government
through a complete term of office. Why should that be? One reason
might be because the independence movement in India was, in fact, a
very pluralistic civil movement, involving different ethnic groupings,
different religious groupings, different segments of class and caste, and
so on, whereas Pakistan was specifically the creation of a monoculture,
an Islamic monoculture. In other words, the creation of Pakistan

actually destroyed, for the Pakistanis, a lot of that civic associationalism that has . . . served India very well."

Perhaps dams and stock exchanges are not the best gift from us to the Third World. We could try Rayner's methods. He says, "I prepared a paper for a conference in Islamabad, and when I presented it to my colleagues in Washington they laughed. What I said was that if we were serious about wanting to advance the development process in countries like Pakistan, we would do well to abandon large, capital-intensive projects and go out and buy Ping-Pong tables and sheet music." His point is that Ping-Pong tables and sheet music and football and sewing clubs bring people together to learn new ways of relating across religious or ethnic lines. They also help them learn how to get things done, even if those things seem trivial, like throwing a parade or raising money for matching jackets. "The people in Washington laughed," says Rayner. "Interestingly, the people in Islamabad said, 'Well, of course!' They understood perfectly well that the key to sustainable development for them is to bring about an end to feudalism. And Pakistan, unfortunately, is very much a feudal society, where, although formally it's illegal, slavery persists to this day. The power of the landlords is very strong. The people who want to change that recognize that . . . the development of this sort of civic layer is the key to dismantling the feudal structure of the state."

Especially in the post-Afghanistan, post-Iraq war period, the industrialized nations, with their one-track mindset — centred on one kind of culture, economics and politics — are not only trying to jump-start other countries with inappropriate massive development, they think they can simply export their political ideologies. Up until the 1980s, the North even used many Third World countries as battlegrounds for the undeclared war between the Soviet Union and the United States. For example, Tewolde Egziather asserts that people died of starvation in Ethiopia in the 1970s and 1980s not because they were unable to grow their own food, but because they were at war. He says, "The disturbances in Africa have their roots in the Cold War. . . . In 1973, Ethiopia was supported by the United States, with Western Europe as a strong ally. In 1974, it was supported by the Soviet Union. In 1973, Somalia was supported by the Soviet Union. In 1974, Somalia was supported by the United States. The switch in a few months was most dramatic . . . . You can't imagine what kind of destabilization that produced."

Terrible images of starving Ethiopians and Somalians in the mid-1980s elicited an outpouring of anguish, money and food from Western nations. But mass starvation had nothing to do with any inherent incapacity of Africans to feed themselves. "In 1984–85, we had one of the biggest food deficits in anybody's history," Tewolde Egziather says. "In 1994, we still had quite a big food deficit. In 1995, we fed ourselves very well, and the prospects looked good. And in 1997, we exported quite a substantive amount of food. If this were a technical problem, in such a small number of years, such a dramatic change wouldn't be possible. The simple reason for the improvement is that the instabilities were removed. People were able to mind their business, to go on with living. And life returned to normal in a very short time."

Of course, there were internal problems as well. War and mismanagement worked together to erode a wise, experienced culture's ability to manage land and food. Tewolde Egziather says, "As a reaction to the external world, we created a strongly centralized government. And when you centralize, your number-one enemy is local community, so you destroy local organization. That meant the disappearance of age-old systems of environmental management. The new centralizing state didn't have the capacity or the experience to replace what it was destroying. So it destroyed environmental management without trying to put anything in its place."

Efforts are now being made to strengthen local organization rather than sacrificing everything to central authority. "People have spontaneously organized themselves to manage their environment much better," Tewolde Egziather says. "My own village, the village I was born in, when I went there in 1973, I was most depressed. There was virtually not a tree left. I went back six years ago, and it looked like a forest, with so many trees! It's the most pleasant transformation I have ever seen, simply because people were allowed to mind their business. If we are allowed to be, we will be."

Many other peoples in the Third World are saying the same thing. Victoria Tauli-Corpus is a member of the Igarot, an indigenous mountain group in the Philippines. She also works with an indigenous coalition and is the liaison officer for the Third World Network in the Philippines. She lives in an area that is being invaded by Canadian gold-mining companies whose vicious dealings abroad were central to the CBC dramatic special *Human Cargo* in early 2004. These companies are being given carte blanche by the Philippine government, which in its quest to follow the First World lead, grants corporations free access to resources in the name of globalization. The government is composed mostly of wealthy people who are as far removed from the daily life of Philippine peasants and indigenous people as are the middle class of the First World. As happened in Somalia, Yugoslavia and Mozambique, the government has bought in to the single vision of prosperity that requires slashed social programs, low wages and increasing advantages to transnational corporations in trying to be globally competitive. But the miners have not been able to get much gold out of the mountains yet, and the people who live there are resisting them, just as the Ogoni did in far-off Nigeria.

Tauli-Corpus, like Oronto Douglas, lives this struggle every day. She says her husband is involved in the resistance and has been threatened by local employees of a Canadian gold-mining company. The Igarot want to control their own lands, but not just for their economic resources. She says the whole concept of this kind of competition has always been discouraged by her people, even though they traditionally pan for gold themselves, because it clashes with their values. She says, "The reason the resources are still up in our mountains is because we have a very simple lifestyle. We haven't squandered them. We really have used them sustainably. We look at them as part of the whole community, because they're not just for us, but for future generations. I think that this whole kind of world-view is really the wealth that we have. And it's why I think the indigenous peoples . . . are on a very high moral ground. We feel that we still have that kind of strength left to defend what we have, despite what others have said in the past — that we were backward, unscientific, pagan, et cetera. Now we feel that the values we have are actually the things that this world needs the most."

Currently, we are moving inexorably towards a system where there is only one accepted, enforced way of doing things. But living in a world that has no variety, no diversity and no adaptability in its human communities and belief systems is very dangerous, just as dangerous as using up all our natural resources. What would it mean to lose the world-views and value systems that groups like Ethiopian farmers, the Maya of

Chiapas, the Igarot, the Ogoni or the Okanagan have to offer? Steve Rayner sums it up in two sentences: "Social and cultural diversity are fully as essential to the welfare of the planet as biodiversity is, and for very much the same reasons. If you violate this law of requisite variety, you have an unsustainable social system." And an unsustainable planet.

CHAPTER 10
COMPLEX PLEASURES

Thought without concrete action is demoralizing.
— RUTH STEPHENS MIDDLETON,
ELIZABETH MAY'S GRANDMOTHER

As chapter 3, "Bigfoot," makes abundantly clear, the facts are in, and the conclusions we have to draw from them are pretty undeniable. We now know how much of the world's forests are being cut down annually, how physical changes in the atmosphere are beginning to affect weather systems worldwide, how toxic pollutants have spread to the farthest corners of the globe, how much we have depleted the oceans, and how our numbers are not only rising, but our demands on the earth for a rapidly shrinking pool of natural resources are growing exponentially. It's pretty clear we're pushing the limits of this planet's ability to support us, its ability to regenerate clean air, water, productive soils and energy faster than we can use them up. It's also clear we have very little time left to come up with new ways of living and new ways of using and sharing what we have. Denying there is a serious problem or, even worse, giving in to a sense of impotence or hopelessness, is to render a catastrophe inevitable.

We have to fight despair. Nelson Mandela must have been sorely tempted to give up during his quarter-century of confinement, and East and West Germans could hardly believe that the Berlin Wall and the border between them would ever come down. Yet Mandela and the East Germans triumphed over apartheid and division because they persisted and never abandoned hope. The ecological crisis now under way demands that we cling to hope and continue to strive to avoid collapse.

Human beings are often at their best when responding to immediate crises — car accidents, house fires, hurricanes. We are less effective in the face of enormous but slow-moving crises such as the loss of biodiversity or climate change. When the crisis is environmental and global, we feel not only frightened and overwhelmed, but also insignificant and helpless. The Sierra Club's Elizabeth May believes that no matter how overwhelming these problems have become, we can still act to mitigate them. She says, "This is not a parade of environmental problems that we can afford to watch as observers. We are not observers. We are no more observers than when you're sitting in a doctor's office and the doctor explains where the cancer is in your body and what it's doing. You are not an observer to that process. You'd better become a participant pretty quickly, and you'd better understand that there are things you can do that will improve your chances."

With major environmental problems, as with a terrifying medical diagnosis, our first reaction may be to feel depressed or to avoid confronting the issue altogether. We may try to find a distraction and pretend it will all go away, or we may simply question the accuracy of the diagnosis. But anyone who has been faced with the challenge of healing knows that such responses lead only to a quicker decline. We're not dead yet, and as May says, we ought to be thinking, "Right, the prognosis may not be good, but I can change the outcome through things that I can do." She says, "We're being told by the global doctors, 'You'd better cut down on the smokestacks. Stop partying beyond your means. Start living within the planet's ecological means. And then you can have a long and healthy life.' We have to recognize that it's our lives, our bodies, our health, our planet that are at risk. We are participants. We have to be involved and engaged at a very active, deep and personal level."

May herself is extremely engaged and active. I've long admired her, and am proud to have her as a friend. Her energy and conviction brought her national attention when, as a university student, she fought pesticide spraying for spruce budworm in Cape Breton. The emulsifying agent used in the pesticide was eventually linked to Reye's syndrome, a deadly swelling of the brain that affects children. The pesticide mix has also been blamed for destroyed runs of salmon in the East. May's actions helped stop the spraying in Nova Scotia, and she went on to become an influential adviser to the federal environment minister in the late 1980s. When he approved a huge dam project on the Rafferty and Alameda rivers in Saskatchewan, May resigned so she could speak out against this decision.

Elizabeth May comes by her activism honestly — she was raised that way. "It comes straight from something my grandmother, Ruth Stephens, used to say: 'Thought without constructive action is demoralizing.' So there is absolutely no point in learning about these issues and thinking about these issues unless you're prepared to take constructive action. I mean, my mom's paralyzing fear for the future was of mutual nuclear annihilation from the arms race. My first press conference, I was sitting on Mother's lap. I was two. She was protesting against atmospheric nuclear weapons testing, which actually was stopped with a test-ban treaty. If you'd asked my mom back then, 'If we don't change, where are we headed?' she'd have said we were headed to a world where the levels of strontium-90, the levels of childhood leukemia, would have been through the roof. As it is, childhood leukemia has increased. Yes, we poisoned the atmosphere for quite a while. But action did occur."

Now that countries rarely explode nuclear bombs above ground, we look back on the debate over atmospheric testing and fallout as something that was solved politically because our leaders gradually understood the danger. We forget what really happened. There was a massive, global movement of ordinary people, most of them women, who said "No!" to atmospheric testing. They staged sit-ins, raised money, formed blockades, marched in parades and held press conferences. "In those days, my mom was just a housewife volunteer," May remembers. "But because people like her got busy, they changed the whole damn world. And we can do it in our generation as well."

May knows all the ways people can become discouraged. She says, "One of the things that we mustn't do in our current moment of struggle is try to assess if we're winning or losing. By any reasonable measurement, we're losing. So let's not waste our time with that. What we have to do is continue the struggle for fundamental change. We do have our moments. We do have our opportunities. And whether we're trying to arrest biotechnology before it contaminates every crop and food source we have; or whether we're trying to reduce the amount of pesticides that we use, knowing that those increase childhood cancers; or whether we're dealing with climate change, and trying to get any level of government to act responsibly and actually create the kind of economic signals that will get society away from fossil fuels — it's all the same thing. I know these changes are doable, but we need the right, clear signals. These things will come from governments, but only when they feel the heat of an enraged population breathing down their necks."

We need to remember that every poll on the subject, every real question that addresses the future, shows that people everywhere still very much want wildlife and clean water and sustainable industries and a future for their children that includes intact and productive natural systems. Even the core of activists working on these issues, as May points out, grows slightly larger all the time. "But we need everybody in the Canadian population. We need everybody to say, 'This is my personal commitment. Even if it's only an hour a week, I'm going to do something to ensure that when my kids are grown, they're not going to be looking at pictures of the African elephant and seeing it the way we look at pictures of the dodo. I don't want my kids to look at a map that shows where the low-lying island states *used* to be before they were inundated by melting ice caps. I don't want my kids growing up in a world where the levels of planetary disruption create a desert in the northern Amazon, which the Hadley Meteorological Centre in England is now predicting unless we deal with greenhouse gas emissions. I don't want my children growing up in a world where we're still fighting wars over fossil fuels that create generations of misery and bitterness in other countries, or drilling for a few months' worth of SUV fodder in national parks.'

"These are human-created problems," May reminds us. "The solutions can be created by humans too. But nothing moves without a very, very motivated, engaged, committed and dedicated public movement." If we applied the same energy to figuring out sustainable technologies and sustainable lifestyles that we do to trying to make the most money in the shortest period of time, creating a new weapon or inventing a new technology, I'm sure we could arrive at true sustainability very quickly. The idea that politicians are too corrupt or distant, and that we ourselves are powerless to do anything that makes a difference, is, as May says, "one of the most pervasive, pernicious and evil myths out there. We're not powerless. We're damn lazy. We have tremendous power in our society. We can, in fact, move the whole world in a different direction. We just have to care enough about it to make it a priority in our own lives."

And a Child Shall Lead Them

If the members of a local community want their community to cohere, to flourish, and to last, here are some of the things they should do: Always include local nature — the land, the water, the air, the native creatures — within the membership of the community; . . . ask how local needs might be supplied from local sources; . . . see that the old and young take care of one another; . . . always be aware of the economic value of neighbourly acts; and always care for your old people, and teach your children.

— WENDELL BERRY,
FARMER AND PHILOSOPHER

In terms of who holds power, perhaps the lowest on the social totem pole are the very young. We pay a lot of lip service to ensuring children's rights, but child mortality and child poverty have increased dramatically under our new economic order, even in the developed world. Children don't have any legal say in our political and social decisions. They can't vote, they're not protected by unions, and when they talk to adults we rarely take them seriously. And yet even they can teach us a thing or two about making a difference. Let's look at an example from just a few years ago.

In West Vancouver in British Columbia, a group of about twenty children and their parents formed Orca, a nature club, one of those Third Sector groups that cut across age, language and religious lines that Steve Rayner believes are so central to creating a functional society. Club members did simple, fun things such as kayaking, hiking and designing projects to help them learn about wild plants and animals or the stars. On a study project, while rooting around a stream, they discovered an unusual creature, a tailed frog.

The first thing they did was call the University of British Columbia to find a scientist who could tell them more about the little creature. They were informed by the astonished biologist that these frogs are very rare and were believed not to occur south of Squamish. The youngsters got excited and worked to help the biologists establish that densities were higher in the suburban stream than up north in the wild, and they realized immediately that the habitat was special. These tiny frogs are the size of your thumb and are the only species in North America to live in clear, fast-running streams. They have other unusual features as well: they breed internally and maintain a tail-like appendage, even as adults. Then the club members discovered that this very stream was due to be developed for a housing project, thereby putting the frog's habitat at serious risk. The children decided, on their own initiative, to make a submission to the West Vancouver city council, because, as they said, "there are no laws to help the frogs." They wanted to try to convince the council members to make the developer change his plans and accommodate this species.

Four of the children wrote speeches and made the presentation. David Maszaros and Kathleen Folley, eleven and twelve years old at the time, were two of the children involved. David says of that first meeting, "I don't think [the councillors] took us very seriously at all." When I asked them if they had been nervous, Kathleen replied, "Yes! Because... well, they're bigger!" But luckily, a television reporter got wind of the story and interviewed the children. Children and frogs make a nice news diversion, and they're visual. So when the Orca kids returned to city hall a second time, there were video cameras present to record it. David recalls, "By the fourth time, when we got there, there were, like, nine, ten TV stations all just filming us. And there were radio broadcasters."

The councillors began to become uncomfortable under such close scrutiny. The upshot was that the Orca members, thanks to their media exposure, were able to persuade the developers to leave a generous buffer zone around the stream for the frogs, divert all waste water away from the creek and refrain from cutting any of the old-growth trees growing along its banks. David says proudly, "We had a huge effect, I think. They started to take us seriously after about the second or third time.... They could see we had a really big point. The frogs are a blue-listed [extremely rare] species!"

Staff members in the ministry of provincial parks told David's father, "These kids have done more to protect our creeks than anybody in the past thirty years." I asked David how that made him and his fellow Orca members feel. He replied, "Well, we feel good! We went up and had a big influence; we saved animals who can't speak for themselves." Of course, you might say the kids were lucky; they live in the affluent North, they're cute and they attracted media attention. But the fact remains that a group of informed and committed children — normally the most powerless people in our society — fought city hall, big business and big money — and won.

Five or Six Very Difficult Things You Can Do to Help Save the World a Little Bit

This is the one possible chink in the armour of the ... monolith represented by the television networks, Disney World, the suburb and the shopping mall, and all the kind of overwhelming material-consumption features of our lives. If those things made us as happy as they claim, then it would not make any difference what environmental arguments they raised. The fact that they don't make us all that happy is their Achilles' heel.

— BILL McKIBBEN, JOURNALIST

One of the most powerful environmental books of the past ten years was *The End of Nature* by Bill McKibben, a journalist who lives in the Adirondacks. *The End of Nature* is a frightening vision of what the world has become and where it is heading. His book has been likened to a modern *Silent Spring,* the seminal work by Rachel Carson that catapulted the environment into everyone's consciousness.

McKibben's book focuses on climate change as a result of global warming. It's one of our most difficult problems, because solving it will involve basic changes in lifestyle: phasing out fossil fuels, changing trade and economic structures so that economies function locally and we don't need so much energy for transport, and much more. McKibben hasn't been afraid to tackle the really big issues that underlie our current problems, and he isn't any more paralyzed by the enormity of the task than were Elizabeth May or the Orca kids. When people ask him what they are supposed to do about such horrendous, systemic problems as over-consumption, overpopulation and global warming, he says, "In practical terms, there are many answers, all of them both simple and difficult.

Someone wrote a book once called *Fifty Simple Things You Can Do to Save the World*. There were some pretty good ideas in that book, but it was a very . . . American notion — that there are going to be fifty things you can do to save the world, and they are all going to be simple. My work has basically been writing many chapters of a long book called *Five or Six Unbelievably Difficult Things You Could Do That Would Help a Little Bit*."

Where do we start? McKibben is the father of one child, a eleven-year-old girl named Sophie. He says, "We've got to address questions of our population. That means we have to address the most difficult and intimate questions of how many children we're each going to have. For those of us in the developed world, it makes real sense to think about and explore single-child families for a couple of generations." And, of course, it's not just a question of numbers of people. McKibben says, "We've got to think about how much we consume." The figures are clear. If we got rid of 80 percent of the world's population tomorrow by eliminating all the poor people, we'd *still* be in the predicament we're in now. That's because those of us who live in the First World, although we make up only 20 percent of the world's population, consume more than 80 per-cent of the planet. We're long past using up all our biological interest and are now destroying natural capital and, with it, our children's future.

Is consumption on this level normal? Does everyone "naturally" want to consume on the scale we do? That certainly seems to be what the popular media would have us believe. Of course, people have always been interested in wealth and material security. But since the Second World War, our culture and our government policies have actually *discouraged* thrift, personal savings or debtlessness, while encouraging and rewarding ever-rising patterns of consumption and debt. We've been exhorted by ads, taxes and our global media culture to believe that having more money in order to buy more things is the real ticket to happiness; that the virtues of the past, like frugality and self-denial, were mistaken, and now being truly happy largely means indulging ourselves. But this entire pat-tern of thought could simply be wrong. "We've been told that the more we have, the happier we'd be," McKibben says. "But a look around us doesn't seem to bear that out."

Since the 1950s, we've almost tripled the amount of material goods in the average North American's possession. Even students without enough cash to eat properly own TVs, computers and cars. However, fifty years of surveys asking people to rate their happiness show absolutely no correlation between the attainment of material goods and personal satisfaction. Once we've got beyond having enough basic food and clothing and a roof over our heads, we're not made much happier by extra consumer goods. In fact, the number of people who say on these survey forms that they're very happy has actually *declined* from its peak back in the 1950s. "Material goods aren't making us happier now," says McKibben. "We've reached some point where we have enough, and the struggle to get more and keep it makes us not more happy, but probably less happy."

A group in Seattle called the New Road Map Foundation is actively looking for ways to enable ordinary, middle-class people to be both happier and less of a burden on the Earth. One of its founders, Vicki Robin, says, "The single most important thing anybody can do now for the environment and for the economy is to save. For the environment, money not spent is resources not consumed at some level. And money saved is putting yourself on a firmer financial footing in a very, very scary and shaky world."

It's interesting how quickly the very suggestion of buying less and saving more, those two notions of sane, stable, responsible people only a generation ago, have come to sound completely radical. That's because we really have come to believe that if you don't do your share by buying a new car or a new kind of soft drink, you'll contribute to job losses and a general economic slide. Vicki Robin says that belief in constant growth and increased consumption is madness. It may have made sense fifty years ago, given our understanding of the world at the time, but now it's just helping us avoid facing ecological limits or the population issue. As Herman Daly put it, it's just helping us to avoid the issue of sharing. Robin concurs. "What constant growth actually does not do is give us the opportunity to develop. It's like a human body. We don't grow forever; we grow until we reach a mature size. And then we develop — we develop our intelligence, our skills, our experiences in life. We become ever more beautiful human beings because of what's going on inside."

As we mentioned earlier, nothing in nature grows forever, except cancer cells; and even they die in the end. "A beautiful tree grows to a certain size and then it can produce fruit," explains Robin. "In fact, if it's pruned, if it faces restrictions and challenges, it will produce more and better fruit. This principle is the same throughout nature." She says that when ideals as sensible as thrift and conservation become radical and abnormal, it reflects an actual disease in society.

The way we live has less to do with individual choices and more to do with the general bent of our society than many of us realize. We forget that we don't *need* everything. As Robin says, if you've noticed that no matter how much stuff you get, you don't feel satisfied, it's probably because material things satisfy human beings only so much.

Eating Right

In Japan it's referred to as "farming with a face on it." Europeans call such arrangements subscription farming. "Linking farmers with consumers" is the phrase used in the United Kingdom . . . . All these terms translate into a dynamic new [global] development in small-scale farming. Community-supported agriculture, or CSA, serves as an umbrella for farmers striving to achieve a number of goals: mutually beneficial relationships with consumers; environmentally sustainable farming practices; and public education on agricultural issues.

— DANIEL IMHOFF, JOURNALIST

There are a lot of ways we can act to conserve the Earth. Like the Orca Club children, we can take an interest in endangered nature and try to help out locally. We can follow Bill McKibben's advice and work on big issues like reducing fossil-fuel consumption and curbing population growth. We can demonstrate and work against all these global trade organizations. Or, like Vicki Robin, we can simply consume less and enjoy ourselves more.

One issue of vital concern to everyone is food. Our disconnection from the Earth is epitomized by our relationship to food. Most urban people associate food with supermarkets but fail to connect it with the land. Brewster Kneen lives in rural British Columbia and publishes a monthly newsletter called the *Ram's Horn*, which deals with agricultural issues. He informs people about how industrial farming, biotechnology and agricultural patenting intrude into their lives, and how they can resist systems that overuse fossil fuels and exhaust the soil. One of his favourite ways to fight back is through community-supported agriculture (CSA). It's a movement that has spread with incredible rapidity over all of North America, and many readers may know about it now because it is available to virtually anyone living near a big city in Canada or the United States. It works by bringing an urban family together with a farmer to share the risks of producing food. The family pays in advance for its vegetables — say, a summer's supply of sweet corn and tomatoes or a winter's supply of potatoes and onions. With the prepayment, the farmer's susceptibility to unexpected pest outbreaks or exceptionally bad weather is removed, because he or she is allowed to substitute, for example, more tomatoes or zucchini if the corn crop fails; the farmer is assured of a market and doesn't have to depend on huge industrial entities, trucking, middlemen, preservatives or crop varieties bred for long shelf life instead of flavour.

As Kneen says, "People can say, 'Look, we have our family doctor. We have our dentist. And now we have our family farmer who looks after us. We have a responsibility to that farmer.' So if it doesn't rain, we don't say, 'Hey, farmer, you failed us.' Food production is a shared thing. He or she had nothing to do with the drought." This means the farmers don't have to go to the bank for loans to get the seed in the ground. They don't have bites taken out of their profits because they have to arrange for middlemen to transport their harvest. "You've begun to attack dependency and distancing in the food supply," explains Kneen. "You eliminate the debt factor." You can also get more variety. You can ask your farmer to grow arugula or bok choi or any vegetable you have trouble getting. And they can introduce you to their favourites. One Japanese CSA farmer presented her clients with different kinds of greens they'd never seen before, and then she taught them how to cook them. Kneen says, "I remember being with them, and those clients were all coming back and saying, 'Hey, that was fantastic! It was so good! I've been eating this iceberg lettuce all these years, and you mean there's all these other kinds of greens we could have?'"

Arrangements like CSA mean that in a bad year for potatoes, people learn to eat cabbage or turnips or whatever other vegetable produced well. They learn that nature is always changing, and that every year is different. Farmers start producing chickens when they learn that chickens can help with the weeding and insect control, and then they benefit from fresh eggs as well. They start growing heirloom vegetables and saving seed. They even start to share. In any given year, there's always a surplus of pumpkins or tomatoes or whatever, and that surplus can be sold cheaply to people who don't have the money to buy into the plan up front, or donated to food banks. "What you're really doing," says Kneen, "is reconstituting a community and a social life around food, which is quite proper." In fact, most CSAs quickly get into having feast days and harvest dances. Kneen says, "You're creating a whole new culture. And I think that's what we have to realize we're doing, and that's what we need."

Getting Control

> *Many, many years ago, culture was ours.... Bit by bit we lost control of our own culture, and suddenly corporations are doing all the talking. They're telling all the stories. And bit by bit, the whole of our media culture has become one large advertisement for a lifestyle of consumption. And now the big job ahead is to start telling our own stories again, to get dominion back over our culture.*
>
> — KALLE LASN, EDITOR OF *ADBUSTERS*

Perhaps you're not that focused on food or the country life; you're urban, young and interested in the Internet, movies and mass culture. So you probably know about *Adbusters,* the glossy magazine that Kalle Lasn produces. Its hip layouts give voice to a growing global movement called culture-jamming. Lasn is an ex-adman who was radicalized when he realized he couldn't go into a television station in North America — even if he had the money — and buy air time to put on a message he believed in. That's because there's a kind of censorship that has outlawed messages that don't conform to the current paradigm of endlessly growing economies and frenzied consumption. Lasn says, "The real problem facing us is our own culture, in all its different aspects, be it our nutritional agenda or our transportation agenda or even the way that we run our society's information delivery systems. Somehow, the whole culture has now become dysfunctional. If you want to fix things, you have to deal with the culture, part by part; but you have to think of it as one huge dysfunctional thing that needs to be jammed back onto a sustainable path. We see ourselves as culture-jammers. We're trying to launch the new social-activist movement of the Information Age. We use anything we can to get our consumer culture to bite its own tail — and to slowly veer it off in the directions we believe it has to go to become sustainable."

Adbusters specializes in grabby graphics and spoof ads: perfect replicas of the style and glossiness of Madison Avenue. One ad, for example, featured Joe Camel getting chemotherapy in a cancer ward; another showed the famous bottle of Absolut vodka drooping like a dysfunctional penis, to remind the viewer that drinking increases impotence. *Adbusters* gets pro bono work from professionals, leaked memos from inside the corporate world and donations from inspired amateurs. To remind people what consumption does to the Earth and to their own lives, the magazine sponsors a global holiday, Buy Nothing Day, on November 26, just as most North Americans are gearing up for the pre-Christmas shopping orgy. Articles share strategies, such as teaching people to turn out the lights on billboards, hang up consumer-disruptive signs in malls or alter signs and logos to expose reality (one wag glued a cartoon balloon over the heads of two Marlboro cowboys so that one was saying, "Bob, I miss my lung"). In short, *Adbusters* is constantly controversial, and the magazine faces lawsuits every day. Lasn writes about the kinds of things most of us are afraid to because of libel chill, SLAPP suits, and the sheer, overwhelming power of corporate money the world over. The staff members of *Adbusters* live on a shoestring, and they are prepared to go under financially and resurface tomorrow. But the magazine is one way urban people, who may not be inclined to wade in streams looking for frogs, can use their creativity and talents and get involved in saving the Earth.

The Media Foundation, which Lasn also runs, tried to sue the CBC, among other broadcasters, for refusing to air its ads against cars and television, ads that criticize globalization and the neo-liberal economic agenda. The foundation had the money to pay for the air time, and its message wasn't racist or pornographic; yet it doesn't have the freedom to put it on the airwaves. The CBC at first signed a contract to air the spots, then changed its mind, Lasn believes, because of pressure from other advertisers. The network then broke the contract. Lasn says the

CBC feels the CRTC allows it to manage its airwaves any way it sees fit, and it doesn't want to run the foundation's ads.

Lasn took the case from the B.C. Supreme Court, through the Court of Appeal; the Supreme Court of Canada refused to hear the case. He then went to the World Court at the Hague, demanding redress under Article 19 of the Universal Declaration of Human Rights. He's fighting for what he calls the right to communicate. He says that since the Magna Carta and the Bill of Rights, people have been fighting to be heard and to have input into the decisions made by the elite groups that govern us. "I think that in this Information Age of ours, we need a new set of human rights," he says. "From television to cyberspace, we have to make sure that the corporations don't run away with the ball; that we, as citizens of the world, have, in our own constitutions and also in the UN charter, a little clause that gives us this new right, the right to communicate."

The right to communicate is one step beyond freedom of speech. "Freedom of speech simply means that you're allowed to stand up in a park on a box and say what you want to people who are passing by," Lasn explains. "But now that there are so many of us, that freedom doesn't give you access to enough people to have an effect on society. The really important battle of our Information Age is the battle for access to larger populations. And if we can win this battle and win the right to communicate legally, then I think we will have a level playing field between citizens and corporations. We would be able to talk back to the corporate image factory. We would be in a position to create alternative futures and to pursue alternative visions, and eventually to create the sort of a world that we need rather than the sort of the world that the corporations are trying to railroad us into at the moment."

How Hard Do We Have to Try?

Of course, it is likely enough, my friends . . . that we are going to our doom: the last march of the Ents. But if we stayed at home and did nothing, doom would find us anyway, sooner or later. That thought has long been growing in our hearts; and that is why we are marching now.

— TREEBEARD THE ENT, IN J.R.R. TOLKIEN,
THE LORD OF THE RINGS

We're not really a special kind of people.... We don't want to die, but... Great Britain is a very small place, and it's a beautiful place. And at the rate that it's being destroyed... there's going to be nothing left. We're fighting for our children now. It's not for us.

— ANDY, A DIGGER PROTESTING THE EXTENSION OF THE
MANCHESTER AIRPORT INTO A FORESTED VALLEY

One of the most popular stories being retold in the early years of the twenty-first century is the J.R.R. Tolkien saga *The Lord of the Rings*. For a fantasy best-seller, this is a remarkably dark and depressing book. Although it comprises three long volumes, it's not only voraciously read by young and old, it has spawned the most popular film trilogy of all time, outgrossing even the iconic *Star Wars* franchise. Largely written by a veteran of the First World War during the days preceding the holocaust of the Second World War, *The Lord of the Rings* can be discussed in many ways. It can be read as a political allegory warning of the gathering militarization and industrialization of the modern world, or as personal psychology, that is, the search for integrity and truth in an immoral and threatening world. The Shire and Mordor, for example, are also often read as metaphors for an idyllic natural world increasingly in the grip of horribly destructive industrial development and, as in most literary analyses, these are only some possibilities. But when it comes to his story of the Ents in volume 2, Tolkien's symbolism brooks few other interpretations. He draws the reader in with a wonderful fantasy: if only trees, those undeniably beautiful, ancient and beneficent forms of life, had some kind of more mobile defenders, like half-tree shepherds who lived among them! How wonderful it would be if they could rise up and fight against the outrageously wasteful and often pointless destruction of the world's forests.

It's not at all too much to say that increasingly today the Ents Tolkien imagined two generations ago have sprung into existence. Regular people from very widespread nationalities, ages and social backgrounds have not only allied themselves with trees so closely that they place their bodies between the forest and the axe, but they have also proved willing to move right into the forest, to live in the wind and the rain and the snow, exactly like the trees. When we first researched this book, the hotbed of their activity was in the United Kingdom. Calling themselves both Diggers and Tree People, these groups of mostly young activists were actually living up in trees for months at a time to prevent them from being cut down for timber or for development. High in the branches, the activists build simple wooden platforms on which they pitch tents. They also construct a system of rope bridges to move from tree to tree. In England, professional climbers are often hired to get them down, and that takes time and money. Their actions are intended to hold up construction and add greatly to the cost of any development. The tactic of the Diggers is to dig tunnels underground and chain themselves to concrete blocks many metres below. That prevents bulldozers and heavy equipment from coming in, because they'd be crushed. It can take even longer to get the Diggers out than it does to remove the Tree People.

When we interviewed him in 1998, one of the Tree People, calling himself Robin Forman-Quercus (Quercus being the Latin name for oak), had joined a protest against the expansion of the Manchester airport in northern England that would put a beautiful, forested river valley under tonnes of concrete. At nineteen years old, he was a veteran, having spent over two years of his young life living in a 500-year-old oak in the south of England in an effort to save it. After a couple of months up on the platforms in Manchester, he and a friend, Fiona, were evicted by the bailiffs. We met them during the mop-up operations. Many of the people who would benefit financially from the expansion were also on the planning board that had approved the new runway. But to the local residents, the trees and river valley were far more precious than the airport expansion. The citizens had, in fact, tried every legal means to save their valley, but they failed. So they had turned to the Tree People and the Diggers, and now, although both groups had resisted bravely, they were being forced out by superior forces.

One of the local citizens, Catherine Mitchell, a public nurse and housewife who had been helping other townspeople smuggle food, water, food and blankets to the activists, said, "We tried for five whole years to stop this the democratic way, but power and greed have won over. We never expected all our methods — petitions, letters, an enquiry — to lose. But the Tree People came, with their funny haircuts and tatty clothes, and their strong ideals and intelligence and integrity and humour. They are very loving, very peaceful. And it's done something to a lot of people." Robin was certainly young, dreadlocked and in tatty clothes. He was also not only a veteran of fighting for nature; he was a veteran of losing. The oak he lived in for more than 900 of his days and nights, in every kind of loneliness, hardship and filthy weather, was finally cut down. "It wasn't just that tree," he says. "There were a lot of others — and crops — that we were trying to save . . . . Now they've chopped all the trees down. It's something we've just got to deal with. I went back to see it, and I sat on the stump of my oak. It was, like, two feet high. And I cried for about four hours."

Pretty hard work, fighting and losing. But they might not have lost as much as they think. The Tree People movement in Britain operated at its height between the early and late 1990s. It was a direct response to an extremely ambitious roadbuilding program put forth for the whole British Isles by the then-Tory government. Twenty-three billion pounds was allocated in 1992, a sum that's been slashed to only a few billion today, with nearly 500 out of 600 of 1989's road schemes scrapped. A construction periodical wrote last year that "the major roadbuilding programme has virtually been destroyed." No one, especially not the government, is saying this was due to the actions of people like Robin Forman-Quercus. But now that the Tree People movement in Britain is quiescent, there have been some serious analyses as to what exactly happened to all that road expansion. Tree People were able to gain so much public support from voters like Mrs. Mitchell, and even more through their media antics, that the government is now very careful to try to placate "both sides" every time it suggests a road. EarthFirst!, one of the major NGOs involved in the issue, calls the scenario, "noisy defeats, quiet victories." It admits that "It's hard to quantify any such direct link [between their actions and the budget-slashing]. However, pro-roads lobbyists and local green activists agree that the Twylford protests were a major factor in the scrapping of the East London River Crossing through Oxleas Wood in 1993; and [actions in] Newbury had an effect on the decision to drop the Salisbury Bypass because of its 'environmental disbenefits' in 1997."

North America is about to find out how well these extreme forms of protest work, in a much more violent regulatory atmosphere. The United Kingdom used to send Tree People over to teach Americans how it's done, and so far the most famous American tree-sitter is Julia

"Butterfly" Hill, whose many months in a 1,000-year-old redwood called "Luna" in northern California won her Internet and media fame. When she finally came down, like many former sitters and Diggers in Britain, she applied her new knowledge and confidence to more mainstream forms of social work, founding a youth NGO called the Circle of Life Foundation. The Ents that inspired her and the British road-resisters are back on the move, however, as never before, due to the Bush administration's policies of allowing commercial timber new kinds of access to public lands. Most tree-sits are out west in British Columbia, California, Oregon, and now Washington and Alaska, but the technique is spreading. There are also tree-sitters at Wachusett Mountain, Massachusetts, protecting a state-owned mountain from a ski hill expansion, as well as abroad in the Otway Range of Australia and the de Efteling park expansion in Holland.

During the summer of 2003, a *Washington Post* front-page article described a Greenpeace and National Forest Protection Alliance camp in Montana that teaches the climbing, non-violence, and general hunkering-down skills of tree-sitting to Americans distressed at the new commercial presence in parks and national forests such as the Tongass National Forest in Alaska. More than a hundred people attended, including David Muller, a fifty-six-year-old bookseller who has never done such things, and Molly Karp, a twenty-one-year-old direct-action veteran studying criminal justice. They'll be trying to slow down what the organizations call "a massive increase of commercial logging and other industrial resource extraction" with tactics that activists, citing the Boston Tea Party, likened to "the most traditional, mom-and-apple-pie way to show your patriotism in America." Mathew Koehler of the Native Forest Network, an NGO more familiar with public hearings and legal appeals than with direct action, says, "We need all these tools. No single form of protest is enough."

Such actions were called "eco-terrorism" long before 9/11, and it's anyone's guess how this form of protest will fare in the courts in an age of eroded civil liberties. Some people may fall for the accusations that have dogged peaceful EarthFirst! protesters like Rod Coronado, the victim of a very expensive, logging-industry-funded media campaign that even includes television ads and full-page inserts in newspapers. But history tells the tale, and so far the only people who have been physically hurt in the eco-battle are the people fighting to save the trees. David Muller says, "If every American could see the Tongass and southeast Alaska, they would be outraged to think the Bush administration wants to log it. They can take away our right to appeal or to file lawsuits or to participate in forest management, but they can't take away non-violent direct action. It's at the very heart of who we are and how far we will go to protect the places we love."

It's easy to become so preoccupied with the details of life — getting up, going to work, dealing with family crises. We don't usually come face-to-face with the opportunity to do anything about these enormous issues, to make a difference in the larger world. Fiona, a twenty-two-year-old student when we met her, had shown up at the Manchester protest just to see what the fuss was about, and ended up spending the next three months living in a tree. She says, "You just have to try one thing, just one small thing. You'll see it work and it'll give you spirit . . . . All I did was come here for the day because I felt an urge to do it, a pull. And now I know what I want to do with my life. I understand what's going on. I understand that we're in an emergency situation and we have to take action, and we're the people to do it . . . . I mean, it really is annoying that you can't get on with your life because the planet is being destroyed. But I, personally, can't just ignore it, because it's part of me. It's part of all of us. I think a lot of people don't see the connections that run through everything. It's as simple as . . . thinking about your children and what they're going to say to you. You know, 'What did *you* do, Mother?' "

Saving the Whales

> *One of my best friends the summer of 1960, when I was nine years old, was a beaver . . . . The next year, when I went back to our camp, I couldn't find him. I found out that trappers had taken all the beaver over the winter. I became pretty angry, . . . and when I was ten, eleven, twelve, I started destroying the trap lines, collecting leg-hold traps and generally being a nuisance in the neighbourhood for disrupting duck hunts and deer hunts.*
>
> — PAUL WATSON,
> HEAD OF THE SEA SHEPHERD SOCIETY

Paul Watson is a Canadian known around the world for his daring efforts to stop the killing of whales and seals by direct intervention. Although he was one of the originators of Greenpeace, he broke away from that group to form the Sea Shepherd Society, which is dedicated to protecting marine mammals. As captain of the ship *Sea Shepherd,* Watson cuts a swashbuckling figure, charging between whales and their hunters. We asked him how he ended up spending his life sinking whaling boats and getting beaten up by enraged sealers. He told us the dramatic and moving story of a day that changed his life.

"In 1975, we had encountered the Russian whaling fleet just off the coast of northern California. We'd come up with this idea to put ourselves in small inflatable boats, to put our lives on the line, reasoning that nobody is going to kill a whale if they've got to risk killing a human being. We were reading a lot of Gandhi at the time and we were quite naive . . . . Anyway, Bob Hunter and I were in a small rubber boat, and we were blocking a Soviet harpoon vessel. This 150-foot steel vessel was bearing down on us at twenty knots. And in front of us, eight magnificent sperm whales were fleeing for their lives. Every time the harpooner swivelled the harpoon, I would manoeuvre the small boat . . . to block his path; this worked for about twenty minutes. Then the captain came down the catwalk and screamed into the ear of the harpooner . . . and we knew we were in trouble. And a few minutes later, the harpooner fired the harpoon over our heads and this 150-pound exploding grenade zoomed over our heads and slammed into the backside of one of the females in this pod of sperm whales. And she screamed. Blood was squirting everywhere. The largest whale in that pod suddenly rose up and dove.

"We'd been told by all of the experts that the whale would attack us, because we were the smallest targets and it would be very angry. As we waited with a great deal of anxiety for fifty tons of very angry animal to come up underneath us, the ocean erupted behind us and we turned in time to see the whale hurl himself from the water straight at the Russian harpooner who was on the bow. But the harpooner was waiting for him, and he very nonchalantly pulled the trigger and sent a second harpoon, at point-blank range, into the head of the whale. The whale screamed and fell back in the water, blood everywhere now. And as he was rolling and thrashing about, I caught his eye. And he saw me. Then he dove.

"I saw this trail of bloody bubbles coming straight towards us, real fast. The whale came up and out of the water at an angle, so it looked like he was just going to come forward and crush us. And then it was almost as if he stopped in mid-air. He was so close I could have reached up and grabbed one of these six-inch teeth. I looked up into this eye, which was the size of my fist, and what I saw in that eye was understanding. That whale really knew what we were trying to do, because the easiest thing for the whale to do was to come forward and crush us or seize us in his jaws, and he did neither. He just very slowly and deliberately and with great effort slid back beneath the waves and died.

"There was something else I saw in that eye, and it was pity. Not for himself, but for us, that we could be engaged in this kind of blasphemy, where we could so wantonly destroy life. And for what? The Russians in this case were killing the whales to provide lubricating oil for ICBM missiles. We were wiping out a species in order to make a weapon meant to destroy human beings. It was absolute insanity.

"And as I sat there in that small rubber boat . . . the water turning red around me, I began to think, 'What is this all about? Why are we doing this?' And I said to myself, 'For the rest of my life, I'm going to pledge to protect as many of his kind as I possibly can.' And personally I don't really care what people think about what I do. My satisfaction comes from living up to that obligation and protecting whales directly by destroying the illegal vessels that are out there destroying them."

Paul Watson's passion for his life commitment is unmistakable — and essential. Much as we'd like politicians to take the burden from the shoulders of individuals and care for the Earth, they haven't. The real force for environmental protection still comes from individuals and thousands of non-governmental organizations around the world. "The only thing that has ever really made a difference in any social movement has been individual initiative, individual passion," Watson argues. "Anthropologist Margaret Mead once put that very distinctly when she said, 'Never depend upon institutions or government to solve any problem. All social movements are founded by, guided by, motivated and seen through by the passion of individuals.'"

Paul Watson has been on the frontlines, protecting wildlife for thirty years, and he speaks from experience. He has many examples of what one person can do. "A few years ago I had a man call me from

Glasgow, Scotland, and he said, 'They're killing grey seals up here. What are you going to do about it?' My answer to him was: 'I'm not going to do anything about it. What are *you* going to do about it? You're Scottish, you do something about it.' And he said, 'Well, what can I do? I'm just one person.' Well, we talked to him and encouraged him, and he set up a Sea Shepherd group in Scotland. And in only three years, they shut that hunt down. He was an architect, so he wasn't a professional conservationist or anything. It just shows you what individuals can do if [they're] given the encouragement to see what kind of power [they] can have."

Complex Pleasures

The story of the twentieth century was finding out just how big and powerful we were. And it turns out that we're big and powerful as all get out. The story of the twenty-first century is going to be finding out if we can figure out ways to get smaller or not. To see if we can summon the will, and then the way, to make ourselves somewhat smaller, and try to fit back into this planet.

— BILL MCKIBBEN

As I've said repeatedly in this book, if we want to learn how to live in balance and harmony with our surroundings, we need input from many different people and cultures. Scientists and researchers can document the state of the planet, but we also need the perspective and knowledge of other cultures, especially those that have always depended on nature for physical and spiritual sustenance. Matthew Mukash, formerly subhead of all of Canada's First Nations and a Cree from northern Quebec, was born in a teepee in 1951, hundreds of kilometres from what we call civilization. Until he was nine, he and his family followed a completely nomadic way of life. Mukash spent his childhood among caribou and bear, wild geese and arctic char, on the huge rivers, muskeg and black spruce barrens of the Far North. But he yearned for a modern education, and eventually ended up studying at McGill University in Montreal.

Mukash recounts: "When I decided to go to school, my father told me, 'You have grown up in the traditional way. You have our traditional knowledge of survival and all that means. Now you have to learn the other way. I don't want you to lose what you already have. You have to keep on practising your way of life. You have to remember the teachings that you received from me and from the elders . . . . Remember what they tell us. Do what you want. Go get your education. But you always have to remember that there's a purpose for you as a Native person. There's a purpose for every nation in the world to be, to exist. That's because each nation has a particular gift. If you put those gifts together, you will find answers for the whole of humanity, for the protection of the Earth and for living a good life . . . . This is what the elders are saying.' "

Just as with Tewolde Egziather describing the importance of Africa in a changing world, Mukash's father was talking about the importance of a variety of beliefs and perspectives held by different peoples for our long-term survival. As circumstances change, one cultural system may be specially placed to offer an adaptive value that will add to our prospects for survival. That's why, at this critical juncture in the relationship between human beings and the Earth, we need elders of every human group who will share their knowledge and ideas.

It's worth our while to pay attention to people like the Cree, who have demonstrably lived in the same place for more than 5,000 years without destroying the ecosystem around them. The Cree know they need to understand us, because our dominant culture so profoundly threatens their way of life. Mukash says, "I had to learn about the history of the world, about modern life. I got a degree in political science. I know how politics works; I know the history of law, the history of philosophy. And for me it helps, because there are certain people who are gifted to . . . be messengers to other nations. And every nation has those people."

Matthew Mukash says the Cree elders think it's important to offer their own insights to other people, to talk to them and share these teachings. "People in modern society, the elders say . . . are doing things without thinking of the sacred. You know, everything is done in the absence of the sacred. But the Cree say that everything is balanced. If any element in a human being is missing, then there are problems. That's why so many people do things to . . . satisfy their own needs, rather than trying to meet the needs of the whole of humanity."

The Cree elders say that our way of life, for all its comforts and pleasures, is missing a key ingredient, and that is human spiritual fulfillment. That doesn't mean that they reject all our knowledge and wisdom, or that they deny the benefits of modern medicine, transportation or information technology. The elders appreciate and accept the advantages and comforts of modern life. They just want them placed into a broader context of a useful and balanced life. Mukash says, "We're here for a reason. We're not here to make money. It's good if you make money. You can get a car; you can have a cottage in the country. But if that spiritual aspect, that spiritual element, is not part of your life, then you're breaking the law. And as an individual, you're off balance."

Mukash says that if we can pull back from our destruction of the natural world, from our fears of poverty, from our dislike of sharing, especially from our wars, we will be able to see what we need to do. "All the problems that we have with regards to the economy, to the environment, all the problems of human conflict in the world — they all have to do with the absence of that balance that should be there. This is the knowledge that our elders are passing on to us, and it's very intense. It's very difficult to understand, but all these things that happen in the world, they're happening because there needs to be a balance."

Bill McKibben also agrees there needs to be balance in life. To find the peace and happiness we seek, he says we need to look at how we're made. "We arrive here with this collection of limbs and muscles and emotions and reason and senses. Clearly, we were designed for something more than reclining on the couch, flicking the remote control. And we were probably designed for something more than obtaining money so we can purchase things with it . . . . We're built for challenge, for things that are difficult, for walking up mountains, for taking on real problems around us. And we're built, I think, for some sense of contact with the divine — however you want to define that — for some sense that there's something larger than us around."

We've come to see the creature comforts our culture provides as the sole purpose of life. When we're very young, our needs are simple — warmth, food, physical comfort, a little stimulation. But as we grow, our pleasures become far more complex. Our current culture seems to have responded by giving us more and more of what made us happy as babies. But a mature organism is not the same as an inexperienced, growing one. Mature organisms flower and fruit; they produce and share; they don't just feed and take. McKibben says, "I set up and helped run a small homeless shelter in New York City when I lived there. And given all the charms and excitements that New York has to offer — fine restaurants and going to the ballet and so on — that should have been only a kind of onerous duty. But, in fact, it was the most fun, in many ways, that I had in my whole time there, because it was real and useful and took full measure of my emotions and abilities, not just my ability to think or to write, but the ability to use my muscles and my hands and my emotions."

Think about the last time you volunteered at a shelter or a bake sale, went to visit someone who was sick or helped a neighbour plant her garden. McKibben says, "We know that once you volunteer to do something or go out to help people, the pleasure is mostly yours. We're clearly built to be of service. That we are able to be convinced otherwise, convinced that our pleasures will come from focusing entirely on ourselves, is just one of several great lies that the twentieth century tried to foist on us."

What the last century should have taught us is to reflect much more deeply before implementing new projects. If we want to overcome child mortality worldwide or increase life expectancy, we will eventually have to place limits on the number of children we can have. If we want to increase agricultural crops, we'd better make sure that this will not negatively affect the natural ecosystem services on which those crops depend. We can increase our material satisfactions with computer games, more television channels and bigger cars and homes, but giving attention to such things uses up much of the time and energy we could be putting into the familial, community and spiritual needs that are so basic to our natures. Matthew Mukash says, "I was always taught that life has to be difficult, because if you're comfortable, there's no room for spiritual growth because all you focus on is your physical needs . . . . That's why our elders today are always on the land. They say, 'You stay in the community, you stay in a house. There's nothing to do! You watch TV. What does that bring you?' They go out on the land. They chop wood. They fish, make things. For them, life has to be difficult."

At first glance, this sounds like more asceticism. But that's too glib a dismissal. The most coddled Yuppie waxes ecstatic about her new gym trainer and how hard he makes her work. We spend our time and money on vacations climbing mountains and shooting rapids. The extreme wealth in industrialized countries has made a cornucopia of food available year-round and machines to perform our labour, and this has resulted in epidemic levels of obesity and heart disease, two things that prove that self-indulgence and ease don't make life better. Down deep, we know that life might not be as easy if we started cutting back the luxuries and began sharing with others and helping to conserve parts of the planet. But we might also discover that life would be a whole lot more fulfilling, even a great deal more fun. As Mukash says, "The three parts of you have to be in balance: your physical health, your mental health and your spiritual health. They all three have to be functioning. If your spiritual side doesn't function, then you become greedy. This is a natural teaching; it's common to all nations. And that is what the problem is. We're way off balance."

As a species, humanity is in its infancy. We appeared very recently in evolutionary time and exploded as an unprecedented force in the twentieth century. We know that we are exceeding the sustainable limits of the planet's carrying capacity. It's not deliberate; it's simply the collective impact of all we do. Both editions of this book have examined some of our most cherished notions about science and technology, information and economics, to reveal how they, too, create an imbalance with our surroundings. The challenge is to restore the balance between our needs and the Earth's capacity to supply those needs and absorb our impact.

The psychiatrist and author Robert Jay Lifton once told me a story about Hiroshima after the atomic bomb was dropped on it. A rumour spread among the survivors that nothing would ever grow again in the city, that the soil had been sterilized forever. Despite the horrific death toll and the terrible suffering, the survivors were anguished even more by the thought that nature itself might have been destroyed by the great weapon. Only when plants began to grow again did the wave of horror and despair subside.

Lifton's poignant story reminds us that we emerged from nature, remain embedded in it, and need it to nourish us physically, socially and spiritually. We may not think about that very often, but as the Hiroshima story reveals, the very thought that we might be capable of destroying the source of all life is too much to bear. The path we are now on leads to the destruction of nature. It is suicidal and too horrific to contemplate. We have to listen to all the voices around us, telling us to harmonize those needs, so that we can restore the balance we so desperately need.

We know we've left the reader with serious challenges, and after the first, very successful edition of this book, many people contacted us. They said they agreed with our arguments, but they didn't know what kind of action to take to address these problems. That's why, within a year of originally writing *From Naked Ape to Superspecies,* we wrote a second book, specifically to address the most difficult word an environmentalist or social reformer ever faces: How? How can we find that sustainable balance in business and in consumption? How can we harmonize our needs with those of the planet? How can we raise our children and prosper without destroying the livelihoods of others or our own future? The book that grew out of these questions is called *Good News for a Change.* Those who agree with much of what we've said here but don't know how to make the systemic changes in their values, way of life and daily practices that can make a real difference to the future can not only start working with the organizations listed at the back of this book, they can go on to read *Good News for a Change.*

REFERENCES

Much of the information contained in this book comes directly from the taped interviews for the radio series on which this book is based. However, we also benefitted from published and Internet information sources, which we list here for each chapter.

INTRODUCTION
- Page 3: Rachel Carson, *Silent Spring* (Boston: Houghton Mifflin, 1962) has been a continual reference and inspiration in a variety of contexts throughout this book. Linda J. Lear's biography of Carson, *Witness for Nature* (New York: Henry Holt, 1997), has served as a fresh reminder of the enormity of Carson's contribution to the environmental movement.

CHAPTER 1
SHARING EACH OTHER'S SKIN
- Page 9: Quoted by Satish Kumar in "Gandhi's Swadeshi: The Economics of Permanence," Jerry Mander and Edward Goldsmith, eds., *The Case Against the Global Economy* (San Francisco: Sierra Club Books, 1996) 418–24.
- Page 10: Nine pristine Eurasian preserves in the former Soviet Union alone "have fallen prey to Putin, the World Bank and ecotourists." See "The Wild, Wild East," *The Ecologist*, February 2003, 48–51.
- Page 10: Cynthia Cockburn, *Machinery of Dominance* (London: Pluto Press, 1985) 255.
- Page 14: Patricia Pearson, "See No Evil, No More," *Globe and Mail*, April 19, 2003.
- Page 18: See "Controversial Transgenic Rice to Be Commercially Grown," *Mainichi Shimbun*, September 2, 2003.
- Page 24: Susan Q. Stranahan, "Monsanto vs. the Milkman," *Mother Jones*, January/February 2004.

- Page 24: Justin Gillis, "Making Way for Designer Insects," *Washington Post,* January 22, 2004.
- Page 25: Oliver Wright, "Scientists' Bid to Combat TB Made It More Virulent," *Times* (London), December 27, 2003.
- Page 25: Joseph Stiglitz, *Globalization and Its Discontents* (New York: W.W. Norton & Co., 2002).
- Page 28: Dena Hoff, "Home-Grown No More," www.TomPaine.com, September 8, 2003.
- Pages 31–32: Maude Barlow, *Profit Is Not the Cure: A Citizen's Guide to Saving Medicare* (Toronto: McClelland & Stewart, 2002) 192–94.
- Page 32: Quoted by Satish Kumar in "Gandhi's Swadeshi: The Economics of Permanence," Jerry Mander and Edward Goldsmith, eds., *The Case Against the Global Economy* (San Francisco: Sierra Club Books, 1996).
- Pages 36–37: Gwynne Dyer, "New Gang in Town," syndicated column, September 17, 2003; see also "The WTO Under Fire," *The Economist,* September 20, 2003.
- Page 39: "Campaign Finance Reform: Your Guide to the Money in U.S. Elections," www.OpenSecrets.org.
- Pages 40–42: Jeanette Armstrong, "Sharing One Skin: Okanagan Community," Jerry Mander and Edward Goldsmith, eds., *The Case Against the Global Economy* (San Francisco: Sierra Club Books, 1996) 460–70.

CHAPTER 2
BUGS Я US

- Pages 43–45: Information on the expectations for and actual results of Biosphere II can be obtained from a large variety of newspaper and periodical sources. A few examples are: H. Westrup, "Biosphere II: Scientists Make Home in a Dome," *Current Science,* March 29, 1991; J.M. Laskas and P. Menzel, "Weird Science," *Life,* August 1991; "Sealed Up," *The New Scientist,* September 21, 1991; P. Heard, "Lost in Space," *The New Republic,* January 21, 1991. Some descriptions of the problems that began to arise can be found in "Bubble Trouble," *Harper's,* February 1992; N. Zeman and L. Howard, "Biosphere or Biobust," *Newsweek,* February 24, 1992; "Lean Times for Glasshouse People," *The New Scientist,* December 5, 1992; Michael Hirschorn, "People in Glass Houses," *Esquire,* August 1992; and "Biosflop?" *Environment,* July/August 1992.
- Pages 47–48: We referred especially to Edward O. Wilson's *The Diversity of Life* (New York: Norton, 1992), and to two volumes Wilson edited for the National Research Council and the Smithsonian Institution: *Biodiversity* (Washington, D.C.: National Academy Press, 1988) and Marjorie Reaka-Kudla and Don Wilson, *Biodiversity II* (Washington, D.C.: Joseph Henry Press,

1997). Another guide to the subject is Paul and Anne Ehrlich, *Extinction: The Causes and Consequences of the Disappearance of Species* (New York: Random House, 1981). Norman Myers, *The Sinking Ark* (Oxford: Pergamon, 1979) is also still useful.

- Pages 51–52: The classic publication on the Gaia hypothesis, now upgraded to a theory, is James Lovelock, *Gaia: A New Look at Life on Earth* (Oxford: Oxford University Press, 1979), and his update, *The Ages of Gaia: A Biography of Our Living Earth* (New York: Norton, 1995). However, there are many others, including works by Lovelock's collaborator, microbiologist Lynn Margulis, such as *Microcosmos,* with Dorion Sagan (New York: Summit, 1986). The understanding of the Earth as a super-organism hasn't diminished since Gaia was written, but has become increasingly supported by research and the hard sciences, such as chemistry and physics, and by the mathematical model called "Daisy World." Stephen Schneider and P. Boston edited the seminal *Scientists on Gaia* (Cambridge, MA: MIT Press, 1991), the first "establishment" scientific conference on the Gaia hypothesis, and the one that established it as legitimate ground for further research.

- Pages 54–55: Tom Reimchen, "Bears, Gardeners of the Forests," *Vancouver Sun,* September 30, 1999, and Tom Reimchen, "Bears: A Vital Link for Salmon, Trees," *Times-Colonist* (Victoria, British Columbia), September 14, 1999. The symbiosis between fish, trees and mammals was also explored in Francis Backhouse, "Fewer Salmon, Fewer Bears, Fewer Trees?" *Canadian Geographic,* September/October 1999. See also David Suzuki and Holly Dressel, *Good News for a Change* (Vancouver: Greystone Books, 2003) 233–41.

- Page 56: Orrie Loucks's studies on forest dynamics and air pollution effects are featured in *Appalachian Tragedy,* a photo volume edited by Harvard Ayers and published by Sierra Club Books. Loucks is also the senior editor of the textbook *Sustainability Perspectives for Resources and Business* (New York: Sierra Sea Press, 1999).

- Page 59: David Pimental et al., "Environmental and Economic Effects of Reducing Pesticide Use," *Bioscience* 41.6 (1991): 402–09. The article states: "Although pesticide use has increased during the past four decades, crop losses have not shown a concurrent decline....The share of crop yields lost to insects has nearly doubled during the last 40 years, despite more than a ten-fold increase in both the amount and the toxicity of synthetic insecticide used."

- Pages 66–70: For more detail on the increasing links between astrophysics, philosophy and religion, see Brian Swimme, *The Hidden Heart of the Cosmos: Humanity and the New Story* (San Francisco: Orbis Books, 1996) and Brian

Swimme and Thomas Berry, *The Universe Story* (San Francisco: Harper, 1994). For a more detailed discussion of the Okanagan approach to life, see Jeanette Armstrong, "Sharing One Skin: Okanagan Community," Jerry Mander and Edward Goldsmith, eds., *The Case Against the Global Economy* (San Francisco: Sierra Club Books, 1996).

- Page 69: Fritjof Capra, *The Web of Life* (New York: Anchor, 1996) is a rich source for those who wish to understand what physics is now offering biology. See also Stephen R. Kellert, *The Value of Life: Biological Diversity and Human Society* (Washington, D.C.: Island Press, 1996) and Stephen R. Kellert and Edward O. Wilson, eds., *The Biophilia Hypothesis* (Washington, D.C.: Island Press, 1993) for insights into the philosophical, ethical and emotional importance of biodiversity to human life.

CHAPTER 3
BIGFOOT

- Pages 72–75: Mathis Wackernagel and William Rees, *Our Ecological Footprint: Reducing Human Impact on the Earth* (Gabriola Island, BC: New Society, 1996) forms the basis for this entire chapter, and we drew from it extensively.
- Pages 82–83: We obtained figures on the distributional change of world income over the past thirty years from UNESCO, Sources 58, May 1994. Along with many other writers, we are indebted to Paul and Anne Ehrlich, *The Population Bomb,* and its update, *The Population Explosion* (New York: Simon & Schuster, 1990). The journal *Population and Development Review* has many pertinent articles on population issues, and we relied as well on figures provided by the Population Institute in Washington, D.C. See Organizations to Contact for further information. See also Kenneth Arrow et al., "Economic Growth, Carrying Capacity and the Environment," *Science,* April 28, 1995, and Norman Myers, "Population Dynamics and Food Security," S.R. Johnson, ed., *Food Security: New Solutions for the 21st Century,* (Des Moines: Iowa State University Press, 1997).
- Page 84: Estimations of the amount of ocean floor destroyed annually by draggers are from *Conservation Biology* 12.6 (December 1998). For recent facts and figures, we used Chris Wood, "Our Dying Seas," and Joseph MacInnis, "Breaking the Bonds," *Maclean's,* October 5, 1998. To learn more about the state of our oceans, refer to Sylvia Earle, *Wild Oceans* (Washington, D.C.: National Geographic Society, 1999), or the same author's *Sea Change: A Message of the Oceans* (New York: G.P. Putnam & Sons, 1995).
- Page 85: Nancy Langston, *Forest Dreams, Forest Nightmares* (Seattle: University of Washington Press, 1995) is a remarkable general reference on the well-intentioned but often disastrous history of professional forestry.

- Page 86: Alan Durning, *How Much Is Enough?* (New York: W.W. Norton, 1992) has been useful in many chapters, beginning with this one, and we made special reference to his time-lapse image of the world's forests, which appears in "Saving the Forests: What Will It Take?" *Worldwatch Paper* 117, December 1993.
- Page 87: The World Resources Institute's groundbreaking *Report on the Frontier Forests of the World* deserves special attention from Canadians, as so much of this issue intersects with our own national interests and the behaviour of our governments. Refer to Organizations to Contact for addresses.
- Pages 88–89: See also Patricia Marchak, Scott Aycock and Deborah Herbert, *Falldown: Forest Policy in British Columbia* (Vancouver: David Suzuki Foundation/Ecotrust Canada, 1999).
- Pages 90–91: Of the more than 200 scientific books, papers and reviews authored by Stephen Schneider, we referred especially to *Laboratory Earth: The Planetary Gamble We Can't Afford to Lose* (New York: Basic Books, 1997). He is also the founder and editor of the interdisciplinary journal *Climatic Change*. See also his *The Coevolution of Climate and Life* (San Francisco: Sierra Club Books, 1984) and *Global Warming: Are We Entering the Greenhouse Century?* (San Francisco: Sierra Club Books, 1990). The last caused a serious wave of concern when it was published, and remains an important work.
- Page 91: Bill McKibben, *Hundred-Dollar Holiday* (New York: Simon & Schuster, 1998) and *Maybe One: A Personal and Environmental Argument for Single-Child Families* (New York: Simon & Schuster, 1998) detail the ideas broached in the text. McKibben's *The End of Nature* (New York: Anchor, 1990) has been a constant reference in this and several other chapters of this book.
- Pages 92–93: See also the many publications of the Intergovernmental Panel on Climatic Change (IPCC), published by Cambridge University Press. For a concise discussion of the "controversy" of global warming and of environmental dangers in general, see Norman Myers and Julian Simon, *Scarcity or Abundance: A Debate on the Environment* (New York: Norton, 1992). An Internet search will provide literally hundreds of newspaper articles on this discussion; the more recent the articles, the more apparent it becomes that global warming is irrefutable and caused by humans.
- Pages 97–98: Figures for the amount of toxic waste in Sydney, Nova Scotia, come from the Sierra Club of Canada. That organization has mounted a major campaign to demand appropriate government action and will provide pertinent data to anyone who is interested. The Center for Health, Environment and Justice in Falls Church, Virginia, just outside of Washington, D.C., has a publications catalogue and also provides statistics and data on landfills, toxic waste dumps, and contaminants like dioxin in the United States, and will gladly provide them upon request. See Organizations to Contact for details.

CHAPTER 4
SEZ WHO?

- Pages 101–02: The "World Scientists' Warning to Humanity" can be obtained in its entirety by contacting the Union of Concerned Scientists. See Organizations to Contact for details.
- Pages 103–05: The chapter entitled "Food Fight" in Sheldon Rampton and John Stauber, *Mad Cow USA* (Monroe, ME: Common Courage Press, 1997) outlines the lawsuit against Oprah Winfrey and Howard Lyman, and the bibliography gives a good rundown of top media stories on the issue.
- Pages 107–10: Lawrence Grossman, *The Electronic Republic: Reshaping Democracy in the Information Age* (New York: Penguin, 1995). See also Neil Postman, *Amusing Ourselves to Death* (New York: Penguin, 1985).
- Pages 115–16: David Shenk, *Data Smog* (New York: HarperCollins, 1997).
- Page 119: Jed Greer and Kenny Bruno, *Greenwash: The Reality Behind Corporate Environmentalism* (Penang, Malaysia: Third World Network, 1996) gives a fascinating history of how corporations use public relations and public perceptions to alter their image and convince us of noble environmental intentions, while pursuing the bottom line as usual.
- Pages 120–21: Joshua Karliner's Internet magazine *Corporate Watch* is an excellent source of information about who is doing what in the corporate world. We also referred to Karliner's *The Corporate Planet: Ecology and Politics in the Age of Globalization* (San Francisco: Sierra Club Books, 1997).
- Pages 121–24: The single most quotable and eye-opening book on the manipulation of information is John Stauber and Sheldon Rampton, *Toxic Sludge Is Good for You: Lies, Damn Lies and the Public-Relations Industry* (Monroe, ME: Common Courage Press, 1995). It uses black humour as it looks at the role of the media and private vested interests in obscuring, perverting or subverting environmental and social issues.
- Page 122: For further discussion of the public-relations wars, see Paul and Anne Ehrlich, *The Betrayal of Science and Reason* (Washington, D.C.: Island Press, 1996), their take on what is termed "brownwashing."
- Pages 127–28: James Ridgeway and Jeffrey St. Clair, *A Pocket Guide to Environmental Bad Guys* (New York: Thunder's Mouth Press, 1999) was a useful reference. Publications from the Environmental Working Group of the Clearinghouse on Environmental Advocacy and Research, especially "Target: Democracy — The Effect of Anti-Environmental Violence on Community Discourse," April 22, 1997, back up EarthFirst! claims and supply more examples of the kind of incidents of violence against peaceful environmental protest that we mention in this book. Other papers list the corporate-funded "astroturf" groups that seek to undermine or weaken environmental legislation. *PR Watch,* a quarterly published by the Center for Media and Democracy in Madison, Wisconsin, gave us help and figures on many issues.

We particularly referred to the periodical's July 1999 story: Carmelo Ruiz, "Burson-Marsteller: PR for the New World Order." See Organizations to Contact for addresses.

*   Pages 130–31: A major source of information and inspiration was Jerry Mander, *Four Arguments for the Elimination of Television* (New York: Quill, 1978). See as well Mander's *In the Absence of the Sacred* (San Francisco: Sierra Club Books, 1991), which also provided food for thought in this and several other chapters.

CHAPTER 5
UNNATURAL SELECTIONS

*   The senior author of this book was trained as a geneticist and has more than thirty years' experience as a practising scientist. The volume of articles and books that contributed to the position he takes in this chapter is far too great to include here. We attempted to pare it down to key articles and books that provide perspective on the debate. Specific facts and figures are as up-to-date as possible and therefore come mostly from journalistic sources. The *Times* (London), *The Guardian* and *The Independent* are excellent English-language sources for the ongoing genetic controversy in Europe. The Reuters news service, which appears in many papers, follows the issue, and the *St. Louis Post Dispatch,* located in Monsanto's hometown, is a rich source of articles on the controversies surrounding genetically engineered agricultural crops (see especially "Ham-Fisted Force-Feeding," April 15, 1999). We wish we could name a Canadian newspaper that is following the issue as diligently; Canadian coverage, though sometimes good, has been sporadic. See the *Province* (Vancouver), "Gene-Tooled Food Going Unchecked," May 14, 1999.

*   Page 135: Mae Wan Ho, *Genetic Engineering: Dreams or Nightmares? The Brave New World of Bad Science and Big Business* (Penang, Malaysia: Third World Network, 1997). See also Mae Wan Ho, *Independent Report on Biosafety,* prepared by the Third World Network in 1997 for the biosafety negotiations that brokered the Biosafety Protocol of 2003 at the United Nations.

*   Page 136: Israel's bomb engineered to affect ethnic Arabs was reported by Uzi Mahnaimi and Marie Colvin in the *Sunday Times* (London), November 15, 1998. For details of the effects of genetically manipulated crops on beneficial organisms like the monarch butterfly, lacewings and ladybugs, see Bob Hortzler's article on the monarchs in *Science,* January 1999, and Angelika Hilbeck's studies on lacewings for the Swiss Federal Research Station for Agroecology and Agriculture, Zurich, Switzerland, published in the *Journal of Environmental Entomology,* February 1997. We also quoted from reports of the International Meeting of Entomologists in Basel, Switzerland, in 1999, where Mark Whalon of Michigan State University said, "This heightened

threat of Bt resistance, coupled with the devastation of the beneficial insect populations that help keep plants in check, could lead to massive crop failures." See also 1999 studies by Nicolas Birch of the Scottish Crop Research Institute on ladybird beetles. Studies at both Cornell and Iowa State on monarchs reveal similar problems.

- Page 137: Neal Talbot, "County Backs Effort to Prevent 'Frankenwheat,'" *Daily Herald-Tribune,* September 4, 2003. See also "Tougher European GMO Legislation Slap in the Face to U.S. Corporate Interests," Greenpeace press release, July 2, 2003; "Bush Administration Says New EU Biotech Laws Are Onerous," Associated Press, July 2, 2003; and Jeffrey Sparshott, "U.S. Sour on EU's Rules for Bio-Foods," *Washington Times,* July 3, 2003.

- Page 140: See the important textbook, Deborah K. Letourneau, Beth E. Burrows, eds., *Genetically Engineered Organisms: Assessing Environmental and Human Health Effects* (New York: CRC Press, 2001) for the pros and cons of biotechnology. See also Jeremy Rifkin, *The Biotechnology Century* (New York: Jeremy P. Tarcher/Putnam, 1998) for an appraisal of the social effects of this new technology. Brewster Kneen, *From Land to Mouth* (Toronto: NC Press, 1993) provides an explanation of the vertical integration of our food supply. See also Kneen's *Farmageddon: Food and the Culture of Biotechnology* (Gabriola Island, BC: New Society, 1999). See as well Nicanor Perlas, *Overcoming Illusions About Biotechnology* (London: Zed Books, 1994), and Miges Baumann, Janet Bell, Florianne Koechlin and Michel Pimbert, *The Life Industry: Biodiversity, People and Profits* (London: Intermediate Technology Publications, 1996).

- Page 142: The deformed cotton plants were reported in many farm journals and news media in the United States and Australia. See Allen Myerson, "Breeding Seeds of Discontent: Cotton Growers Say Strain Cuts Yield," *New York Times,* November 9, 1997; Bill Lamprecht, "Many Farmers Finding Altered Cotton Lacking," *St. Louis Post Dispatch,* April 12, 1998; and Robert Steyer, "Monsanto Refuses to Pay $1.94 Million to Farmers," *St. Louis Post Dispatch,* June 20, 1998. Probably the single most useful overview of the controversies surrounding biotechnology for the layman is *The Ecologist,* September/October 1998, an entire issue devoted to biotechnology and subtitled "The Monsanto Files." The articles and references in this issue pretty much cover the topic.

- Page 144: "Bt Cotton Winning a Battle, but Losing the War," *Gharat Textile,* India, July 4, 2003; Geoffrey Lean, "Insects Thrive on GM 'Pest-Killing' Crops," *The Independent,* March 30, 2003. See also the abstract of the Venezuelan study by Ali H. Sayyed, Hugo Cerda and Denis J. Wright, "Could Bt Transgenic Crops Have Nutritionally Favorable Effects on Resistant Insects?" *Ecology Letters* 6.3 (March 2003): 167, http//:nrs.Harvard.edu/ura-3:hul.eresource:ecology1.

- Page 153: Poison-free methods of dealing with weeds are legion and widely reported in the literature. A recent example is Fred Pearce, "Farmer's Friend," *Science,* October 24, 1998.
- Pages 153–54: The toxicity of Roundup is detailed in two heavily researched papers by Caroline Cox, "Glyphosate, Part l: Toxicology," *Journal of Pesticide Reform* 15.3 (Fall 1995), and "Glyphosate, Part 2: Human Exposure and Ecological Effects," *Journal of Pesticide Reform* 15:4 (Winter 1995). Other papers we consulted included William S. Pease et al., "Preventing Pesticide-Related Illness in California Agriculture: Strategies and Priorities," for the Center for Occupational and Environmental Health, the School of Public Health, University of California at Berkeley, 1993. It should be noted that Roundup is so generally toxic that researchers are even investigating its potential as an antimicrobial. See F. Roberts et al., *Nature* 393 (1999): 801–05. Many related articles are available through PAN, the Pesticide Action Network. See Organizations to Contact for details.
- Pages 153–58: Physicist Vandana Shiva is ubiquitous in the areas of both biotechnology and food security. We met with her four times, and also consulted some of her scores of articles, books and pamphlets on these subjects, notably *Monocultures of the Mind* (Penang, Malaysia: Third World Network, 1993) and *The Violence of the Green Revolution: Third World Agriculture, Ecology and Politics* (London: Zed Books, 1993).
- Page 154: Anders Legarth Schmidt, "Poisonous Spray on a Course Towards Drinking Water," *Politken,* Denmark, May 10, 2003. See also *The Danish Pesticide Leaching Assessment Programme Monitoring Results May 1999–June 2002,* June 2003; the Danish EPA at www.mst.dk/, tel.: 45 32 66 01 00; and industry statements at www.environmentdaily.com/docs/ 30605a.doc.
- Page 158: Peter Rosset, "The Parable of the Golden Snail," *The Nation,* December 27, 1999.
- Page 159: J. Madeleine Nash, "Grains of Hope," *Time,* July 31, 2000. See also "New Rice Could Save Lives," Reuters, January 13, 2001, at www.abcnews.go.com, and Shane Wright, "GM Opponents Should Stand Trial — Golden Rice Inventor," Australian Associated Press, November 24, 2003.
- Page 160: K.T. Arasu, "U.S. Farmers Reach $110 Million StarLink Settlement," Reuters, February 10, 2003.
- Page 161: Murray Lyons, "Roundup Ready Wheat Could Prove Costly, NFU Told," *StarPhoenix,* Saskatoon, November 23, 2002; Monsanto Canada spokesperson Trish Jordan, quoted in Murray Lyons, "Roundup Ready Wheat Could Prove Costly, NFU Told," *StarPhoenix,* Saskatoon, November 23, 2002.
- Page 162: Justin Gillis, "Biotech Firm Mishandled Corn in Iowa," *Washington Post,* November 14, 2002; Randy Fabi, "USDA Probes Nebraska Biotech Crop Contamination," Reuters, November 15, 2002.

- Page 163–64: "Consumers Want Mandatory Labelling of Genetically Modified Foods," Consumers' Association of Canada press release, December 3, 2003.
- Page 164: "GM Foods with an 'ID Card' Debut in Beijing," *China Youth Daily,* posted by People's Daily Online, July 21, 2003. Regarding the fears expressed by Christine von Weizsaecker, see J.A. Nordlee et al., "Identification of a Brazil-Nut Allergen in Transgenic Soybeans," *The New England Journal of Medicine,* March 14, 1996. Problems with synthetic insulin were detailed in a series of articles in *The Guardian,* beginning with the March 9, 1999, article "Diabetics Not Told of Insulin Risk." Another human health issue is detailed in "Meningitis Fears from GE Crops," a story reported both on BBC News and the *Sunday Times* (London), April 19, 1999. The *Times* article states: "The UK's Advisory Committee on Novel Foods and Processes fears antibiotic-resistant genes will be picked up by commonly carried bacteria, such as those for meningitis…"
- Page 165: Mark Schapiro, "Sowing Disaster?" *The Nation,* October 28, 2002. Prior to the tragedy of the contaminated Mexican land races of corn, there were many established incidences of biological pollution via wind, pollen, and insects as well as fears about microorganism contamination expressed by many researchers. See Nick Nuttall, "Bees Spread Genes from GM Crops," *Times* (London), April 15, 1999, and M. Brookes, "Running Wild," *The New Scientist,* October 31, 1998. Monsanto's own lawsuits against farmers have raised the contention of uncontrolled pollen contamination in fields adjacent to or even some kilometres away from GM crops. They have been widely reported; we used an article by Marie Woolf that appeared in *The Independent,* March 14, 1999. See also Jane Rissler and Margaret Mellon, "Perils Amidst the Promise: Ecological Risks of Transgenic Crops in a Global Market," 1993, available through the Union of Concerned Scientists.
- Page 166: See "*Nature* Refuses to Publish Mexican Government Report Confirming Contamination of the Mexican Maize Genome by GMOs," Food First press release, October 24, 2002.
- Page 169: Andrew Rowell, "The Sinister Sacking of the World's Leading GM Expert — and the Trail That Leads to Tony Blair and the White House," *Daily Mail,* July 7, 2003.
- Page 170: John Vidal, "GM Genes Found in Human Gut," *The Guardian,* July 17, 2002.
- Page 170: David Scott, "Crop Trials Must Stop," *The Scotsman,* November 19, 2002.
- Page 171: Some of the biotech industry's increasing regulatory problems affecting profit margins are listed in Andrea Ahles, "Biotech Companies Seek Funds from Non-Wall Street," *Philadelphia Inquirer,* June 7, 1999. Regarding the failings of regulatory bodies in this matter, see Mae Wan Ho

and Ricarda Steinbrecher, "Fatal Flaws in Food Safety Assessment: Critique of the Joint FAO/WHO Biotechnology and Food Safety Report," *Environmental and Nutritional Interactions* 2 (1998). We are particularly indebted to Richard Strohman, "The Coming Kuhnian Revolution in Biology," *Nature Biotechnology,* March 1997.

- Pages 171–72: Richard Strohman's charge that commercial concerns are twisting academic research has been noted for some years. See Brian Tokar, "The False Promise of Biotechnology," *Z Magazine,* February 1992, which states: "Twenty-four leading research universities all have ten or more faculty members with…direct corporate affiliations in biotechnology; Harvard alone has 69, Stanford has 40…in the most striking case, almost one-third of the MIT biology department is tied to biotech companies." The overall picture has grown much worse since that article was written. Regarding ethics in the scientific profession, we were informed by Theodore Roszak, "The Soul of Science," *Resurgence,* September/October 1997. The opposite approach — that the ethical issues surrounding cloning and other genetic interference can be dealt with through PR spin and habituation — is outlined in Oliver Morton, "Overcoming Yuk: It May Be Unnatural, but Encouraging Genetic Choice in Humans Is Not Bad (It's Also Inevitable)," *Wired,* November 1998.

CHAPTER 6
YOUR MONEY OR YOUR LIFE FORMS

- Page 177: See "Cashing in at Public Expense," *Boston Sunday Globe,* April 5, 1998, which states, "Taxpayers subsidize an industry awash in profits, then pay onerous prices for drugs their tax dollars helped create." See also Andy Kimbrell, *The Human Body Shop: The Engineering and Marketing of Life* (San Francisco: HarperCollins, 1993).
- Page 178: On the patent scramble for neem, mustard, ginger, castor, basmati and more, see especially Vandana Shiva, "Biopiracy: The Need to Change Western IPR Systems," *The Hindu,* July 28, 1999.
- Page 179: The Staphylococcus controversy was reported in Marlene Cimons and Paul Jacobs, "Biotech Battlefield: Profits vs. Public," *Los Angeles Times,* February 21, 1999.
- Pages 180–81: Mark Schapiro, "Sowing Disaster?" *The Nation,* October 28, 2002.
- Page 181: "U.S. Withdraws Genetically Engineered Animal Feed Donation After Bosnia's Hesitation," Agence France Presse, January 30, 2001.
- Page 181: George Monbiot, "The Covert Biotech War: The Battle to Put a Corporate GM Padlock on Our Food Chain," *The Guardian,* November 19, 2002.
- Page 182: Chandrika Mago, "Government Takes Firm Stand on GM Food," *Times* of India, March 8, 2003. For details of the organized rural resistance to

GM crops in India, known journalistically as the "Cremate Monsanto" movement, see "Farmers Group Storms Mahyco Office in Hyderabad," *Indian Express,* Bangalore Edition, December 2, 1998, or "KRRS 'Cremates' Cotton in Bellary," *Times* of India, December 4, 1998.

- Page 183: Vandana Shiva's published works are even more extensive in the area of biopiracy and farmers' rights than they are in food systems and biotechnology. We specifically recommend *Biopiracy: The Plunder of Nature and Knowledge* (Boston: South End Press, 1997) and *Staying Alive: Women, Ecology and Development* (London: Zed Books, 1988). A very useful overview is Ruth Hubbard and Elijah Wald, *Exploding the Gene Myth: How Genetic Information Is Produced and Manipulated by Scientists, Physicians, Employers, Insurance Companies, Educators and Law Enforcers* (Boston: Beacon Press, 1993).

- Page 184: The Vandana Shiva article cited above for page 178 also explains the remarkable fact that transport of a living thing from one country to another is the only current requirement for ownership. Shiva points out that section 102 of the 1952 U.S. Patent Act doesn't consider use of any item, even an invention, in another country to be "prior art" that would disallow patenting. In other words, until it's patented in the United States, it's free game.

- Page 188: "Tougher European GMO Legislation Slap in the Face to U.S. Corporate Interests," Greenpeace press release, July 2, 2003, at www.greenpeace.org/news/details?item%5fid=290260.

- Page 188: Evidence of a link between glyphosate and a dangerous cancer, non-Hodgkin's lymphoma, are detailed in a paper by oncologists Lennart Hardell and Mikael Eriksson in the *Journal of the American Cancer Society,* March 15, 1999.

- Pages 188–89: The corn in question is Bt176, grown in Spain by Syngenta. Early investigations reveal no common diseases or errors in feeding, but the Robert Koch Institute, responsible for licensing the feed in Germany, has refused to respond to calls for further investigation. Today, the large majority of the livestock eaten in North America are being raised on diets of genetically engineered grain even as researchers are learning the Bt substance "becomes degraded more slowly than had been anticipated," appearing in stomachs, intestines and even excrement. See "Cows Die Mysteriously on Farm in Hesse, Germany," Greenpeace Germany press release, December 8, 2003.

- Page 190: Regarding liability, see Justine Thornton, "The Price of Disaster Will Not Be Paid by Those Responsible," *The Independent,* February 28, 1999.

- Pages 191–92: The "revolving door" list on page 191 is only a small part of the picture. In the biotech special edition of *The Ecologist,* September/October 1998, Jennifer Ferrara and others provide many more examples in their section entitled "Revolving Doors: Monsanto and the Regulators." For allegations of bribery in Canada and Britain, see Nick Cohen, "All Monsanto's Men?" *The Observer,* February 21, 1999, which says, "Dr. Margaret Haydon told the Canadian Senate that she and her superior in the Human Safety Division, one Dr. Drennan, [in October of last year] had met representatives of Monsanto, the hormone's (rBGH) manufacturers. 'An offer of one to two million dollars was made,' she said." On April 28, 1999, CBC Radio News reported on Agriculture Canada's refusal to provide any details about Monsanto research being conducted at the department's facilities throughout Canada. Agriculture Canada also refused to disclose how much money it had received from the chemical and biotech company over the past fifteen years. The story was also reported in "Giant Food Companies Control Standards: Critics," *Toronto Star,* April 28, 1999, and "Federal Government Invested in GM Wheat," CBC Radio, November 29, 2003. See also "Farmers, Citizens and NGOs Protest Genetically Modified Wheat at Agriculture Minster's Office," AGNET Canada Press Release, December 9, 2003, and Charles Abbott, "Monsanto Sees U.S. Bio-Wheat Approval Within 3 Years," Reuters, December 8, 2003.

- Page 192: The description of Monsanto's behaviour as a corporate citizen is well documented. The best popular source is Brian Tokar, "Monsanto: A Checkered History," *The Ecologist,* September/October 1998. More detailed sources would include: Cate Jenkins, "Criminal Investigation of Monsanto Corporation — Cover-up of Dioxin Contamination in Products — Falsification of Dioxin Health Studies," USEPA Regulatory Development Branch, November 1990; Peter Schuck, *Agent Orange on Trial: Mass Toxic Disasters in the Courts* (Cambridge, MA: Harvard University Press, 1987); "Case of Mislabeled Herbicide Results in $225,000 Penalty," *Wall Street Journal,* March 25, 1998; Erik Millstone, "Increasing Brain Tumour Rates: Is There a Link to [the Monsanto Product] Aspartame?" University of Sussex Science Policy Research Unit, October 1996. See references for Joseph Cummins, "PCBs — Can the World's Sea Mammals Survive Them?" *The Ecologist,* September/October 1998. And see especially Samuel S. Epstein, "Unlabelled Milk from Cows Treated with Biosynthetic Growth Hormones: A Case of Regulatory Abdication," *International Journal of Health Services* 26.1 (1996). For excellent overviews of the regulatory history of Monsanto and other polluting chemical companies such as Dow and DuPont, see Jed Greer and Kenny Bruno, *Greenwash: The Reality Behind Corporate Environmentalism* (Penang, Malaysia: Third World Network, 1996).

- Page 193: For the early story of Percy Schmeiser and Edward Zielinski, two Saskatchewan farmers being sued by Monsanto, we used Marie Woolf, "Two Farmers Being Pursued by Monsanto for Growing Their Seed Without a Licence in Landmark Cases," *The Independent.* For later developments, we used E. Ann Clark, "A Fanciful Tale on the Appeal of the Percy Schmeiser

Decision," posted in 2002 at www.percyschmeiser.com, a good source for updates on the story.

- Page 196: Alex Binkley, "Farmers Should Be Able to Save GE Seed," *Manitoba Co-Operator,* June 13, 2002.
- Page 197: "Terminate the Terminators," *ISIS Report,* July 12, 2001; Dr. Martha Crouch, "How the Terminator Terminates," available on many web sites, including www.greens,org/s-r/18 or www.theedmondsinstitute.org.
- Page 198: See Ricarda Steinbrecher and Pat Mooney, "Terminator Technology: The Threat to World Food Security," *The Ecologist,* September/ October 1998. For a critique of hybrid seed, see Richard Lewontin and Jean-Pierre Berlan, "The Political Economy of Agricultural Research: The Case of Hybrid Corn," Ronald Carroll et al., eds., *Agroecology* (New York: McGraw Hill, 1989). In terms of the history of seed control, we consulted Richard Lewontin and Jean-Pierre Berlan, "Genetically Modified Food: It's Business as Usual," *The Guardian,* February 22, 1999. The Terminator seed actually contains a poison that is set off in the late stages of production by exposure to the antibiotic tetracycline or to the other chemicals mentioned. Technically speaking, temporary "gene silencing" of the poison gene or failed activation of the Terminator countdown could cause infection of wild relatives with sterility. Obviously, tetracycline itself could cause problems if released in such amounts into the environment. The Terminator's potential as a weapon is well-known to the military and, in theory, crop diseases could also be triggered by seeds with a hidden feature that wouldn't kick in until activated by specific chemicals or conditions.
- Page 199: Robert Hartwig, quoted in Kristen Philipekoski, "Food Biotech Is Risky Business," *Wired,* December 1, 2003.
- Page 202: Don Westfall, quoted in "Five Reasons to Keep Britain GM-Free," *The Ecologist,* online edition, June 22, 2003.
- Page 204: Joe Cummins, "Transgenic Meat Scandal," Institute for Science in Society, www.i-sis.org.uk/TransgenicMeat.php.
- Page 204: Kendall Powell, "U.S. Academia Held Accountable for GM Products," *Nature Biotechnology* 21.7 (July 2003): 720–21.
- Page 207: Jeremy Bigwood, "Health: Scientists Link GM Crop Weed Killer to Powerful Fungus," IPS article, Washington D.C., August 20, 2003.
- Page 208: The controversy surrounding the experiments by Dr. Arpad Pusztai (described on pages 168–69) occupied the British media for weeks. The players, the chronology and the controversy are laid out in Michael Sean Gillard, Laurie Flynn and Andy Rowell, "Food Scandal Exposed," *The Guardian,* February 21, 1998. The authors write: "[Dr. Pusztai's] team found that the liver and heart sizes of the rats were decreasing — worse still, the brains were getting smaller.... 'It is very unfair to use our fellow citizens as guinea pigs,'

said Dr. Pusztai. Two days later he was summarily suspended." We, of course, consulted the primary data, including the Rowett Research Institute audit report by Philip James, which disagreed with Pusztai's findings, and the October 1998 experiments by pathologist Stanley Ewen of Aberdeen University Medical School, which supported Pusztai's position. Pusztai's findings have never been duplicated because no one has tried to, which means they still remain unrefuted. The main point, supported by twenty prominent international scientists in a public statement in February 1998, was that such preliminary findings in a government-funded agency should have triggered serious health-effects research, not a forced resignation and a cover-up.

- Page 208: Andy Coghlan, "Weedkiller May Boost Toxic Fungi," *The New Scientist,* August 14, 2003.
- Pages 208–09: Randy Boswell, "Roundup May Harm Wheat," *Leader-Post*/CanWest News Service, August 19, 2003.
- Pages 211–13: In the bioweapons section, documentation acquired by RAFI via a Freedom of Information application, "Biotechnology Workshop 20/20," May 29–30, 1996, held at the U.S. Army War College, outlines a wide range of military uses for biotechnology that the authors believe to be feasible by the year 2020.
- Pages 214–15: Gaia Vince, "Pig-Human Chimeras Contain Cell Surprise," *The New Scientist,* January 13, 2004.
- Page 216: For action strategies, see Melanie McDonagh, "How Women Took on the Supermarkets — and Won," *Evening Standard* (London), March 24, 1999. See also "GM Food Ban Boosts Iceland Supermarket Chain," *Business: The Company File,* March 23, 1999. See as well Organizations to Contact.

CHAPTER 7
FOLLOW THE MONEY

- Page 218: Robert Constanza et al., "The Value of the World's Ecosystem Services and Natural Capital," *Nature* 387: 1997. See also the following books on economics by Hazel Henderson: *Building a Win-Win World: Life Beyond Global Economic Warfare* (San Francisco: Berrett-Koehler, 1996), *Creating Alternative Futures: The End of Economics* (New York: G. P. Putnam's Sons, 1978), and *Paradigms in Progress: Life Beyond Economics* (San Francisco: Berrett-Koehler, 1995).
- Page 222: The fascinating statistics on this page and the Barber Conable quote at the bottom came from Clifford Cobb and Ted Halsted, *The Genuine Progress Indicator: Summary of Data and Methodology* (San Francisco: Redefining Progress Institute, 1994). See also Organizations to Contact.
- Pages 223–24: Herman Daly's work in ecological economics is unsurpassed. A few of his books that we recommend are: *Steady-State Economics* (Boston:

Beacon Press, 1997), *For the Common Good,* with John Cobb (Boston: Beacon Press, 1989), *Valuing the Earth, Economics, Ecology and Ethics,* with Kenneth Townsend (Boston: MIT Press, 1993), and *Beyond Growth: The Economics of Sustainable Development* (Boston: Beacon Press, 1996).

- Page 226: See the article on *Pfiesteria* by Joanne Burkholder in *Scientific American,* August 1999.
- Pages 228–29: See the summary of a conference held in Saskatchewan on ILOs, "Beyond Factory Farming," in November 2002 at www.scc.ca that quotes Joann Jaffe, associate professor of sociology from the University of Regina; Bill Weida, an economist from Colorado College, Denver; and John Keen, farmer and economist, at www.dnpp.ca.
- Page 229: The Cindy Watson story was reported in Bill Moyers's television special *Free Speech for Sale,* broadcast on PBS, July 1999.
- Pages 230–34: Articles such as "The Failures of Bretton Woods" in Jerry Mander and Edward Goldsmith, eds., *The Case Against the Global Economy* (San Francisco: Sierra Club Books, 1996), and "Toxic Empire: The World Bank, Free Trade and the Migration of Hazardous Industry" in Joshua Karliner, *The Corporate Planet: Ecology and Politics in the Age of Globalization* (San Francisco: Sierra Club Books, 1997), back up Jane Ann Morris and Harry Glaesbeck's charges in our section entitled "Corporate Morality"; the latter piece also details Union Carbide's behaviour in Bhopal.
- Pages 234–35: David Korten's books, *When Corporations Rule the World* (New York: Kumarian Press, 1995) and *The Post-Corporate World: Life After Capitalism* (New York: Berrett-Koehler and Kumarian Press, 1999), provide more background to his positions. He also founded *YES! Magazine,* an organization looking for positive futures for the world. See Organizations to Contact for more details.
- Pages 236–37: The statistics on these pages are taken from World Bank, World Development Report 1990, Poverty, Washington, D.C., 1990, and Michel Chossudovsky, *The Globalisation of Poverty: Impacts of IMF and World Bank Reforms* (Penang, Malaysia: Third World Network, 1997). We obtained figures on the distributional change of world income over the past thirty years from UNESCO, Sources 58, May 1994.
- Page 239: John Cassidy's figures on middle-income decline were published in "Who Killed the Middle Class?" *The New Yorker,* October 16, 1995.
- Page 240: For a further discussion of perverse subsidies, see Alan Durning, *How Much Is Enough?* (New York: W.W. Norton, 1992). Also see Durning's booklet *The Car and the City.*
- Page 245: An indispensable book for anyone interested in the historical evaluation of modern culture is John Ralston Saul, *Voltaire's Bastards* (Toronto: Penguin, 1992). See also Saul's *The Unconscious Civilization* (Toronto: House of Anansi Press, 1995).

- Page 246: For more details about "natural capitalism," see Paul Hawken, *The Ecology of Commerce: A Declaration of Sustainability* (New York: HarperBusiness, 1993), and any of Hawken's other writings.

CHAPTER 8
GLOBALIZATION BLUES

- Page 249: The most indispensable book for this chapter and a layperson's overview of the entire globalization controversy is Jerry Mander and Edward Goldsmith, eds., *The Case Against the Global Economy* (San Francisco: Sierra Club Books, 1996), which includes articles by many of the people quoted in other chapters of this book (Ralph Nader, Vandana Shiva, David Korten, Andy Kimbrell, William Greider, et cetera). It's a good idea for anyone to read or reread the touted father of modern economics, who actually put things differently than we are led to believe: Adam Smith, *The Wealth of Nations: Books I–III* (Harmondsworth, England: Penguin, 1978). Joshua Karliner, *The Corporate Planet: Ecology and Politics in the Age of Globalization* (San Francisco: Sierra Club Books, 1997) was very useful here. Also see, of course, John Cavanaugh and Richard J. Barnet, *Global Dreams: Imperial Corporations and the New World Order* (New York: Simon & Schuster, 1994).
- Page 254: Maude Barlow and Tony Clarke, *Making the Links: A Citizen's Guide to the World Trade Organization and the Free Trade Area of the Americas,* a pamphlet published by the Council of Canadians, 1-800-387-7177, www.canadians.org.
- Page 255: For a detailed analysis of the on-the-ground effects of NAFTA in Canada, the United States and Mexico, see Bruce Campbell et al., *Pulling Apart: The Deterioration of Employment and Income in North America Under Free Trade* (Ottawa: The Canadian Centre for Policy Alternatives, 1999).
- Page 262: To learn more about the global economy's input into recent famine, genocide and war, see Michel Chossudovsky, *The Globalisation of Poverty: Impacts of IMF and World Bank Reforms* (Penang, Malaysia: Third World Network, 1997), or Chossudovsky's article for the *Journal of International Affairs,* "Global Poverty in the Late 20th Century" (Columbia University, Fall 1998). See also "Control of the World's Food Supply," Jerry Mander and Edward Goldsmith, eds., *The Case Against the Global Economy* (San Francisco: Sierra Club Books, 1996), and Vandana Shiva, Afsar Jafri and Gitanjali Bedi, *Ecological Cost of Economic Globalisation: The Indian Experience* (New Delhi: Research Foundation for Science, Technology & Ecology [RFSTE], 1995).
- Page 270: David Korten, *When Corporations Rule the World* (San Francisco: Kumarian Press, 1996). See specifically the chapters "Decline of Democratic Pluralism" and "Agenda for Change."

- Page 270: Edward Goldsmith, "Development as Colonialism," Jerry Mander and Edward Goldsmith, eds., *The Case Against the Global Economy* (San Francisco: Sierra Club Books, 1996) 253.
- Page 272: Alan Durning, *How Much Is Enough?* (New York: W.W. Norton, 1992) 56.
- Pages 272–73: The MMT controversy was widely covered in the Canadian press and by the CBC. See "Fuel Additive Debate Recharged" or "Ottawa Reintroduces MMT Bill: Proposed Ban on U.S.-Made Substance Could Start Trade War," *Globe and Mail,* March 25, 1996; "Banning MMT Prudent and Responsible," *The Financial Post,* April 3, 1997; and "Lawsuit Looms over MMT," *Sarnia Observer,* January 29, 1997.
- Pages 279–80: A typical local article on the Daishowa suit is Debbie Parkes, "Silencing Citizens," *Montreal Gazette,* December 7, 1996.
- Page 285: Ralph Estes, *The Tyranny of the Bottom Line* (San Francisco: Berrett-Koehler, 1996).
- Page 290: Tony Clarke and Maude Barlow, *MAI: The Multilateral Agreement on Investment and the Threat to Canadian Sovereignty* (Toronto: Stoddart, 1997) is a must-read for every Canadian. There are many additional publications about NAFTA and free trade available from the Council of Canadians. See Organizations to Contact for details.
- Page 290: For a fascinating overview of how one corporation is trying to corner the global food market, see Brewster Kneen, *Invisible Giant: Cargill and Its Transnational Strategies* (Halifax, NS: Fernwood Publishing, 1995).

CHAPTER 9

THE OTHER WORLD

- Pages 297–301: The accusations against Shell by the two people interviewed, Oronto Douglas of Environmental Rights Action, Nigeria, and Danny Kennedy of Project Underground, are backed up by the following published sources. On general background information: Stephen Kretzmann and Shannon Wright, *Drilling to the Ends of the Earth: The Ecological, Social, and Climate Imperative for Ending Petroleum Exploration* (Project Underground and Rainforest Action Network, 1998); Stephen Kretzmann and Shannon Wright, *Independent Annual Report: Human Rights and Environmental Operations Information on the Royal Dutch/Shell Group of Companies 1996–1997* (Project Underground, Rainforest Action Network and Oilwatch, 1997); *People and the Environment, Shell Petroleum Development Company of Nigeria Ltd. Annual Report,* 1997, put out by Shell International. Specific information provided in interviews is backed up by World Bank figures, which point out that millions of people have flocked to the Niger Delta in hopes of development employment, but Shell still only accounts for 1.3 percent of modern sector employment in the entire country. See World Bank, "Defining an Environmental Development Strategy for the Niger Delta," volume 1, 1995. Also see the World Council of Churches, "Ogoni: The Struggle Continues" (Geneva, 1996).

- Pages 298–99: Project Underground undertook the research detailed on these pages in April 1997; the samples were analyzed by Citizens Environmental Laboratory (CEL) Cambridge, Massachusetts. For the dangers of flaring and hydrocarbons, see EC Directive 75/440/EEC and Greenpeace Oil Briefing #7, "Human Health Impacts of Oil," Greenpeace United Kingdom, January 1993. The *New York Times* covered the Smithsonian Tropical Research Institute's work in Panama, where the levels of spilled oil reached similar levels as in the Niger Delta. These levels were termed, in a comparable tropical mangrove system, "catastrophic," and amounted to a "chronic oil spill" every time the rainy season arrived (December 21, 1993).

- Page 300: Shell's admissions of payments to the military are based in part on a letter to Paul Brown and Andy Rowell from Eric Nickson, head of media relations for Shell International (June 11, 1996). This letter also admits to providing transport for security forces on two occasions that involved loss of Ogoni life. However, as Stephen Kretzmann and Shannon Wright note in Project Underground's *Independent Annual Report: Human Rights and Environmental Operations Information on the Royal Dutch/Shell Group of Companies 1996–1997,* "The company maintains that they were merely protecting their interests and [were] unconnected to the military activity." Memos that seem to point to "the existence of a coordinated effort between Shell and the military to suppress dissent" are also detailed in the *Independent Annual Report.* Note that there are at least eleven statements on record from Shell denying military connections, and there are as well many other charges of wrongdoing and intimidation from the Ogoni. Owens Wiwa, Ken's brother, says that Brian Anderson of Shell Nigeria said he could free Ken, then imprisoned by the Nigerian military, but only if the international protest against Shell was called off and the environmental damage disallowed. As the *Independent Annual Report* states, "Shell disputes Dr. Wiwa's account of the meetings, but gives no plausible alternate explanation for the reason for the meetings."

- Page 302: Gwynne Dyer, "São Tomé and the Curse of Oil," syndicated column, July 20, 2003.

- Pages 311–12: Vandana Shiva, Afsar Jafri and Gitanjali Bedi, *Ecological Cost of Economic Globalisation: The Indian Experience* (New Delhi: Research Foundation for Science, Technology & Ecology [RFSTE], 1995).

- Page 312: Refer to sources on the effects of the Green Revolution in chapter 5. See also Francis Moore Lappé, Joseph Collins, Peter Rosset and Luis

Esparza, *World Hunger: 12 Myths* (New York: Grove/Atlantic and Food First Books, 1998), and Andy Kimbrell, "Why Biotechnology and High-Tech Agriculture Cannot Feed the World," *The Ecologist,* September/October 1998. For famine, food production and population figures, see any number of the Food and Agriculture Organization of the United Nations (FAO) reports on its web site www.fao.org — for example, "Food Supply Situation and Crop Prospects in Sub-Saharan Africa, Special Report No. 1," April 1993.

- Page 314: See Sir Albert Howard, *An Agricultural Testament* (Mapusa, Goa, India: The Other India Press: 1956) 9–15.
- Page 315: Mariam Mayet, "Africa — the New Frontier for the GE Industry," African Centre for Biosafety, January 2004. The centre reports that USAID is funding various GE projects "in a bid to take control of African agricultural research." USAID is also working in Zimbabwe and elsewhere to "promote weak and ineffective biosafety legal regimes and redirect capacity building towards genetic engineering rather than biosafety."
- Pages 318–19: The analysis of lack of satisfaction with government in northern and southern Italy comes from Robert D. Putnam et al., *Making Democracy Work: Civic Traditions in Modern Italy* (Princeton, NJ: Princeton University Press, 1994).
- Pages 324–25: A visual map showing the human and environmental cost of gold and other metal mining in the Third World was published in *Harper's,* October 1999.

CHAPTER 10
COMPLEX PLEASURES

- Page 335: Get a copy of Vicki Robin and Joe Dominguez, *Your Money or Your Life* (New York: Viking Penguin, 1992).
- Page 337: Check the Internet under Community Supported Agriculture to find farmers near you.
- Page 343: See www.eco-action.org/dod/no7/1-4.html and the U.K. Department of Environment, Transport and the Regions News Release, No. 76/Transport, 28/7/97.
- Page 344: Mark Matthews, "Learning a New Platform for Protest," *Washington Post,* July 18, 2003; Sherry Devlin, "Learning the Ropes," *The Missoulian,* July 25, 2003, posted at www.missoulian.com; Bob Doran, "Eco-Terrorism in Humboldt? PL Ads Spark Controversy," *Northwest Journal,* Humboldt County, California, April 24, 2003, posted at www.northwestjournal.com.

ORGANIZATIONS TO CONTACT

If the issues raised in this book have concerned you, you should know that many elected officials in Canada are actively supporting policies that will make matters even worse in terms of genetic engineering, global warming, species protection, food safety and phasing out of toxins. Elected officials respond to pressure, and they get a lot of it from vested interests. If you want to exert some pressure of your own, send letters to your Member of Parliament or to the relevant ministry at the following address (the main address, after the ministry, is always the same, and the postage is free):

> House of Commons
> Ottawa, ON K1A 0A6

You can also visit the web site www.net-efx.com/faxfeds and fax the Feds!

For biotech, food safety, human health as related to toxins and global warming, write:
> Health Canada (www.hc-sc.gc.ca)
> Canadian Food Inspection Agency
>   (www.canada.com/health; www.inspection.gc.ca)
>   Ottawa: (613) 225-2342 Hull: (819) 997-2919

For global warming issues, globalization issues, biotech and industrial toxins, write:
> Industry Canada (www.ic.gc.ca)

For climate change, forests, toxins and biotech, write:
> Natural Resources Canada (www.nr.can.gc.ca)

For globalization issues, the Tobin Tax, structural adjustment inequalities, or to demand forgiveness of Third World debt and democratization of the global economy, write:
> Department of Finance Canada (www.fin.gc.ca)

For globalization issues, water export and more, write:
> Department of Foreign Affairs and International Trade
>   (www.dfait-maeci.gc.ca)

To push for tougher legislation on endangered species, to speak out against exporting our water and to protect biodiversity, write:
   Environment Canada (www.ec.gc.ca)

If you're concerned about consumption issues, write especially to the Department of Foreign Affairs and International Trade and the Department of Finance Canada. If you have questions about a specific issue, contact the Sierra Club (see address below) or fax the Feds at the web site on the previous page.

The following is a list of effective organizations that are concerned with concrete action about the issues raised by this book. It begins with some organizations that deal with overall environmental issues, and then groups others by their topics of focus. We hope you'll be inspired to go further and learn more about what we've discussed here.

GENERAL ENVIRONMENTAL ISSUES

   The Corner House
   PO Box 3137
   Station Road
   Sturminster Newton
   Dorset, UK DT10 1YJ
   Phone: 44 1258-473795
   Fax: 44 1258-473748
   Email: cornerhouse@gn.apc.org
   Web site: http://cornerhouse.icaap.org
Publishes monthly briefings on subjects such as climate, nuclear contamination, biotechnology, globalization and other issues.

For Web-based NGOs working on democratic campaign reform, try:
   United States: www.MoveOn.org; www.progressiveportal.org;
      www.truemajority.org; www.commoncause.org
   Canada: www.democracywatch.org

   The Ecologist
   Unit 18, Chelsea Wharf
   15 Lots Road
   London, UK SW10 0QJ
   Phone: 44 171 351-3578
   Fax: 44 171 351-3617
   Email: ecologist@gn.apc.org
   Web site: www.theecologist.org

The Ecologist is a long-established bimonthly of the highest status and credibility, published by Edward Goldsmith. It features some of the finest experts in the world contributing articles investigating every aspect of environmental issues — scientific, cultural and economic.

Friends of the Earth (FOE) U.S.
1025 Vermont Avenue NW
Washington, D.C. 20005
Phone: (202) 738-7400
Fax: (202) 783-0444
Email: foe@foe.org
Web site: www.foe.org

FOE International
PO Box 19199
1000 GD Amsterdam, Netherlands
Phone: 31 20 622 1369
Email: foe@foei.org
Web site: www.foei.org

FOE UK
26–28 Underwood Street
London, UK N17JQ
Email:susdev@foe.co.uk
Web site:www.foe.co.uk.org

Friends of the Earth is an international environmental advocacy organization with member groups in sixty countries that often involve themselves in direct action. FOE is democratically based, with individual groups varying widely in their campaigns. FOE attempts to mobilize grass-roots citizen efforts in order to rescue the planet from environmental disasters.

Greenpeace Canada
250 Dundas Street West, Suite 605
Toronto, ON M5T 2Z5
Phone: (416) 597-8408, ext. 3030
Web site: www.greenpeace.ca

Greenpeace International
Keizersgracht 176
1016 DW Amsterdam, Netherlands
Phone: 31 20 523 6222

Fax: 31 20 523 6200
Active in food-related issues, especially biotech.

Population Institute
107 Second Street NE
Washington, D.C. 20002
Phone: (202) 544-3300
Email: web@populationinstitute.org
Web site: www.populationinstitute.org
The Population Institute studies and provides figures on population, fertility, food security and other related issues.

Sierra Club of Canada
1 Nicholas Street, Suite 412
Ottawa, ON K1N 7B7
Phone: (613) 241-4611
Toll-free: 1-888-810-4204
Email: sierra@web.net
Web site: www.sierraclub.ca
Arguably the most active and effective general-issue environmental organization in the country, the Sierra Club of Canada spearheads campaigns on issues such as biotechnology, toxic waste, pesticide reduction, forest and endangered species conservation, global corporate rule, and initiatives against human-rights violations. The SCC is very different from its U.S. counterpart, operating on a shoe-string and tackling issues in a more direct manner.

Third World Network, 121-S
Jalan Utama, 10450
Penang, Malaysia
Phone: 60 4 2266728/2266159
Fax: 60 4 2264505
Email: twnet@po.jaring.my
Web site: www.twnside.org.sg
The Third World Network is an independent network of organizations and individuals working on issues related to development and the North-South dichotomy. It researches and publishes helpful books and articles on issues pertaining to struggles in the Third World: a daily out of Switzerland, the *SUNS Bulletin;* a twice-monthly magazine focusing on GATT, the WTO and the IMF, *Third World Economics;* and a monthly illustrated magazine, *Third World Resurgence.* The secretariat is in Penang, but the organization has offices in Delhi, Montevideo, Geneva, London and Accra.

Union of Concerned Scientists
Two Brattle Square
Cambridge, MA 02238-9105
Phone: (617) 547-5552
Web site: www.ucsusa.org
This is a conservative group of hundreds of scientists who have banded together
to investigate and publicize unimpeachable statements regarding global warm-
ing, deforestation and other issues.

CONSUMER ISSUES

Adbusters and Media Foundation
Kalle Lasn
1243 West 7th Avenue
Vancouver, BC V6H 1B7
Phone: (604) 736-9401 and 1-800-663-1243
Fax: (604) 737-6021
Email: info@adbusters.org
Web site: www.adbusters.org
The Media Foundation works to help people step back and analyze the role of
public relations and advertising in the world. It exposes abuses and brainwashing
and especially tries to fight the unregulated spread of a corporate consumerism
that is wasting the life-support systems of the planet. *Adbusters* is the glossy
monthly magazine of the culture-jammer movement. Subscriptions don't fund
any corporations.

Alternative media web sites we recommend for thoughtful, unusual and free-of-
corporate influence information, most of which were founded after 1999 (all
begin with www.):
The Guerrilla News Network (guerrillanews.com)
GlobalPublicMedia.com
Disinfo.com
Indymedia.org
TomPaine.com
Straightgoods.com
Undercurrents.org

New Road Map Foundation
PO Box 15981
Seattle, WA 98115
Phone: (206) 527-0437
Fax: (206) 528-1120
Web site: www.newroadmap.org

The New Road Map Foundation is an all-volunteer, educational and charitable organization concerned with the role of personal responsibility and personal initiative in effecting positive global changes. They are involved with the voluntary simplicity movement, offering audiocassette courses and workbooks to people wishing to gain control of their financial lives, with all proceeds donated to non-profit organizations.

Northwest Environment Watch (NEW)
Alan Durning, Director
1402 3rd Avenue, Suite 500
Seattle, WA 98101
Phone: (206) 447-1880
Fax: (206) 447-2270
Email: new@northwestwatch.org

Northwest Environment Watch is a non-profit research centre looking to find sustainable, long-term ways to live in harmony with the ecological area extending between northern California and southern British Columbia and bounded by the Rockies. They publish health checkups for the region, as well as long reports like *The Car and the City* and *What Can the Earth Afford? Beyond the Consumer Society in the Pacific Northwest.*

Redefining Progress
1904 Franklin Street, 6th floor
Oakland, CA 94612
Phone: (510) 444-3041
Fax: (510) 444-3191
Email: info@redefiningprogress.org
Web site: www.rprogress.org

The people at Redefining Progress are pioneers in the effort to develop a new measure for national well-being to counteract the inadequate gross domestic product (GDP). Their method, the genuine progress indicator (GPI), is an attempt to measure more than just the exchange of money in judging a society's well-being. Redefining Progress believes that a world of true abundance lies not in more financial transactions, but in a high quality of life for all, sustained by the networks of nature and community and passed on from generation to generation.

FOOD ISSUES AND BIOTECHNOLOGY (ABROAD)

Australian Gene Ethics Network
60 Leicester Square, Floor 1
Carlton 3053
Victoria, Australia
Phone: 03 9347 450 (Int. +613)

Fax: 03 9345 1166
Email: info@geneethics.org
Web site: www.geneethics.org
This grass-roots, non-profit, public-interest group monitors and tries to protect consumer rights and to demand labelling of genetically engineered products.

Ecoropa
President, Christine von Weizsaecker
Postfach 130 165
53061 Bonn, Germany
Phone: 49 228 9181 033
Fax: 49 228 9181 034
Email: cvw@isd.de
Ecoropa is a twenty-year-old European environmental group specializing in biotechnology, agricultural issues, toxics and food-safety issues.

NAVDANYA
(National Program for Conservation of Native Seed Varieties)
Director, Vandana Shiva
A-60, Hauz Khas, New Delhi 110 016
Phone: 91 11 696 8077 or 651 5003
Fax: 91 11 685 6795; 696 2589
NAVDANYA is involved with actively saving the seeds of rural India from patenting and industrial development.

Women's Environmental Network
PO Box 30626
London, UK E1 1TZ
Phone: +44 (0) 20 7481 9004
Fax: +44 (0) 20 7481 9144
Email: info@wen.org.uk
Web site: www.wen.org.uk
Involved in both direct action and education against biotechnology in England and Europe, WEN also has a unique research and campaign team, four members of which are biologists or geneticists. Their publications are used all across the United Kingdom, Europe, Canada and South America.

For a constant stream of up-to-date biotechnology information that has been rigorously screened by a microbiologist, write to Genet@agoranet.be.

FOOD ISSUES AND BIOTECHNOLOGY (CANADA)

Action Group on Erosion, Technology and Concentration
(ETC Group, formerly RAFI, Rural Advancement Foundation International)
478 River Avenue, Suite 200
Winnipeg, MB R3L 0C8
Phone: (204) 453-5259
Fax: (204) 284-7871
Email: etc@etcgroup.org
Web site: www.etcgroup.org
The ETC Group is headquartered in Winnipeg, Manitoba, Canada, with affiliate offices in Europe, the Third World and the United States. The organization is dedicated to the conservation and sustainable improvement of agricultural biodiversity, and to the socially responsible development of technologies useful to rural societies. It is currently very active in the controversies surrounding genetically engineered crops and technology.

Beyond Factory Farming Coalition
Room 420, 230 Avenue R South
Saskatoon, SK S7M 0Z9
Phone: (306) 955-6454
Fax: (306) 955-6455
Toll-free: 1-877-955-6454
Email: choltslander@canadians.org
Web site: www.beyondfactoryfarming.org
A national coalition of the many provincial groups fighting factory farming. The organization sponsors a very useful list-serve to keep people appraised of ILO activity, and it promotes livestock production that supports food sovereignty and ecological, human and animal health, as well as sustainability, community viability and informed citizen/consumer choice.

Canadian Organic Growers
125 South Knowlesville Road
Knowlesville, NB E7L 1B1
Phone: (506) 375-7383
Toll-free: 1-888-375-7383
Fax: (506) 375-4221
Email: office@cog.ca
Web site: www.cog.ca
The national office of organic farmers, gardeners and consumers in Canada.

Equiterre
2177 rue Masson, bureau 317
Montreal, QC H2H 1B1
Phone: (514) 522-2000
Email: info@equiterre.org
Web site: www.equiterre.org
Quebec's major NGO puts consumers together with CSA farmers and works for diversity in farming and for fair trade and fossil-fuel-use reduction.

Hogwatch Manitoba
2–70 Albert Street
Winnipeg, MB R3B 1E7
Phone: (204) 926-1914
Email: info@hogwatchmanitoba.org
Web site: www.hogwatchmanitoba.org
One of the oldest local NGOs fighting industrial farming, Hogwatch remains an excellent source of information about how to work for sustainable farms and rural areas.

The Ram's Horn
Brewster Kneen, Editor
S6 C27 RR1 Sorrento, BC
Phone/Fax: (250) 675-4866
Email: ramshorn@ramshorn.bc.ca
Web site: www.ramshorn.bc.ca
Brewster and Cathleen Kneen's monthly newsletter keeps people abreast of crucial food issues in Canada and abroad.

FOOD ISSUES AND BIOTECHNOLOGY (USA)

Edmonds Institute
Beth Burrows, Director
20319–92nd Avenue West
Edmonds, WA 98020
Phone: (425) 775-5383
Fax: (425) 670-8410
Email: beb@igc.org
Web site: www.edmonds-institute.org
The Edmonds Institute is a public-interest, non-profit environmental think tank; its current emphasis is on biosafety, intellectual property rights and just policies for the protection of biodiversity. The institute encourages pro bono research and policy analysis by scientists and scholars, and its special talent is creating alliances and coalitions with like-minded organizations and individuals.

Foundation on Economic Trends
1660 L Street NW, Suite 216
Washington, D.C. 20036
Phone: (202) 466-2823
Fax: (202) 429-9602
Email: campaign@uol.com or jrifkin@foet.org
Web site: www.biotechnology.org

The Foundation on Economic Trends (FET) is a non-profit organization whose mission is to examine emerging trends in science and technology and their effects on the environment, economy, culture and society. Jeremy Rifkin is its president and founder.

Humane Society of the United States
Farm Animals and Sustainable Agriculture Campaign
2100 L Street NW
Washington, D.C. 20037
Phone: (202) 452-1100
Toll-free: 1-800-444-8359
Web site: www.hsus.org

The Humane Society of the United States fights for safer and more humane conditions for food animals, in an effort to also help prevent the spread of E.coli, salmonella and BSE (mad cow disease) outbreaks in humans, which the organization contends are a natural concomitant of present industrial agriculture.

Institute for Agriculture and Trade Policy (IATP)
Trade and Agriculture Program
2105 First Avenue South
Minneapolis, MN 55404
Phone: (612) 870-0453
Fax: (612) 870-4846
Email: iatp@iatp.org
Web site: www.iatp.org

IATP works to create environmentally and economically sustainable communities and regions through sound agriculture and trade policy. The data it gathers are distributed as educational materials to policy makers and the public at large.

International Center for Technology Assessment
Center for Food Safety
310 D Street NE
Washington, D.C. 20002
Phone: (202) 547-9359

Email: info@icta.org
Web site: www.icta.org
The International Center for Technology Assessment examines the economic, ethical, social, environmental and political effects of the applications of technology, and carries on lawsuits against government entities and corporations where necessary. The Center for Food Safety addresses the impact of our food production system on human health, animal welfare and the environment. It fights for testing, labelling and regulation of genetically engineered foods, preserving strict national organic food standards, and preventing potential health crises caused by food-borne illness, including mad cow disease. Both organizations are energetic sources of information and of legal action.

Organic Consumers Association
Ronnie Cummings
6101 Cliff Estate Road
Little Marais, MN 55614
Phone: (218) 353-7652
Fax: (218) 226-4164
Web site: www.organicconsumers.org
A public-interest organization that deals with issues of food safety and genetic engineering. Campaign tactics include public education, litigation, boycotts, media work and direct action. The Organic Consumers Association also publishes a free monthly electronic newsletter, *Organic Bytes.*

RAFI USA
PO Box 640
Pittsboro, NC 27312
Phone: (919) 542-1396
Fax: (919) 542-0069
Email: info@rafiusa.org
Web site: www.rafiusa.org
RAFI USA, originally an offshoot of RAFI Canada, is now an independent U.S. NGO dedicated to community, equality and diversity in agriculture.

Seed Savers Exchange
3076 North Winn Road
Decorah, IA 52101
Phone: (563) 382-5990
Fax: (563) 382-5872
This group grows, saves, distributes and sells organic and heritage seed. It also funds workshops for Third World participants to increase their knowledge of sustainable food cropping.

SoilFoodWeb Inc.
Elaine Ingham, President, SoilFoodWeb Inc.
1128 NE 2nd Street, Suite 120
Corvallis, OR 97330
Phone: (541) 752-5066
Fax: (541) 752-5142
Email: info@soilfoodweb.com
Web site: www.soilfoodweb.com

This organization advises, educates and provides the means for farmers to wean their soils from agricultural chemicals and learn how to build soil for sustainable agriculture.

Waterkeepers' Alliance
828 South Broadway, Suite 100
Tarrytown, NY 10591
Phone: (914) 674-0622
Email: info@waterkeepers.org
Web site: www.waterkeepers.org

A nine-year old coalition of 110 river-watch groups run by Robert Kennedy, Jr., Waterkeepers not only defends rivers and waterways in the United States, Canada and elsewhere from pollution and agricultural run-off, it attacks the causes of that pollution, like factories and industrial farming.

Food-safety-issues information can also be obtained in the United States by calling: Food First: (510) 654-4400 or Consumer's Union: (914) 378-2000.

GLOBALIZATION ISSUES

Council of Canadians
502–151 Slater Street
Ottawa, ON K1P 5H3
Phone: (613) 233-2773
Fax: (613) 233-6776
Email: inquiries@canadians.org
Web site: www.canadians.org

The Council of Canadians is a non-profit, non-partisan public-interest organization supported by more than 100,000 members. It serves as a government watchdog, critic of globalization and catalyst for grass-roots organizing. The council works to protect Canadian sovereignty and promote democratic rights. It tackles such issues as unregulated, genetically engineered food, corporate control of the media, and the commodification of our water supply.

International Forum on Globalization
1009 General Kennedy Avenue, Suite 2
San Francisco, CA 94129
Phone: (415) 561-7650
Fax: (415) 561-7651
Email: ifg@ifg.org
Web site: www.ifg.org

The IFG is an alliance of leading activists, scholars, economists, researchers and
writers from twenty-five countries, formed to stimulate joint activity and public
education in response to the global economy. It has organized teach-ins that
have preceded major anti-globalization demonstrations across the United States.

Institute for Policy Studies
733 15th Street, NW, Suite 1020
Washington, D.C. 20005
Phone: (202) 234-9382
Fax: (202) 387-7915
Email: a-quinn@mindspring.com
Web site: www.ips-dc.org

The IPS is a Washington think tank established more than thirty-five years ago.
These days it is concerned primarily with democracy and justice issues as they
relate to trade.

PR Watch
Center for Media and Democracy
520 University Avenue, Suite 310
Madison, WI 53703
Phone: (608) 260-9713
Fax: (608) 260-9714
Email: editor@prwatch.org
Web site: www.prwatch.org

The Center for Media and Democracy, founded by Sheldon Rampton and John
Stauber, the two crusaders against green- and brownwashing described in chapter 4, keeps an eye on transnationals and their media manipulations. The organization publicizes issues such as SLAPP suits, corporate censorship and marketing
to children. Its newsletter *PR Watch* is published quarterly.

Transnational Resource and Action Center (TRAC)
Corporate Watch, Internet magazine
2288 Fulton Street, Suite 103
Berkeley, CA 94704
Phone: (510) 849-2423

Email: corpwatch@corpwatch.org
Web site: www.corpwatch.org
This organization keeps track of corporate power and analyzes the dangers to democracy and to human rights. TRAC's Internet magazine is a rich source of information about who is doing what and how to combat it.

GLOBAL WARMING AND FOREST ISSUES

David Suzuki Foundation
2211 West Fourth Avenue, Suite 219
Vancouver, BC V6K 4S2
Phone: (604) 732-4228
Fax: (604) 732-0752
Email: solutions@davidsuzuki.org
Web site: www.davidsuzuki.org
The David Suzuki Foundation is a non-profit research group that tries to study the underlying structures and systems that cause environmental crises, in order to bring about fundamental change. The foundation commissions research, supports the implementation of ecologically sustainable models, works to educate, and urges decision makers to adopt policies that encourage individuals and businesses to act within nature's constraints.

(Note: if your request or question involves David's CBC Television series *The Nature of Things,* contact it directly by email at tnot@toronto.cbc.ca, or at PO Box 500, Station A, Toronto, ON M5W 1E6; phone: [416] 205-6888; fax: [416] 205-3579.)

Greenpeace U.S.A.
1436 U Street NW
Washington, D.C. 20009
Toll-free: 1 (800) 326-0959
Phone: (202) 462-1177
Fax: (202) 462-4507
Web site: www.greenpeaceusa.org
Greenpeace is the leading independent, international organization that uses peaceful and creative activism to protect the global environment. Its major U.S. campaign at the moment is against global warming.

Rainforest Action Network
221 Pine Street, No. 500
San Francisco, CA 94702
Phone: (415) 398-4404
Fax: (415) 398-2732

Email: rainforest@ran.org
Web site: www.ran.org
Rainforest Action Network works to protect the Earth's rain forests and support the rights of their inhabitants through education, grass-roots organizing, and non-violent direct action.

World Resources Institute
10 G Street NE, Suite 800
Washington, D.C. 20002
Phone: (202) 729-7600
Fax: (202) 729-7610
Email: front@wri.org
Web site: www.wri.org
The World Resources Institute is a public and grant-supported educational organization that believes a healthy environment and a healthy economy can coexist. Current areas of research include economics, forests, biodiversity, climate change, energy, sustainable agriculture, resource and environmental information, trade, technology and health. Its groundbreaking report, *The Last Frontier Forests: Ecosystems and Economies on the Edge,* is a must-read.

NON-VIOLENT DIRECT-ACTION GROUPS

EarthFirst! California
PO Box 83
Canyon, CA 94516
Phone: (510) 848-8724

EarthFirst! Journal
PO Box 3023
Tucson, AZ 85702
Phone: (520) 620-6900
Fax: (413) 254-0057
Email: collections@earthfirstjournal.org
Web site: www.earthfirstjournal.org
The activities of EarthFirst! are detailed in chapter 4. EarthFirst! works mostly on the West Coast of the United States, but also has offices in Boston. Although its members are most famous for non-violent direct action (blockades, demos, sit-ins and the like), it also helps organize communities, petition governments and educate the public, like most democratic groups.

Sea Shepherd Conservation Society
PO Box 2670

Malibu, CA 90265
Phone: (360) 370-5650
Fax: (360) 370-5651
To contribute money, call: 1-888-WHALE-22
Web site: www.seashepherd.org

Paul Watson is captain of the *Sea Shepherd* vessel and a direct-action environ-mental activist. For the past twenty-five years, he and members of the Sea Shepherd Conservation Society have tried to protect sea mammals by putting themselves between harpooners and whales, and by sinking whaling ships. He has been beaten by sealers in Newfoundland, jailed in Iceland for sinking ships and lionized by many as one of the only lines of defence left between the survival and the extinction of a constantly shrinking marine animal population.

Other non-violent action groups include Friends of the Earth and Greenpeace, whose addresses are listed under "General Environmental Issues."

POLLUTION AND TOXICS

Center for Health, Environment and Justice, CHEJ
Lois Gibbs, Executive Director
PO Box 6806
Falls Church, VA 22040
Phone: (703) 237-2249
Fax: (703) 237-8389
Email: chej@chej.org
Web site: www.chej.org

The Center for Health, Environment and Justice helps families protect their chil-dren from chemical poisons and their communities from toxic waste dumps. It was founded, as the Citizen's Clearing House on Hazardous Waste, by Lois Gibbs, the leader of the first successful community campaign against toxics, Love Canal. CHEJ works for, and is supported primarily by, blue-collar workers, farm-ers, low-income families and people of colour. Victims of suspected toxics can turn to their Science and Technical Assistance Program, led by Harvard-trained toxicologist Stephen Lester. CHEJ publishes more than 130 guides and fact-packs and three periodicals, *Everyone's Backyard, Environmental Health Monthly* and *Dioxin Digest*. It spends only 14 percent of its budget on fund-raising and administration, and is a true grass-roots organization.

Environmental Rights Action (ERA)
Friends of the Earth
PO Box 10577
Ugbowo, Benin City

Nigeria
Phone and Fax: 234 52 600 165
Email: eraction@infoweb.abs.net
Web site: www.essentialaction.org/shell/era/era.html

Oronto Douglas, lawyer and friend of Ken Saro-Wiwa, the Nobel Prize–winning poet executed by Nigerian dictators in 1995, is deputy director of ERA. This organization attempts to disseminate news of the human and environmental abuses by multinational oil companies, like Shell Oil and Chevron, in alliance with paramilitary and government groups in the Niger Delta. They also help coordinate the worldwide Shell boycott, and can provide information on the activities of these corporations in the Third World.

Pesticide Action Network
49 Powell Street, Suite 500
San Francisco, CA 94102
Phone: (415) 981-1721
Fax: (415) 981-1991
Email: panna@panna.org
Web site: www.panna.org

The Pesticide Action Network is a fast and scientifically reliable source of information. It knows what compounds our food contains, and it can tell exactly how dangerous and common they are. PAN can also show how to organize to limit or eliminate pesticide use in a given area.

Project Underground
1611 Telegraph Avenue, Suite 702
Oakland, CA 94612
Phone: (510) 271-8081
Fax: (510) 271-8083
Email: project_underground@moles.org
Web site: www.moles.org

Project Underground fights the toxic pollution and human-rights violations of the current worldwide gold rush on metals, led in large part by Canadian corporations. It provides information, legal help and outside research to people suffering the effects of the global land grab for minerals.

INDEX

THE DAVID SUZUKI FOUNDATION:
WORKING TOGETHER FOR A SUSTAINABLE FUTURE

The David Suzuki Foundation was established to work for a world of hope in which our species thrives in balance with the productive capacity of the Earth.

Our mission is to find solutions to the root causes of our most threatening environmental problems. Then, we work with our supporters and their communities to implement those solutions for a sustainable future.

Our mandate is broad, ranging from projects on climate change, air, soil, water, fisheries, forestry, energy, and liveable cities, to defining the foundations of sustainability, how social change occurs, and the potential of new economic models.

We can only accomplish this with the support of concerned citizens who care about the environment. We invite your help.

JOIN OUR PARTNERSHIP . . . JOIN THE FOUNDATION!

NAME   ADDRESS

CITY/PROVINCE   POSTAL CODE   TELEPHONE

Here is my donation of: ❑ $30.   ❑ $50   ❑ $100
                          ❑ $500 E.O. Wilson Circle   ❑ $5000 Patron

❑ Cheque   ❑ Money Order   ❑ Visa   ❑ MasterCard

CARD NUMBER   EXPIRY DATE

SIGNATURE

Yes, I'll become a Friend of the Foundation. I authorize the Foundation to receive the following amount from my chequing account on a monthly basis. I understand I'll receive a tax credit and the benefits of becoming a Foundation Supporter.

❑ $10 a month   ❑ $15 a month   ❑ $25 a month   ❑ $_____ a month

I understand I can change or cancel my pledge at any time.
I enclose a sample cheque marked VOID for bank coding.

SIGNATURE

Please return this reply memo with your tax-creditable donation. Cheques can be made payable to The David Suzuki Foundation. Thank you very much for your support.

 THE DAVID SUZUKI FOUNDATION
219–2211 WEST FOURTH AVENUE
VANCOUVER, B.C., CANADA V6K 4S2
Phone (604) 732-4228; Fax (604) 732-0752
Toll free: 1-800-453-1533

Charitable Registration Number: BN 12775 6716 RR0001